AF352804

Casting and Solidification of
Light Alloys

Casting and Solidification of Light Alloys

Editor

Alexander Vorozhtsov

MDPI • Basel • Beijing • Wuhan • Barcelona • Belgrade • Manchester • Tokyo • Cluj • Tianjin

Editor
Alexander Vorozhtsov
Tomsk State University
Russia

Editorial Office
MDPI
St. Alban-Anlage 66
4052 Basel, Switzerland

This is a reprint of articles from the Special Issue published online in the open access journal *Metals* (ISSN 2075-4701) (available at: https://www.mdpi.com/journal/metals/special_issues/casting_solidification).

For citation purposes, cite each article independently as indicated on the article page online and as indicated below:

LastName, A.A.; LastName, B.B.; LastName, C.C. Article Title. *Journal Name* **Year**, *Volume Number*, Page Range.

ISBN 978-3-03943-737-5 (Hbk)
ISBN 978-3-03943-738-2 (PDF)

Contents

About the Editor

Alexander Vorozhtsov, Doctor of Physical and Mathematical Sciences, Professor, Head of High Energetic and Special Materials Laboratory of Tomsk State University, Russian Federation.

Preface to "Casting and Solidification of Light Alloys"

The processes of casting and crystallization define the development of the final properties of light alloys, and optimization of these processes allows for improving efficiency of the use and applications of light alloys. The analysis and design of such processes requires special attention. The aim of this book is to present the latest achievements in technology, including advanced and modern approaches, structure development, and properties of the light alloys. Included are investigations of the effect of casting and crystallization on the structure and properties of the resulting light alloys (based on aluminum and magnesium to begin with) and, in particular, research connected with detailed analysis of the microstructure of light alloys obtained using various external influences of ultrasonic, vibration, magnetic, and mechanical processing on the casting and crystallization. The use of modern methods to study the properties of alloys in order to assess the effect of structure on the mechanical and functional properties of light alloys is also covered as are studies on the introduction of additives (modifiers, reinforcers, including nanosized ones, etc.) into the melt during the crystallization process, the technological properties of casting (fluidity, segregation, shrinkage, etc.), the structure and physicomechanical properties of light alloys. Finally, this book includes papers that focus on the relations of physicomechanical properties with the defective structure of light alloys and mathematical modeling of plastic deformation of dispersion-strengthening materials.

Alexander Vorozhtsov
Editor

Editorial

Casting and Solidification of Light Alloys

Alexander Vorozhtsov

Laboratory of High Energetic and Special Materials, National Research Tomsk State University, Lenin av. 36, 634050 Tomsk, Russia; abv1953@mail.ru; Tel.: +7-3822-526-120

Received: 9 October 2020; Accepted: 15 October 2020; Published: 23 October 2020

1. Introduction and Scope

At present, light alloys based on aluminum, magnesium, titanium, etc. have been intensively studied. Improving the properties of these alloys by modifying the structure and reinforcement makes it possible to solve the technological problem of increasing the working properties of mechanical engineering products while simultaneously reducing their weight. Concurrent to the study and development of methods for strengthening light alloys, casting technologies also require improvement since traditional casting technologies do not allow for the improvement of alloy properties when hardening particles are introduced into the liquid phase of the base metal. Modern methods of processing light alloys during casting make it possible to increase the physical and, ultimately, the operational properties of the obtained alloys by degassing, reducing the average grain size, increasing the uniformity of the alloy composition, reducing the amount of agglomerates and impurities at grain boundaries, increasing wettability, etc. The development of unconventional casting methods and the study of properties of light alloys are reported in the articles of this special issue.

2. Contributions

This special issue consists of one review and 11 original research articles. Of these 12 articles, 10 articles are devoted to aluminum alloys, one article is devoted to titanium alloys and the review article is devoted to aluminum, titanium, copper and magnesium alloys.

The review article [1] provides an overview of recent advances in light alloys (aluminum, copper, titanium and magnesium alloys) modified using friction stir processing (FSP). The general mechanisms of the formation of subsurface gradient structures in metal alloys processed by FSP under various conditions are described. It is shown that FSP can be used to produce light alloys with subsurface gradient structure, composite subsurface gradient structure and "in-situ" composites.

In the first research article [2], aluminum alloys of the Al-Mg system with titanium diboride particles of different dispersion were obtained. The introduction of titanium diboride particles using ultrasonic treatment of the melt made it possible to significantly reduce the average grain size of the alloy AA5056. The greatest effect of structure refinement was obtained using a master alloy containing titanium diboride particles with a size of 1 μm. It was also found that the introduction of titanium diboride particles led to an increase in the yield strength, tensile strength and plasticity from 57 to 71 MPa, from 155 to 201 MPa, and from 11.5 to 18.8%, respectively.

The effect of aluminum oxide nanoparticles introduced into the melt using ultrasonic treatment of the melt on the structure, and the properties of pure aluminum were studied in [3]. It was found that the introduction of nanoparticles of aluminum oxide using ultrasonic treatment of the melt into commercially pure aluminum allowed microstructure refining and reduced the average grain size from 200 to 69 μm. The introduction of nanoparticles made it possible to increase the hardness from 19 to 22 HB, the yield strength from 12 to 27 MPa, and the tensile strength from 48 to 79 MPa in commercially pure aluminum.

In [4], the effect of porosity on the properties of aluminum composites reinforced with SiC particles was studied using 2D models and finite elements analysis. The authors found that when the particle content did not exceed 11%, and the theoretical density of the material was reached, the shape of the particles (circular or square) did not affect the mechanical properties of the composite. With SiC particle content of more than 11%, the angular particles were more effective in improving the mechanical properties. Furthermore, it was shown that the particles transferred the stress to the soft matrix even in the presence of pores in the composite.

The analysis of the effect of processing the Al-Mg-Si solution on the transformation of β-AlFeSi particles into α-(FeMn) Si and the aging of the Al 6063 alloy were presented in [5]. It was found that the maximum hardness of 104 HV was reached following a solution treatment at a temperature of 600 °C for 2 h and aging at a temperature of 160 °C for 12 h.

In work [6], the Al-Si-Cu-Mg alloy was cast at casting speeds of 1, 2, 3 and 4 mm/s by the Ohno continuous casting (OCC) process. It was shown that the spacing between secondary dendrite arms of α-Al dendrites in the samples decreased significantly with an increase in the casting speed. Moreover, an increase in the casting speed to 4 mm/s allowed the authors to increase the tensile strength of the alloy, which was not possible with the alloy obtained at the casting speed of 1 mm/s.

In work [7], the authors obtained the $Al_8Zn_7Ni_3Mg$ hypereutectic alloy, combining an Al-8% Zn-3% Mg matrix reinforced by an Al_3Ni intermetallic compound. The specified structure of the alloy was obtained by changing the cooling rate in the range from 0.1 K/s to 2.3×10^5 K/s. It was shown that an increase in the cooling rate made it possible to grind intermetallic compounds in the alloy, the size of which was 50 nm, at a cooling rate of 105 K/s. At the same time, the hardness of the alloy increased to 220 HV.

In work [8], the aluminum alloy containing 0.6 wt.% Zr, 0.4% Fe and 0.4% Si was obtained by electromagnetic casting at a high cooling rate (140 K/s). The subsequent procedure of direct cold drawing was used to synthesize Al_3Zr nanoparticles in the alloy with a uniform distribution in the matrix and an average size not exceeding 10 nm. The composite wire sample obtained by the authors had a tensile strength and electrical conductivity of 234 MPa and 55.6 IACS, respectively.

A study of the process of treating commercially pure titanium by friction with stirring using a tool made of a nickel-based superalloy ZhS6U was presented in [9]. It was found that the transfer layer in titanium contained chemical elements of the ZhS6U alloy. After such treatment, the tensile strength of commercially pure titanium increased by 25%.

In work [10], an experiment was carried out on friction stir welding of a sheet made of a fine-grain Al-Mg-Sc-Zr alloy to study the peculiarities of the plasticized metal flow and microstructural evolution. It was shown that the stir zone macrostructure might contain either a single or many nugget zones, depending on sheet thickness and the seam length. Despite the finer-grained structure, the hardness of the welded seam was lower than the hardness of the base alloy, and the tensile strength of the welded seam was comparable to the tensile strength of the base alloy.

The structural phase characteristics of the friction-treated Al-Cu alloy zone were studied in [11]. The presence of Al_2Cu, Al_2Cu_3, $AlCu_3$, Al_2MgCu, etc. intermetallic phases with a nonuniform distribution over volume and size was revealed.

In work [12], the structural and phase states of the TiAl system alloyed with rare earth metals were studied. Tantalum, yttrium and dysprosium were used as additives. It was found that the studied systems Ti(49 at.%)–Al(49 at.%) with additions of Ta, Y and Dy contained intermetallic compounds consisting of $AlTi_3$, TiAl in hexagonal, tetragonal and triclinic forms.

3. Conclusions and Outlook

Future development of new casting methods and techniques for local control of crystallization of light alloys will significantly change the approach to industrial production of high-tech and efficient products. This will make it possible to create light alloys for specific structural elements of automobiles, airplanes, etc., where a set of specific operational properties is required. There is also great interest in joining new light alloys, including those with a composite structure, which also indicates the widespread use of new light alloys.

Implementation of this concept will create a strong demand for innovations in the field of new light alloys and their casting and joining technologies, which are one of the leading topics in this special issue. As a guest editor of this special issue, I am very pleased with the final result and hope that these articles will be useful to both researchers and technologists working in the production and processing of light alloys. I would like to express my gratitude to the authors for their contributions and the reviewers who helped with the review process. I would also like to extend a special thanks to the Metals Editor Office staff for working on the special issue.

Conflicts of Interest: The author declares no conflict of interest.

References

1. Zykova, A.; Tarasov, S.; Chumaevskiy, A.; Kolubaev, E. A Review of Friction Stir Processing of Structural Metallic Materials: Process, Properties, and Methods. *Metals* **2020**, *10*, 772. [CrossRef]
2. Khrustalyov, A.; Kozulin, A.; Zhukov, I.; Khmeleva, M.; Vorozhtsov, A.; Eskin, D.; Chankitmunkong, S.; Platov, V.; Vasilyev, S. Influence of Titanium Diboride Particle Size on Structure and Mechanical Properties of an Al–Mg Alloy. *Metals* **2019**, *9*, 1030. [CrossRef]
3. Zhukov, I.; Kozulin, A.; Khrustalyov, A.; Kahidze, N.; Khmeleva, M.; Moskvichev, E.; Lychagin, D.; Vorozhtsov, A. Pure Aluminum Structure and Mechanical Properties Modified by Al_2O_3 Nanoparticles and Ultrasonic Treatment. *Metals* **2019**, *9*, 1199. [CrossRef]
4. Rivera-Salinas, J.; Gregorio-Jáuregui, K.; Romero-Serrano, J.; Cruz-Ramírez, A.; Hernández-Hernández, E.; Miranda-Pérez, A.; Gutierréz-Pérez, V. Simulation on the Effect of Porosity in the Elastic Modulus of SiC Particle Reinforced Al Matrix Composites. *Metals* **2020**, *10*, 391. [CrossRef]
5. Alvarez-Antolin, F.; Asensio-Lozano, J.; Cofiño-Villar, A.; Gonzalez-Pociño, A. Analysis of Different Solution Treatments in the Transformation of β-AlFeSi Particles into α-(FeMn)Si and Their Influence on Different Ageing Treatments in Al–Mg–Si Alloys. *Metals* **2020**, *10*, 620. [CrossRef]
6. Lee, Y.; Makino, Y.; Nitta, J.; Lee, E. Influence of Continuous Casting Speeds on Cast Microstructure and Mechanical Properties of an ADC14 Alloy. *Metals* **2020**, *10*, 625. [CrossRef]
7. Shurkin, P.; Akopyan, T.; Korotkova, N.; Prosviryakov, A.; Bazlov, A.; Komissarov, A.; Moskovskikh, D. Microstructure and Hardness Evolution of Al8Zn7Ni3Mg Alloy after Casting at very Different Cooling Rates. *Metals* **2020**, *10*, 762. [CrossRef]
8. Belov, N.; Murashkin, M.; Korotkova, N.; Akopyan, T.; Timofeev, V. Structure and Properties of Al–0.6wt.%Zr Wire Alloy Manufactured by Direct Drawing of Electromagnetically Cast Wire Rod. *Metals* **2020**, *10*, 769. [CrossRef]
9. Amirov, A.; Eliseev, A.; Kolubaev, E.; Filippov, A.; Rubtsov, V. Wear of ZhS6U Nickel Superalloy Tool in Friction Stir Processing on Commercially Pure Titanium. *Metals* **2020**, *10*, 799. [CrossRef]
10. Kalashnikova, T.; Chumaevskii, A.; Kalashnikov, K.; Fortuna, S.; Kolubaev, E.; Tarasov, S. Microstructural Analysis of Friction Stir Butt Welded Al-Mg-Sc-Zr Alloy Heavy Gauge Sheets. *Metals* **2020**, *10*, 806. [CrossRef]
11. Zykova, A.; Chumaevskii, A.; Gusarova, A.; Kalashnikova, T.; Fortuna, S.; Savchenko, N.; Kolubaev, E.; Tarasov, S. Microstructure of In-Situ Friction Stir Processed Al-Cu Transition Zone. *Metals* **2020**, *10*, 818. [CrossRef]
12. Belgibayeva, A.; Abzaev, Y.; Karakchieva, N.; Erkasov, R.; Sachkov, V.; Kurzina, I. The Structural and Phase State of the TiAl System Alloyed with Rare-Earth Metals of the Controlled Composition Synthesized by the "Hydride Technology". *Metals* **2020**, *10*, 859. [CrossRef]

Publisher's Note: MDPI stays neutral with regard to jurisdictional claims in published maps and institutional affiliations.

 *metals* 

Article

The Structural and Phase State of the TiAl System Alloyed with Rare-Earth Metals of the Controlled Composition Synthesized by the "Hydride Technology"

Akbayan Belgibayeva [1,2,*], Yuri Abzaev [1,3], Natalia Karakchieva [1], Rakhmetulla Erkasov [2], Victor Sachkov [1] and Irina Kurzina [1]

[1] Chemical Technology Laboratory, National Research Tomsk State University, Lenin 36, 634050 Tomsk, Russia; abzaev2010@yandex.ru (Y.A.); kosovanatalia@yandex.ru (N.K.); itc@spti.tsu.ru (V.S.); kurzina99@mail.ru (I.K.)

[2] Department of Chemistry, Eurasian National University, Kazhymukan 13, Nur-Sultan 010008, Kazakhstan; erkass@mail.ru

[3] Material Research Centre for collective use, Tomsk State University of Architecture and Building, Solyanaya 2, 634003 Tomsk, Russia

[*] Correspondence: bayan_05.06@mail.ru; Tel.: +8-775-255-8611

Received: 30 May 2020; Accepted: 22 June 2020; Published: 29 June 2020

Abstract: The structural state and the quantitative phase analysis of the TiAl system, alloyed with rare-earth metals synthesized using hydride technology, were studied in this work. Using the Rietveld method, the content of the major phases in the initial system Ti(50 at.%)–Al(50 at.%), as well as Ti(49 at.%)–Al(49 at.%), with alloying additions Ta, Y and Dy having a high accuracy was determined. The methods of scanning electron microscopy, transmission electron microscope and X-ray spectral microanalysis of the local areas of the structure for studying the distribution of alloying elements were used. The energies of lattices of separate phases were also determined after the full-profile specification. All the lattices of the identified structures (about 30) turned out to be stable. It was established that in the Ti(49 at.%)–Al(49 at.%) systems under study with alloying additions of metals Ta, Y and Dy, there were intermetallides composed of $AlTi_3$, TiAl in the hexagonal, tetragonal and triclinic units. It is known that after microalloying alloys by Y and Dy metals, the mass fraction of TiAl phases increases significantly (>70%).

Keywords: intermetallides; phase composition; microstructure; hydrides; TiAl system

1. Introduction

Intermetallic alloys based on γ-TiAl are a good example of the way basic and applied research along with industrial development can lead to obtaining a new innovative class of advanced structural materials [1–4]. Nowadays, intermetallic alloys based on the γ-TiAl phase are promising materials for application in aeronautical engineering, owing to their attractive properties: high specific strength, stiffness, creep resistance at temperatures of $T = 600$–800 °C, oxidation resistance and burn resistance at temperatures up to $T = 900$ °C. In the temperature range of $T = 20$–800 °C, the specific modulus of elasticity of these alloys is higher than that of the nickel by 30–50% [2]. It is supposed that in the gas-turbine engine, light γ-TiAl alloys ($\rho \approx 4$ g/cm^3) will partially replace heat-resisting heavy nickel alloys ($\rho = 8$–8.5 g/cm^3), which will allow significant increases in its specific power characteristics during a simultaneous decrease in the fuel consumption, carbon dioxide emissions and nosiness [4].

To achieve a certain combination of properties, it is necessary to optimize the chemical composition and the microstructure [4–8]. Therefore, in the past two decades, the increased focus of researchers of γ-TiAl alloys has been on achieving an optimal combination of mechanical properties by varying the elemental composition and the microstructure with various sizes of columns/grains and thickness of plates. For that, a detailed work on the optimization of the composition and conditions of thermomechanical/thermal treatment of the alloys is being performed.

In modern materials science, a new method of synthesis of binary and multicomponent alloys is of interest, which is called a "hydride" technology (HT) [9]. Modern technologies of production of alloys (mechanical alloying, arc smelting, powder metallurgy, etc.,) are associated with notable labor intensity and hardware difficulties (application of a deep vacuum and creation of inert environment at high temperatures, duration and multiplicity of processes, etc.). The powder metallurgy technique is characterized by a special duration because the result of the interaction of metals in the initial mixtures is mainly determined by solid diffusion rates. The specific difficulties of obtaining quality alloys are also associated with the presence of a dense oxide film on the surfaces of particles of refractory metals, which prevents the passage of mutual diffusion. The HT method allows the avoidance of the majority of them (in particular, it excludes melting). HT is a high-tech method based on the combination of self-propagating high-temperature synthesis SHS of transition metal hydrides and heat treatment of a mixture of hydrides, resulting in alloy formation [10–12]. In the HT method, to obtain high-strength alloys based on transition metals, the powders of hydrides of refractory metals and alloys are used as initial materials. The essence of HT consists of the successive use of processes of hydrides synthesis, their combined compacting and dehydration. The advantage of HT is that the alloy is formed at a relatively low temperature (from 900 to 1150 °C) during rapid exposition (from 1 to 2 h). It is important that metals with different values of melting temperature and density should alloy without melting [10–12]. Other advantages of the method are relative cheapness, the use of refractory materials, as well as obtaining materials with high purity. It is also known that the formation of metal alloys can be carried out easier from hydrides than from metals themselves, since chemical bonds in hydrides are less strong than bonds in metal structures [11,12].

It is well known that the compaction of the alloy structure is one of the most effective ways of improvement of strength and plasticity of the materials. Alloying was demonstrated as a viable approach to enhance the mechanical properties of TiAl alloys at room temperature through improving the materials microstructure [13,14].

To improve the mechanical properties of γ-TiAl alloys, their alloying with rare-earth elements is of interest. It is known that [15–17] the introduction of rare-earth elements (La, Er and Dy) into titanium and intermetallic alloys based on γ-TiAl can lead to the improvement in machinability in the as-cast state, owing to refinement of the structure, to increased heat-resistance and refractoriness, and in the γ-TiAl alloys, in some cases, to the enhancement of the technological plasticity at elevated temperatures.

Previous studies supposed that the addition of yttrium [18] and gadolinium [19–22] could significantly reduce the grain sizes and lamellar spaces of TiAl alloys, enhance strength and plasticity at room temperature, as well as creep resistance at high temperatures. However, the mechanisms of microstructure refinement in the rare-earth metals-modified (REM-modified) TiAl alloys were studied insufficiently because of the complex succession of solidification and solid-phase transformations of TiAl during thermal treatment and cooling [23]. Systematic data on the influence of additions of rare-earth elements on the structure and mechanical properties of γ-TiAl alloys in the literature are practically not presented.

Thus, the development of the methods for producing high-strength alloys of the TiAl system, the study of the dependence of the structural phase state on the synthesis parameters, and the study of the effect of various modifying additives on the physical and mechanical properties is an actual task.

The purpose of the present work is the development of the basics of hydride technology for the production of TiAl alloys and the study of the influence of alloying additions Ta, Y and Dy on the microstructure and the phase composition of the titanium–aluminum system obtained by the HT method. The Ti(50 at.%)–Al(50 at.%) system has been accepted as a basis of the composite material with REM additions of no more than 2 at.%.

2. Materials and Methods

2.1. Obtaining Alloys

Samples with the following selected atomic compositions were prepared TA (TiAl); TAT (TiAlTa); TAY (TiAlY); TAD (TiAlDy). Sample TA was prepared as follows: a weighed amount of the titanium was placed in a quartz boat and heated in a furnace (Nabertherm RS 120/750/13, Lilienthal, Germany) in a stream of the hydrogen. The heating rate of the furnace was 10 °C/min to 450 °C, with a hydrogen volume flow of 500 mL/min.

This sample was cured for 3 h at this temperature; after that, it was cooled down to room temperature. The obtained metal hydrides were mixed with a nanodispersed aluminum powder (the average size of particles was 115 ± 10 nm, specific surface area—19.4 ± 3 m^2/g, loading of active aluminum—$80.8 \pm 0.6\%$). Then, the mixture was pressed into a round plate and was formed (collapsible compression mold, Lab Tools, diameter of 13 mm, thickness of 2 mm), pressing load was under a pressure of 5.3 tons/sm^2, bulk density was 3 g/sm^3 (Lab Tools PAH-20, 2019, St. Petersburg, Russia)). The equation for obtaining the TA sample can be written as:

$$Ti + H_2 = TiH_2 + 27.3 \text{ kcal/mole} \tag{1}$$

The TAT sample was prepared as follows: a weighed amount of the tantalum was placed in a quartz boat and heated in a furnace (Nabertherm RS 120/750/13, Lilienthal, Germany) in a stream of the hydrogen. The heating rate of the furnace was 10 °C/min to 550 °C, with a hydrogen volume flow of 500 mL/min. This sample was cured for 3 h at this temperature; after that, it was cooled down to room temperature. The equation for the production of titanium hydride can be written as:

$$2Ta + H_2 = 2TaH + 56 \text{ kcal/mole} \tag{2}$$

Titanium hydride was obtained as in the case of sample TA. Then, the obtained titanium hydride and tantalum hydride were mixed with a nanodispersed aluminum powder (according to the manufacturer, the average size of the particles was 115 ± 10 nm, specific surface area—19.4 ± 3 m^2/g, loading of active aluminum—$80.8 \pm 0.6\%$). Then, the mixture is pressed similar to TA.

Samples of TAY and TAD were obtained similarly to the method of obtaining TAT. The temperature for yttrium hydride and dysprosium hydride was 420 °C. The equation for obtaining a TAY sample can be written as:

$$Y + H_2 = YH_2 + 56 \text{ kcal/mole} \tag{3}$$

$$2Y + 3H_2 = 2YH_3 + 21.5 \text{ kcal/mole} \tag{4}$$

The equation for the preparation of dysprosium hydride can be written as:

$$2Dy + 3H_2 = 2DyH_3 \tag{5}$$

The obtained sample blanks were obtained in a vacuum unit and heated to a temperature of 1150 °C, with a heating rate of 2 °C/min. The vacuum value was 10^{-4} atm. They were maintained under these conditions for 3 h and cooled at a rate of 5 °C/min.

The equations for obtaining samples TA, TAT, TAY and TAD can be written as:

$$TiH_2 + Al = TiAl + 2H_2\uparrow \tag{6}$$

$$2TiH_2 + 2Al + 2TaH = 2TiAlTa + 3H_2\uparrow \tag{7}$$

$$TiH_2 + Al + YH_2 = TiAlY + 2H_2\uparrow \tag{8}$$

$$2TiH_2 + 2Al + 2DyH_3 = 2TiAlDy + 5H_2\uparrow \tag{9}$$

2.2. Research Methods

The structural state and the quantitative phase analysis of the system TiAl-alloying metal (TA-REM), synthesized by HT, were studied in this work by the Rietveld method and scanning electron microscopy (SEM) [24,25]. Ta, Y and Dy metals were used as alloying additions. The X-ray diffraction studies of the TA-REM system were undertaken using DRON4-07 (Bourevestnik, St. Petersburg, Russia), which was modified for the digital processing of the signal. Spectra were made using copper radiation (K_a) according to the Bragg–Brentano scheme with an increment of 0.02°, exposure time at a point of 1 sec and in the angle range of 10–86°. The voltage on the X-ray tube was 30 kV, beam current—25 mA. The structural state and the quantitative content of the phases were identified by the Rietveld method by means of reflex [26–29]. As the standard lattices, the crystallographic data of the COD base [26] and the model structures of the TiAl system, predicted by the program code USPEX with the interface shell SIESTA [30], were used. In connection with the promising physicochemical properties of the TiAl systems, they were alloyed with metal additions [24–34]. In this work, the search for standards of the TiAl system was supplemented by the USPEX program with the outer shell SIESTA [31,35,36].

To study the distribution of alloying elements in the structure at the local level, the quantitative X-ray spectral microanalysis of the near-surface layer, accompanied by the analysis of the microstructure and the morphology of the surface by the method of scanning electron microscopy, was carried out using the X-ray fluorescence spectral analyzer S4 Pioneer (Bruker, Karlsruhe, Germany) and the raster electron microscope "LEOEVO 50" (Zeiss, Oberkochen, Germany).

Electron microscope studies of the surface microstructure of TA-REM samples were conducted using the transmission electron microscope "JEM-2100F" (JEOL, Tokyo, Japan) with accelerating voltage of 200 kV using the attachment "JEOL" (JEOL, Tokyo, Japan), intended for energy-dispersive spectral analysis. The phases were identified applying well-known procedures using schemes of microdiffraction patterns calculated by the table values of parameters of crystalline lattices.

3. Results and Discussion

The USPEX-SIESTA program by means of the evolution code can predict stable structures of the known elemental composition, space group, among which the structures with the global minimum of enthalpy of the system are predicted. In the present work, the TiAl structures were used for the quantitative phase analysis. For the discovered structures (about 30) based on complete structural information, the energies of lattices were calculated proceeding from the first principles. All the lattices turned out to be stable and they were used at the next stage for qualitative and quantitative phase analysis by the Rietveld method. In Tables 1–5, there are standards in the initial state, obtained from the COD base (AlTi$_3$, AlTi), as well as the structure predicted by means of the USPEX code (TiAl-Struct2, AlTi3-2768). For the alloy with the conventional designation TiAl-Struct2, AlTi3-2768, the energy was calculated in the framework of the electron density functional by the gradient pseudo potential of electron density (GGA). The details of the code are given in the work of reference [26]. The total energy of the lattices was determined at 0 K. Wave functions of valence electrons of TiAl-Struct2 atoms were analyzed on the basis of flat waves with a cutoff radius of kinetic energy of 330 eV. In this case, the convergence of the total energy was $\sim 0.5 \times 10^{-6}$ eV/atom. It was established that the lattice energy after geometric optimization turned out to be equal to $E_{\text{TiAl-Struct11}} = -4978.555$ eV, and in the initial state -4976.099 eV. The final enthalpy of the TiAl-Struct2 structure in both cases is evidence of its stability. TiAl lattices (AlTi$_3$-2768) with Y, Ta and Dy additives (Tables 2–4) also turn out to be stable, and the lattice energies are substantially negative. In this work, based on the results of the qualitative phase analysis of the TA-REM system, it was suggested that the REM nanoadditions were embedded into the interstitial site [0.5 0.5 0.5] of the lattice of the AlTi$_3$ alloy. Quantum chemical calculations of the AlTi$_3$ energy in the initial state, as well as with the embedded nanoadditions in the mentioned site, were made. It was established that: $E_{\text{AlTi3}} = -19{,}318.285$ eV; $E_{\text{AlTi3-Y}} = -19{,}712.792$ eV; $E_{\text{AlTi3-Ta}} = -19{,}613.620$ eV; $E_{\text{AlTi3-Dy}} = -31{,}227.561$ eV.

Table 1. Structural parameters of the lattices and the share of phases, and the convergence criterion of the TA alloy.

Phase	State	a, Å	b, Å	c, Å	Alpha	Beta	Gamma	V, A^3	Space Group	Share, %	E, eV	Rwp, %
AlTi$_3$-2768	Init.	5.764	5.764	4.664	90.00	90.00	120.00	132.56	P6/mmm, Hexagonal	22.33	−19,317.484	
	Spec.	5.763	5.763	4.645	90.00	90.00	120.00	131.996				
TiAl-Struct2 -GeomOpt	Init.	6.339	4.150	4.234	113.36	93.36	92.52	132.56	P1, Triclinic	49.80	−4810.6263	7.193
	Spec.	6.129	4.237	4.017	113.62	88.24	92.42	133.711				
TiAl-Struct2	Init.	6.339	4.150	4.234	113.36	93.36	92.52	101.791	P1, Triclinic	24.08	−4975.776	
	Spec.	6.194	4.119	4.215	113.07	92.88	91.97	98.64				

Table 2. Structural parameters of the lattices and the share of phases, and the convergence criterion of the TAY alloy.

Phase	State	a, Å	b, Å	c, Å	Alpha	Beta	Gamma	V, A^3	Space Group	Share %	E, eV	Rwp, %
AlTi$_3$-2768	Init.	5.764	5.764	4.664	90.00	90.00	120.00	132.56	P6/mmm, Hexagonal	26.99	−19,603.151	
	Spec.	5.736	5.736	4.626	90.00	90.00	120.00	131.828				
AlTi-2770	Init.	2.837	2.837	4.059	90.00	90.00	90.00	32.677	P4/mmm, Tetragonal	41.04	−1660.340	6.317
	Spec.	2.824	2.824	4.070	90.00	90.00	90.00	32.466				
TiAl-Struct2	Init.	6.339	4.150	4.234	113.36	93.36	92.52	101.791	P1, Triclinic	27.63	−4954.073	
	Spec.	6.245	4.128	4.319	114.67	91.27	93.84	100.802				

Table 3. Structural parameters of the lattices and the share of phases, and the convergence criterion of the TAT alloy.

Phase	State	a, Å	b, Å	c, Å	Alpha	Beta	Gamma	V, A^3	Space Group	Share, %	E, eV	Rwp, %
AlTi$_3$-2768	Init.	5.764	5.764	4.664	90.00	90.00	120.00	132.56	P6/mmm, Hexagonal	16.66	−19,705.872	
	Spec.	5.791	5.791	4.736	90.00	90.00	120.00	137.565				
AlTi-2770	Init.	2.837	2.837	4.059	90.00	90.00	90.00	32.677	P4/mmm, Tetragonal	50.61	−1660.341	6.701
	Spec.	2.831	2.831	4.067	90.00	90.00	90.00	32.609				
TiAl-Struct2	Init.	6.339	4.150	4.234	113.36	93.36	92.52	101.791	P1, Triclinic	25.79	−4978.726	
	Spec.	6.568	4.133	4.093	160.27	95.85	92.64	33.821				

Table 4. Structural parameters of the lattices and the share of phases, and the convergence criterion of the TAD alloy.

Phase	State	a, Å	b, Å	c, Å	Alpha	Beta	Gamma	V, Å^3	Space Group	Share, %	E, eV	Rwp, %
AlTi$_3$-2768	Init.	5.764	5.764	4.664	90.00	90.00	120.00	132.56	P6/mmm, Hexagonal	11.20	−31,228.526	
	Spec.	5.771	5.771	4.657	90.00	90.00	120.00	134.34				6.504
AlTi-2770	Init.	2.837	2.837	4.059	90.00	90.00	90.00	32.677	P4/mmm, Tetragonal	65.04	−1660.341	
	Spec.	2.826	2.826	4.074	90.00	90.00	90.00	32.537				
TiAl-Struct2 -GeomOpt	Init.	6.339	4.145	4.234	113.36	93.36	92.52	101.79	P1, Triclinic	16.88	−4978.606	
	Spec.	6.294	4.139	4.260	115.23	92.72	91.94	100.09				

Table 5. Relative coordinates of the atoms in the TiAl-Struct2 lattice.

Symbol of the Atom	x	y	z	Displacement Parameters, (U_iso)	Occupancy
Ti1	0.310	−0.359	−0.678	0.0127	1.0
Ti2	−0.500	0.334	0.667	0.0127	1.0
Ti3	−0.311	0.027	0.012	0.0127	1.0
Ti4	0.227	−0.017	−0.036	0.0127	1.0
Al5	−0.000	0.334	−0.333	0.0127	1.0
Al6	−0.227	−0.316	0.370	0.0127	1.0

The calculations are evidence of the fact that the introduction of the alloying metals into the indicated interstitial site is possible and it leads to the significant stabilizing effect of the lattices in each TA-REM system without exception. In fact, the binding energy of ($|E_{AlTi3}|$) atoms in the lattices accompanied by the addition of the alloying metal increased significantly. It is interesting to note that the increase in the binding energy is accompanied by significant polarization of the Millikan charges [37,38]. The Millikan charges in the metal additions are equal: [(−5.96)Y], [(−4.42)Ta], [(+0.73)Dy], but on the atoms of the main elements—[(−0.16)Al, (+1.540)Ti], [(−0.35)Al, (+1.22)Ti], [(−0.13)Al, (−0.12)Ti], [(−0.17)Al, (−0.13)Ti] in ternary compounds TAT, TAY and TAD, respectively. The analysis of distances between atoms of the main elements and additions showed that these distances were significantly less than the sum of the covalent radii of free elements Ti and Al, and Ta, Y and Dy, which are equal, respectively, to 1.6, 1.21, 1.7, 1.90, 1.92Å [39]. For instance, in the TAT system, the lengths of bonds TiTa and AlTa are equal to 2.030 and 2.031 Å, respectively. Polarization of the charges is indicative of the growth in the share of the covalent bond as a result of introduction of the alloying metals.

The results of the quantitative content of phases in the studied systems are given in Tables 1–4. The tables show the phases, the structural state of the phases before (Init) and after (Spec) the Rietveld refinement of the experimental diffractograms, the corresponding lattice parameters, the lattice volume (V), the space group, the fraction of the individual phases (Share) and the lattice energy after refinement by the Rietveld method (E). The tables also show the fraction of the calculated integrated intensity (Rwp) in the experimental diffractogram. The analysis of the contributions into the integral intensity of separate phases (Table 1, Figure 1) showed that to a high degree of reliability ($R_{wp} < 7.2\%$), the major phases were intermetallides AlTi$_3$, AlTi$_3$-REM, AlTi and TA-Struct2 in the initial and optimized states. The experimental diffraction patterns of the TA, TAT, TAY and TAD systems can be approximated well by the calculated integral intensity; the difference between them is an insignificant value (Figure 1). However, as Tables 1–4 show, the contributions of separate phases differ in different systems. Using the Rietveld method, it is established that in the TA system, the contribution of the triclinic TiAl system (over 73%) predominates. Both lattices differ by parameters, atoms coordinates, as well as the volume (Table 1). The number of the main systems also includes the hexagonal lattice of AlTi$_3$ (22.33%). The introduction of an insignificant amount of Ta, Y and Dy leads to the redistribution of the contributions of the hexagonal and triclinic systems. In the TA-REM systems, the contribution of the geometrically optimized lattice (TA-Struct2-GeomOpt) disappears, and the phase with the AlTi tetragonal lattice appears (Tables 2–4). The introduction of Ta into the AlTi$_3$ lattice significantly increases its binding energy and stimulates the contribution to the integral intensity up to 26.99%. On the contrary, in the TAY and TAD systems, additions Y and Dy lead to the reduction in the share of AlTi$_3$ to 16.66, 4.78 and 11.20%, respectively. The share of phases and the energy of the lattices allow evaluation of the effective energy of the TA, TAT, TAY and TAD systems according to the formula $\alpha E_1 + \beta E_2 + \gamma E_3$, where α, β, γ—the share of phases; E_1, E_2, E_3—the energy of the lattices of separate phases. It was established that the energies of the mentioned systems were equal to −7907.453, −5407.31, −7341.104, −21,188.219 and −15,920.023 eV, respectively.

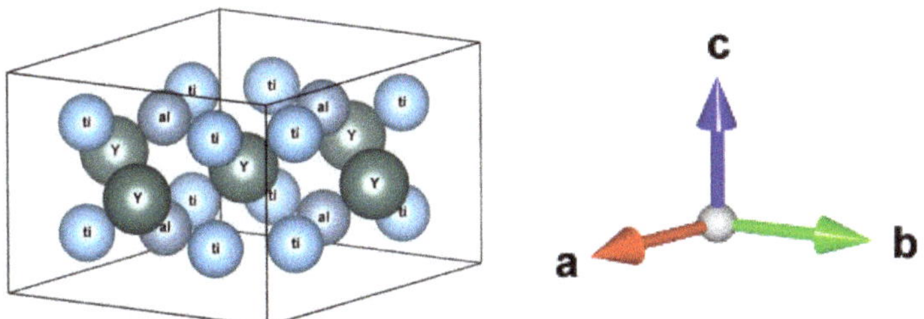

Figure 1. 3D lattice of the AlTi$_3$ alloy with the Y atom embedded in the site [0.5 0.5 0.5].

The relative coordinates of the atoms and isotropic parameters of displacements are given in Tables 5 and 6). The lattice parameters and space group are shown in Tables 1–4.

Table 6. Relative atomic coordinates in the AlTi lattice (AlTi$_3$-2768) with an embedded atom Dy, Ta and Y.

Element	x	y	z	U_iso	Occupancy
Ti	0.833	0.167	0.25	0.0127	1
Al	0.333	0.667	0.25	0.0127	1
Dy; Ta; Y	0.5	0.5	0.5	0.0127	1

Figure 2 shows the results of the observed and calculated intensities of the initial alloy and alloys obtained by doping with rare-earth metals. X-ray phase analysis of the samples obtained by doping with rare-earth metals showed that they have a complex multiphase structure: the initial components (Ti, Al) and the new phases—intermetallic phases AlTi$_3$, AlTi$_3$–REM, AlTi, TiAl-Struct2—in the initial and optimized states were identified.

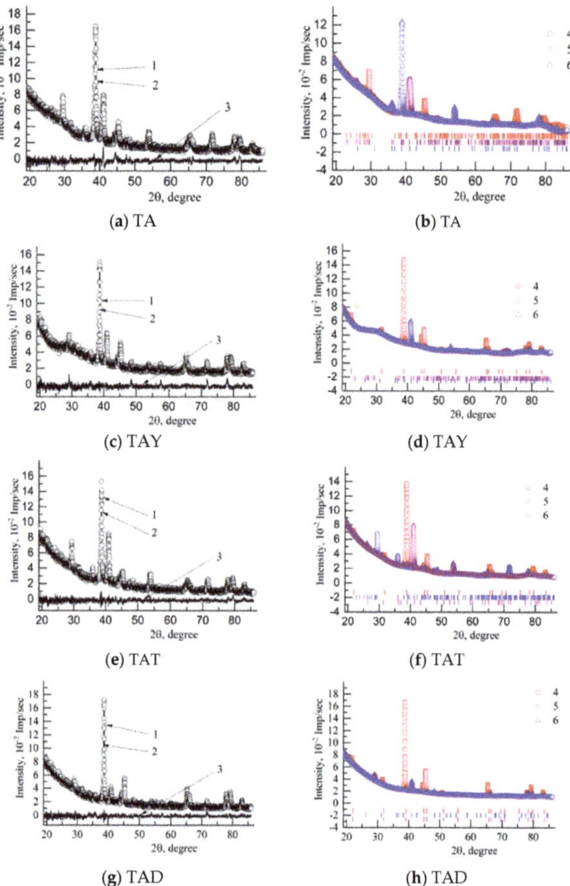

Figure 2. Diffraction patterns of alloys TA (**a,b**); TAY (**c,d**); TAT (**e,f**); TAD (**g,h**): 1—experiment, 2—summary model intensity, 3—difference between intensities, 4–6—contributions to the integral intensity of separate phases.

The study of the microstructure showed that the base of the material was the alloy of titanium and aluminum with a small quantity of alloying metals (no more than 2 at.%). The ratio between the atomic concentrations of Al and Ti is close to the initial 1:1 (Figure 3). The spectra of characteristic X-ray radiation are also given in Figure 3.

As a result of the quantitative analysis, it was established that the matrix of the TAY alloy on average had a composition in mass percent: 29.60% of Al and 68.35% of Ti, which corresponds to the intermetallide phase α_2-Ti$_3$Al according to the stoichiometric ratio (Figure 3a). The elemental analysis of the TAT and TAD alloys (Figure 3b,c) also revealed the presence of titanium aluminide α_2-Ti$_3$Al. This agrees with the results obtained in the works [40,41]. It was established that the microstructure of the rest of the alloys was also characterized by the formation of the α_2-Ti$_3$Al intermetallide.

The choice of these metals as alloying microadditions in the alloys under study is conditioned by a number of positive effects. In particular, metals as an effective modifier of the cast structure [15–22] possess an increased affinity with oxygen, which in turn leads to a substantial decrease in the amount of oxygen in the alloy and, as a consequence, to an increase in low-temperature plasticity because of the reduction in the number of barriers in the form of oxygen atoms, which decelerate the dislocations motion during deformation.

The X-ray phase analysis results of the initial intermetallic alloy TA showed that the basic thermodynamically stable phases were intermetallide compounds Ti$_3$Al, TiAl, TiAl$_2$ and a solid aluminum solution in titanium α-Ti (Table 7).

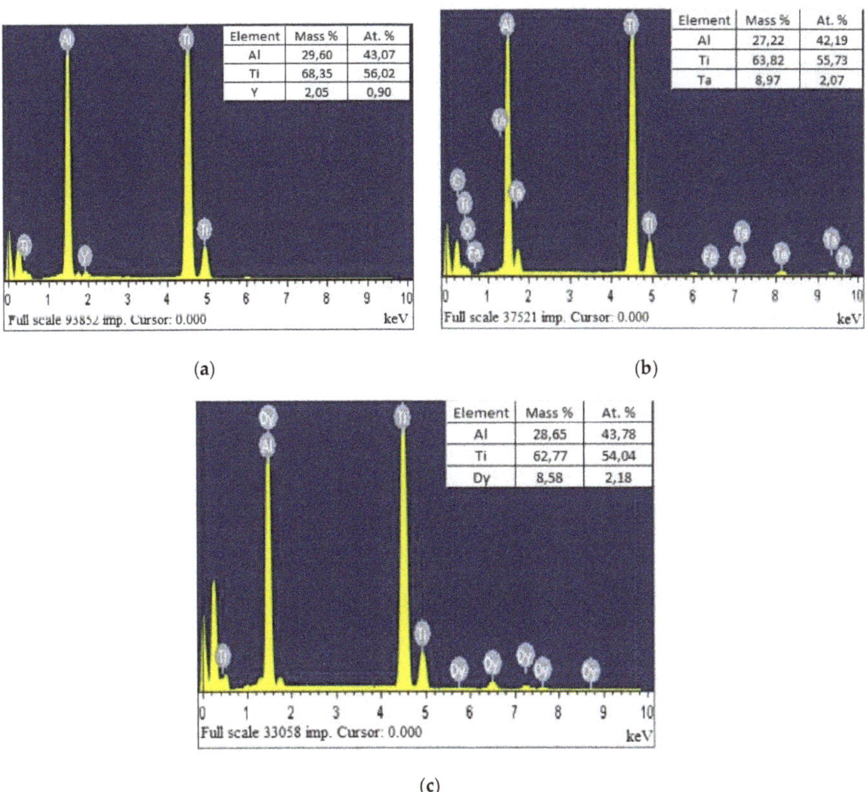

(a)　(b)

(c)

Figure 3. EDX-spectra and elemental composition of the TAY (**a**), TAT (**b**) and TAD (**c**) alloys.

Table 7. Crystallographic data of phases in the system TA.

Composition	Space Group	Syngony	CSR Volume, Å^3	Weight Fraction, %	Lattice Parameters, Å		
					a	b	c
TiAl	*P4/mmm*	tetragonal	33 ± 5	38.4	2.8234	2.8234	4.0768
Ti$_3$Al	*P63/mmc*	tetragonal	134 ± 5	25.2	5.7671	5.7671	4.64646
α-Ti	*C6/mmc*	hexagonal	31 ± 5	18.0	2.9253	2.9253	4.6184
TiAl$_2$	*C/mmm*	rhombic	195 ± 5	11.8	12.0187	4.0232	4.0253
Ti$_5$Al$_{11}$	*P/mmm*	rhombic	261 ± 5	8.3	3.9522	3.9522	17.3119
Ti$_3$Al$_5$	*P/mmm*	rhombic	61 ± 5	2.9	3.8675	3.8212	4.1445
Ti$_2$Al$_5$	*P4/mmm*	tetragonal	440 ± 5	2.5	3.8164	3.8164	30.1785

A small fraction of the sample consists of the following intermediate phases having practically constant chemical compositions: Ti$_5$Al$_{11}$, Ti$_3$Al$_5$, TiAl$_2$, Ti$_2$Al$_5$. The X-ray phase analysis (XPA) results are confirmed by TEM studies. The TEM studies showed that, before the introduction of rare-earth metals, in the initial state, the TA alloy contained phases Ti$_3$Al, TiAl, TiAl$_2$, α-Ti, Ti$_5$Al$_{11}$ and Ti$_3$Al$_5$ (Figure 4e). Consequently, the obtained TA alloy has a complex multiphase structure, as it contains a number of phases with different crystal lattices. This fact confirms a possibility of obtaining intermetallic alloys by hydride technology.

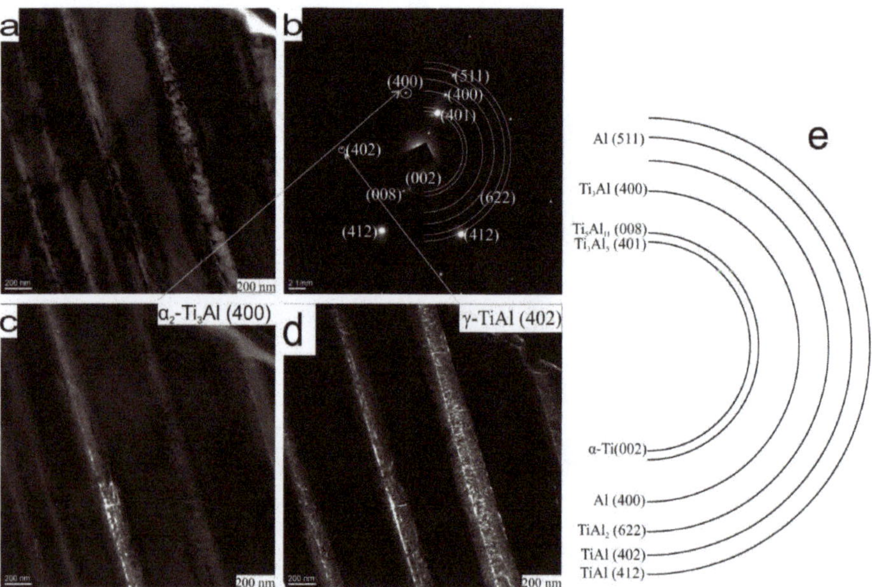

Figure 4. Electron microscope images of the matrix of the initial alloy TA: (**a**) bright-field image; (**b**) microdiffraction pattern; (**c**), (**d**) dark-field images in reflexes, marked by arrows; (**e**) microdiffraction pattern identification.

First of all, the obtained studies of the microstructure confirmed the formation of Ti$_3$Al and TiAl phases. The Ti$_3$Al phase is an ordered phase with a D019 superstructure having an HCP crystal lattice; its space group is *P63/mmc*. Ti$_3$Al is formed as parallel lamellar precipitates (Figure 4a). The intermetallide Ti$_3$Al and TiAl phases under formation are microdimensional, which is evident owing to the dark-field images (Figure 4c,d). The presence of Ti$_3$Al and TiAl is associated with thermodynamics of the phases' formations. The formation of these phases is characterized by a minimum of Gibbs energy [42]. The existence of these intermetallide phases must result in significant alloy strengthening [6,43,44].

A layer-by-layer location of intermetallide phases—that is, alternation by the composition—is possible. Thus, using hydride technology allows for obtaining of complex sandwich structures. Owing to alternation of intermetallide phases, it becomes possible to obtain super strong materials, where each layer strengthens the previous one.

To predict the composition of the formed phases when microalloying the TiAl alloys with rare-earth metals, ternary diagrams of the corresponding ternary systems were considered. The analysis of the TiAlY ternary diagram of the partial isothermal cross-section at 1000 °C shows that the content of the Ti:Al:Y = 49:49:2 components corresponds to the highlighted area of formation of γ-TiAl + YAl$_2$ phases (Figure 5). The ternary diagram of the TiAlTa system at 1100 °C implies that at a ratio of Ti:Al:Ta = 49:49:2 components the TiAl phase is formed (Figure 5c). According to the ternary diagram, TiAlDy at a ratio of components Ti:Al:Dy = 49:49:2, the formation of the area, where DyAl$_2$ + TiAl phases are present, can be expected (Figure 5b).

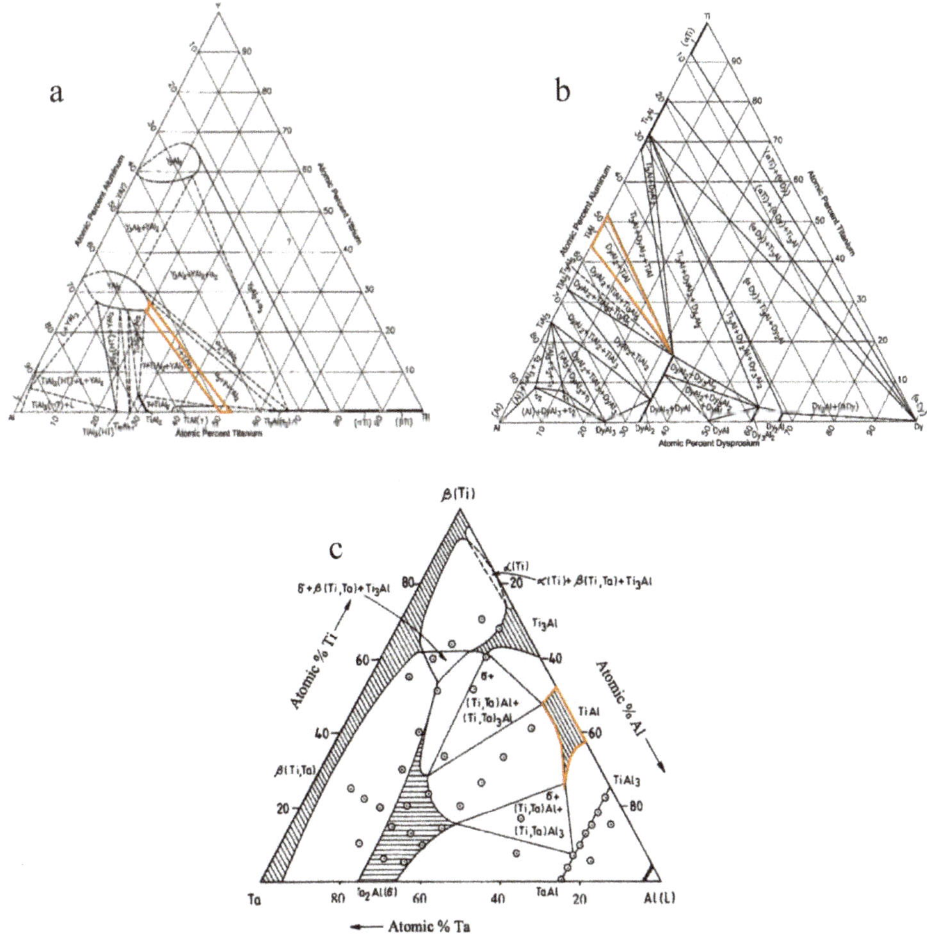

Figure 5. Isothermal cross-sections of the TiAlY system at 1000 °C [45] (**a**), of the TiAlDy system at 500 °C [46] (**b**), of the TiAlTa system at 1100 °C [47] (**c**).

The X-ray phase analysis of the samples, obtained during alloying with yttrium, is given in Table 8.

Table 8. Crystallographic data of phases in the TAY system.

Composition	Space Group	Syngony	CSR Volume, Å^3	Weight Fraction, %	Lattice Parameters, Å		
					a	b	c
Ti$_3$Al$_5$	*P/mmm*	rhombic	65 ± 5	70.3	4.0040	4.0049	4.0710
Ti$_3$Al	P63/mmc	tetragonal	134 ± 5	18.3	5.7661	5.7661	4.6371
Al	Fm-3m	cubic	66 ± 5	8.6	4.0311	4.0311	4.0311
α-Ti	*P63/mmc*	hexagonal	31 ± 5	1.4	2.9186	2.9186	4.6006
TiAl	*P4/mmm*	tetragonal	40 ± 5	1.2	2.7453	2.7453	5.3402
Y	P63/mmc	hexagonal	67± 5	0.3	3.6689	3.6689	5.7302

According to the XPA data, the TAY alloy contains the basic Ti$_3$Al$_5$ phase. The Ti$_3$Al$_5$ phase has a rhombic crystal lattice and the space group P/mmm. Along with the Ti$_3$Al$_5$ grains, the structure of the alloy contains a small number of grains of the Ti$_3$Al phase, having an HCP crystal lattice and the space group P63/mmc. A small fraction belongs to phases TiAl, Al, α-Ti and Y.

Figure 6 presents electron microscope images of the TAY alloy. On the microdiffraction pattern (Figure 6b), there are reflexes belonging to planes (110) and (201) TiAl; (101), (200), (302), (411) Ti$_3$Al; (400) Ti$_3$Al$_5$; (311) YAl$_2$; (100) α−Ti; (200) Al. Thus, the XPA results are confirmed by TEM.

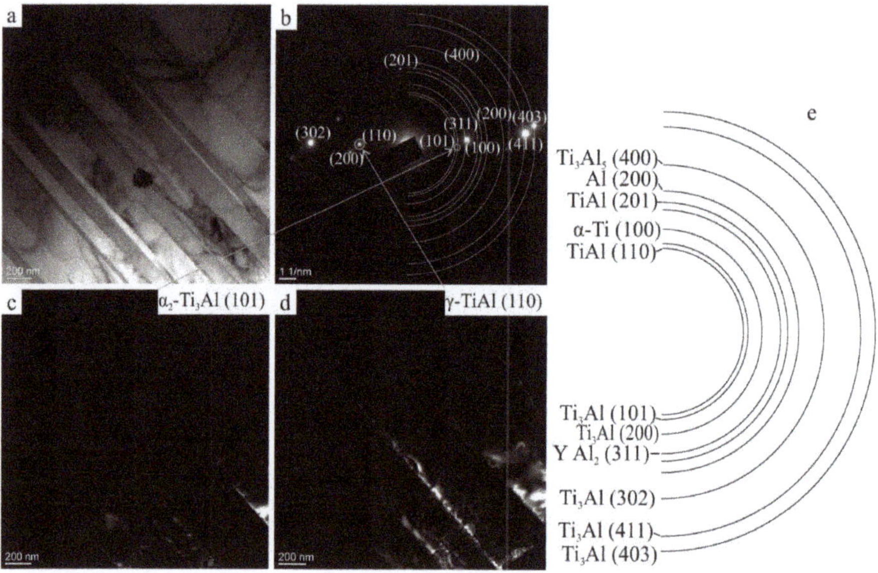

Figure 6. Electron microscope images of the TAY alloy: (**a**) bright-field image; (**b**) microdiffraction pattern; (**c**,**d**) dark-field images in reflexes, marked by arrows; (**e**) microdiffraction pattern identification.

Figure 7 presents TEM images of the depositions of yttrium particles of different shapes inside the grain of titanium aluminide. According to the TEM results, yttrium and Al form the YAl$_2$ intermetallide in intermetallide phases and are located on dislocations or on grain boundaries.

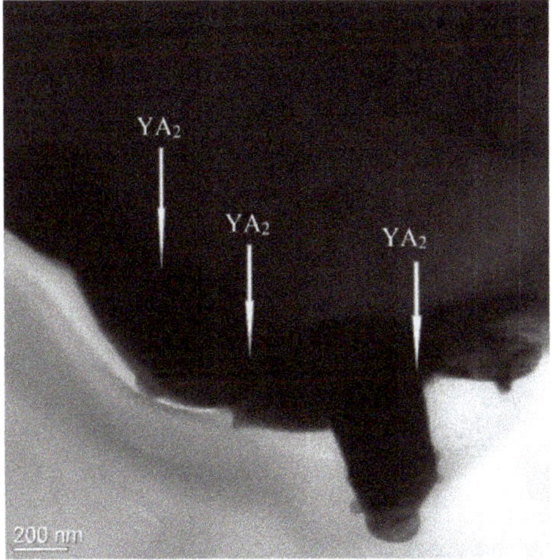

Figure 7. TAY alloy microstructure.

The X-ray phase analysis of the samples obtained when alloying with dysprosium (TAD) and tantalum (TAT) showed that these alloys had a complex multiphase structure: initial components Al, α-Ti and new phases—intermetallide phases TiAl, Ti$_3$Al, Ti$_5$Al$_{11}$, Ti3Al5—were identified (Table 9). The basic phase of the TAD alloy is the TiAl phase, which has an ordered tetragonal-distorted face-centered structure and the P4/mmm space group.

Table 9. Crystallographic data of phases in the TAD and TAT systems.

Alloy	Composition	Space Group	Syngony	Weight Fraction, %	Alloy	Composition	Space Group	Syngony	Weight Fraction, %
TAD	TiAl	*P4/mmm*	tetragonal	74.3	TAT	TiAl	*P4/mmm*	tetragonal	30.1
	Ti$_3$Al$_5$	*P/mmm*	rhombic	10.9		Ti$_3$Al	P63/mmc	tetragonal	22.9
	TiAl$_2$	*C/mmm*	rhombic	5.0		Ti$_3$Al$_5$	*P/mmm*	rhombic	26.4
	Al	Fm-3m	cubic	3.9		Ti$_5$Al$_{11}$	*P/mmm*	rhombic	3.9
	Ti$_5$Al$_{11}$	*I4/mmm*	rhombic	1.4		α-Ti	*P63/mmc*	hexagonal	3.9
	α-Ti	*P63/mmc*	hexagonal	1.4		TaAl$_3$	F-43m	cubic	2.8
	Dy	*P63/mmc*	hexagonal	1,6		Ti$_2$Al$_5$	P4/mmm	tetragonal	1.9

The results of the X-ray phase analysis of the TAT alloy showed that in the alloy, the main share belonged to intermetallide phases: 30.1% of titanium aluminide TiAl of tetragonal syngony with the space group P4/mmm; 22.9% of the Ti$_3$Al phase, having a space-centered lattice of tetragonal syngony with the space group P63/mmc; 26.4% of the Ti$_3$Al$_5$ phase with the space group P/mmm (Table 9). The smallest share belongs to 1.9% of Ti$_2$Al$_5$ with the space group P4/mmm and 3.9% of Ti$_5$Al$_{11}$ with the space group P/mmm. In addition, the intermetallide of tantalum and aluminum (Ta161$_{,8}$Al$_{282,2}$) is present in the sample.

The results of the X-ray phase analysis of the TiAlDy alloy are confirmed by the TEM results (Figure 8). On the microdiffraction pattern there are reflexes belonging to planes (002), (004), (402), (413) TiAl; (600), (203) and (423) Ti$_3$Al; (312) TiAl$_2$; (008) Ti$_5$Al$_{11}$; (511) Al; (300) α-Ti; (551) DyAl$_2$.

Figure 9 shows the electron microscope images of the TAT alloy. According to the microdiffraction analysis in the layers of the TAT alloy, there are the following intermetallide phases with reflexes belonging to planes (312), (412) TiA; (110), (402) and (210) Ti$_3$Al; (311), (440) Ti$_3$Al$_5$; (314) Ti$_5$Al$_{11}$; (103) α-Ti; (008) and (215) TaAl$_3$.

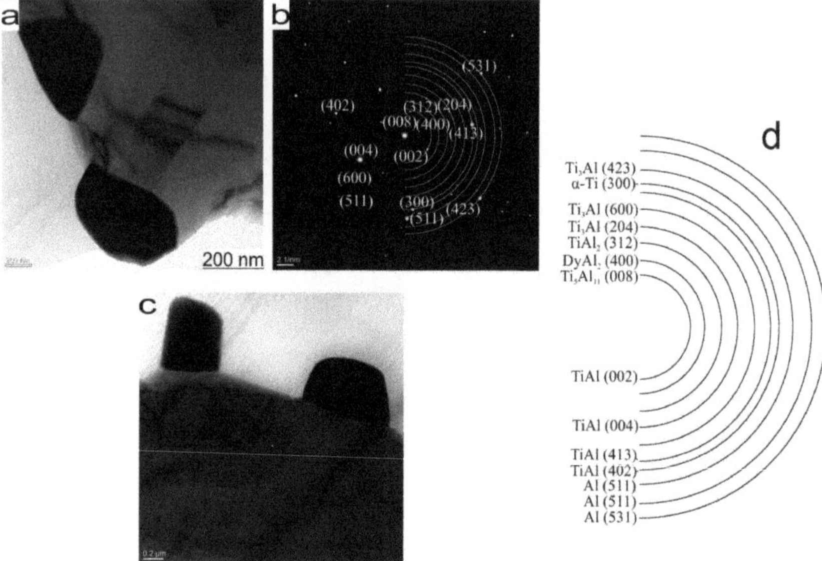

Figure 8. Electron microscope images of the TAD alloy: (**a**) bright-field image; (**b**) microdiffraction pattern; (**c**) dark-field images in the reflex; (**d**) microdiffraction pattern identification.

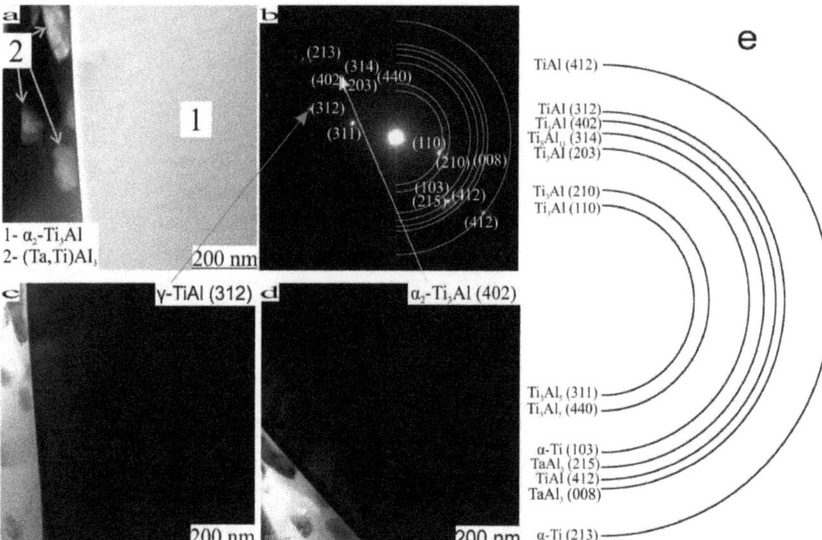

Figure 9. Electron microscope images of the TAT alloy: (**a**) bright-field image; (**b**) microdiffraction pattern; (**c**) dark-field image in reflex; (**d**) microdiffraction pattern identification.

Thus, the TEM studies confirmed the formation of a number of phases of titanium aluminide. The tantalum compounds indicated by the XPA method probably remain in the volume of the sample, but only TaAl$_3$ is present in the film structure.

As Ti and Ta have significant solubility in Al, Ta can form a solid substitutional solution in titanium and contribute to formation of the Widmanstatten microstructure. In addition, it is well known that the TaAl system also dissolves Ti (Ti aluminide) by as much as 60 at.%. Aluminides TiAl and Ti$_3$Al dissolve Ta (Ta aluminide) up to 10 and 15 at.%, respectively [47–49]. The TEM research results comply with the data of the X-ray phase analysis, which imply that an insignificant amount of Ta dissolves in the crystals of TiAl and Ti$_3$Al and are present in the (Ti,Ta)Al$_3$ phase.

Thus, the XPA and TEM studies have shown that the formed solid solution based on titanium and the alloys under study have particles of the new phase in their structure. It is also well known that after microalloying of the alloys with metals Y and Dy, the mass fraction of the TiAl phases increases significantly (>70%). A distribution of tantalum in the titanium–aluminum matrix is evidence of the possibility of the formation of the three component system (Ti,Ta)Al$_3$ [47–49]. Alloying metals, Y and Dy, form a compound with aluminum; they are applied as a dispersion phase and strengthen the alloy structure by the introduction of the second phase into the metal matrix in the form of a high-melting compound. If YAl$_2$ is located on the dislocations or on the grain boundaries, DyAl$_2$ is chaotically distributed in the grain volume.

4. Conclusions

The composite materials based on gamma-aluminides of titanium TiAl-REM were obtained by a new method—"hydride technology". The content of the major phases in the initial TA system, as well as TA-REM, containing microadditions Ta, Y and Dy, was determined by the Rietveld method with high reliability.

1. It was established that in the initial TA system and in the TA-REM system, containing microadditions, TiAl predominate in tetragonal and triclinic units. It was established that the used initial standard lattices, as well as the structures after the full-profile specification, were in the stable states.

2. The results of X-ray phase analysis are confirmed by the results of TEM. It is known that after microalloying alloys by Y and Dy metals, the mass fraction of TiAl phases increases significantly (>70%).

3. From the results of X-ray phase analysis and TEM studies, it follows that an insignificant amount of Ta dissolves in the crystals of TiAl and Ti$_3$Al and are present in the (Ti,Ta)Al$_3$ phase. This fact indicates the possibility of the formation of a three component system (Ti, Ta)Al$_3$.

Author Contributions: Y.A. studied structural state and determined the energies of lattices of separate phases and fraction of quantitative phase, and wrote the paper; N.K. conducted the microstructure research of samples and analyzed the data; I.K. conducted the X-ray phase analyses of samples; A.B. wrote the introduction and conducted the synthesis of samples; V.S. and R.E. analyzed data. All authors have read and agreed to the published version of the manuscript.

Funding: This work was carried out with the support of the Program of Competitive Recovery of TSU, NU 8.2.10.2018 L. and project No. 10.3031.2017/4.6.

Acknowledgments: The authors are grateful to Mark Kalashnikov for fruitful cooperation.

References

1.	Clemens, H.; Chladil, H.; Wallgram, W.; Block, B.; Kremmer, S. *Structural Aluminides for Elevated Temperature Applications*; Kim, Y.W., Morris, D., Yang, R., Leyens, C., Eds.; The Minerals, Metals and Materials Society (TMS): Warrendale, MI, USA, 2008; pp. 217–228.

2.	Clemens, H.; Smarsly, W. Light-Weight Intermetallic Titanium Aluminides—Status of Research and Development. *Adv. Mater. Res.* **2011**, *278*, 551–556. [CrossRef]

3.	Appel, F.; Paul, J.D.; Oehring, M. *Science and Technology*, 1st ed.; Wiley-VCH: Weinheim, Germany, 2011.

4.	Appel, F.; Clemens, H.; Fischer, F.D. Modeling concepts for intermetallic titanium aluminides. *Prog. Mater. Sci.* **2016**, *81*, 55–124. [CrossRef]

5.   Kurzina, I.A.; Kozlov, E.V.; Popova, N.A.; Kalinnikov, M.P.; Nikonenko, E.L.; Savkin, K.P.; Oks, E.M.; Sharkeev, Y.P. Modifying the Structural Phase State of Fine Grained Titanium under Conditions of Ion Irradiation. *Bull. Russ. Acad. Sci. Phys.* **2012**, *76*, 1238–1245. [CrossRef]

6.   Kurzina, I.A. Physical Base of the Metallic Gradient Surface Layers of Titanium Alloys Formed under Ion Implantation. *Adv. Mater. Res.* **2014**, *872*, 184–190. [CrossRef]

7.   Zykova, A.P.; Kazantseva, L.A.; Kurzina, I.A.; Dammer, V.K.; Chumaevaskii, A.V. Influence of the Modifying Ability of Various Compositions on the Microstructure and Properties of the AK7ch Alloy. *Russ. J. Non-Ferr. Met.* **2015**, *56*, 593–598. [CrossRef]

8.   Kurzina, I.; Nikonenko, A.; Popova, N.; Nikonenko, E.; Kalashnikov, M. Fine structure and phase composition of Fe-14Mn-1.2C steel: Influe, nce of the modified mixture based on refractory metals. *Int. J. Miner.* **2017**, *24*, 523–529. [CrossRef]

9.   Kazantseva, L.A.; Kurzina, I.A.; Kosova, N.I.; Pichugina, A.A.; Sachkov, V.I.; Vladimirov, A.A.; Sachkova, A.S. Synthesis of titanium hydrides and obtaining alloys based on them. *Bull. Tomsk. State Univ. Chem.* **2015**, *2*, 69–75.

10.   Dolukhanyan, S.K.; Aleksanyan, A.G.; Ter-Galstyan, O.P.; Mailyan, D.G.; Shekhtman, V.S.; Sakharov, M.K. *Carbon Nanomaterials in Clean Energy Hydrogen Systems*; Baranovski, B., Zaginaichenko, S., Schur, D., Skorokhod, V., Veziroglu, A., Eds.; Springer: New York, NY, USA, 2008; pp. 733–741.

11.   Hakobyan, H.; Aleksanyan, A.; Dolukhanyan, S.; Mnatsakanyan, N.; Shekhtman, V. *Solid State Phenomena*; Bobet, J.L., Ed.; Trans Tech Publications: Stäfa, Switzerland, 2011; p. 354.

12.   Aleksanyan, A.G.; Dolukhanyan, S.K.; Shekhtman, V.S.; Khasanov, S.S.; Ter-Galstyan, O.P.; Martirosyan, M.V. Formation of alloys in the TieNb system by hydride cycle method and synthesis of their hydrides in self-propagating high-temperature synthesis. *Int. J. Hydrog. Energy* **2012**, *37*, 14234–14239. [CrossRef]

13.   Wallgram, W.; Schmolzer, T.; Cha, L.; Das, G.; Guther, V.; Clemens, H. Technology and mechanical properties of advanced γ-TiAl based alloys. *Int. J. Mater. Res.* **2009**, *100*, 1021–1030. [CrossRef]

14.   Klein, T.; Rashkova, B.; Holec, D.; Clemens, H.; Mayer, S. Advancement of Compositional and Microstructural Design of Intermetallic γ-TiAl Based Alloys Determined by Atom Probe Tomography. *Acta Mater.* **2016**, *110*, 236–245. [CrossRef]

15.   Hadi, M.; Shafyei, A.; Meratian, M. A comparative study of microstructure and high temperature mechanical properties of a β-stabilized TiAl alloy modified by lanthanum and erbium. *Mater. Sci. Eng.* **2015**, *624*, 1–8. [CrossRef]

16.   Bulanova, M.; Fartushna, I.; Meleshevich, K.; Samelyuk, A. 16. M. Bulanova, I. Fartushna, K. and A. Meleshevich, Samelyuk, Isothermal section at 850°C of the Ti–Dy–Al system in the Ti–TiAl–DyAl$_2$–Dyregion. *J. Alloy. Compd.* **2014**, *598*, 61–67. [CrossRef]

17.   Hadi, M.; Shafyei, A.; Meratian, M.; Bayat, O.; Ebrahimzadeh, I. Oxidation Properties of a Beta-Stabilized TiAl Alloy. *Oxid. Met.* **2018**, *90*, 421–434. [CrossRef]

18.   Liu, Z.Z.G.; Chai, L.H.; Chen, Y.Y.; Kong, F.T.; Davies, H.A.; Figueroa, I.A. Microstructure evolution in rapidly solidified Y added TiAl ribbons. *Intermetallics* **2011**, *19*, 160–164. [CrossRef]

19.   Wang, X.; Luo, R.; Liu, F.; Zhu, F.; Song, S.; Chen, B.; Zhang, X.; Zhang, J.; Chen, M. Characterization of Gd-rich precipitates in a fully lamellar TiAl alloy. *Scr. Mater.* **2017**, *137*, 50–54. [CrossRef]

20.   Li, W.; Xia, K. Kinetics of the α grain growth in a binary Ti–44Al alloy and a ternary Ti–44Al-0.15Gd alloy. *Mater. Sci. Eng. A* **2002**, *329–331*, 430–434. [CrossRef]

21.   Xia, K.; Wu, X.; Song, D. Effects of Gd addition, lamellar spacing and loading direction on creep behavior of a fully lamellar Ti–44Al–1Mn–2.5Nb alloy. *Acta Mater.* **2004**, *52*, 841–849. [CrossRef]

22.   Li, W.; Inkson, B.; Horita, Z.; Xia, K. Microstructure observations in rare earth element Gd-modified Ti-44 at%Al. *Intermetallics* **2000**, *8*, 519–523. [CrossRef]

23.   Schwaighofer, E.; Clemens, H.; Lindemann, J.; Mayer, A.S. Hot-working behavior of an advanced intermetallic multi-phase γ-TiAl based alloy. *Mater. Sci. Eng. A* **2014**, *77*, 297–310. [CrossRef]

24.   Belgibaeva, A.A.; Erkasov, R.S.; Kurzina, I.A.; Karakchieva, N.I.; Sachkov, V.I.; Abzaev, Y.A. Obtainment of high-strength alloys of the Ti-Al system using "hydride technology". In Proceedings of the Conference New Materials and Technologies, Barnaul, Russia, 2018; pp. 62–66.

25.   Kosova, N.; Sachkov, V.; Kurzina, I.; Pichugina, A.; Vladimirov, A.; Kazantseva, L.; Sachkova, A. The preparation of the Ti-Al alloys based on intermetallic phases. *IOP Conf. Ser. Mater. Sci. Eng.* **2016**, *112*, 012039. [CrossRef]

26. Abzaev, Y.A.; Syzrantsev, V.V.; Bardakhanov, S.P. Simulation of the Structural State of Amorphous Phases in Nanoscale SiO_2 Synthesized via Different Methods. *Phys. Solid State* **2017**, *59*, 1874–1878. [CrossRef]

27. Pecharsky, V.; Zavalij, P. *Fundamentals of Powder Diffraction and Structural Characterization of Materials*, 2nd ed.; Springer: New York, NY, USA, 2005; p. 713.

28. Toby, B.H. R factors in Rietveld analysis: How good is good enough? *Powder Diffr.* **2006**, *21*, 67–70. [CrossRef]

29. Young, R.A. *The Ritveld Method*; Oxford University Press: Oxford, UK, 1996.

30. Abdel-hamid, A.A. Influence of Ta, Zr, V and Mo on the growth-morphology of Ti-aluminide crystals. *Z. Fur Met.* **1991**, *82*, 383–386.

31. Crystallography Open Database. Available online: http://www.crystallography.net/cod/search.html (accessed on 13 March 2019).

32. Braun, J.; Ellner, M.; Predel, B.; Splat, Z. Splat-cooling investigations in the binary-system Ti-Al. *Z. Fuer Met.* **1994**, *82*, 355–362.

33. Xie, Y.Q.; Peng, H.J.; Liu, X.B.; Peng, K. Atomic states, potential energies, volumes, stability and brittleness of ordered FCC Ti_3Al-type alloys. *Phys. B Condens. Matter* **2005**, *366*, 1–17. [CrossRef]

34. Wang, H.Y.; Hu, Q.K.; Yang, W.P.; Li, X.S. Influence of metal element doping on the mechanical properties of TiAl alloy. *Acta Phys. Sin.* **2016**, *65*, 176101. [CrossRef]

35. Lyakhov, A.O.; Oganov, A.R.; Stokes, H.T.; Zhu, Q. New developments in evolutionary structure prediction algorithm USPEX. *Comput. Phys. Commun.* **2013**, *184*, 1172–1182. [CrossRef]

36. Abzaev, Y.A.; Starostenkov, M.D.; Klopotov, A.I. First-principles calculations of the concentration dependence of elastic modules in monocrystals $Ni_3(Ge_{1-x},Al_x)$. *Fundam. Probl. Mod. Mater. Sci.* **2014**, *11*, 56.

37. Segall, M.D.; Pickard, C.J.; Shah, R.; Payne, M.C. Population analysis in plane wave electronic structure calculations. *Mol. Phys.* **2010**, *89*, 571–577. [CrossRef]

38. Segall, M.D.; Shah, R.; Pickard, C.J.; Payne, M.C. Population analysis of plane-wave electronic structure calculations of bulk materials". *Phys. Rev. B* **1996**, *54*, 16317–16320. [CrossRef]

39. Oganov, A.R.; Glass, C.W.; Lyakhov, A.O.; Zhu, Q.; Qian, G.R.; Stokes, H.T.; Bushlanov, P.Z.; Allahyari, S. Lepeshkin. Universal Structure Predictor: Evolutionary Xtallography. Manual. 9.7. Appendices.P.108. Electronic access. Available online: http://han.ess.sunysb.edu/ (accessed on 13 March 2019).

40. Sereda, B.; Zherebtsov, A.; Belokon', Y. *The Modeling and Processes Research of Titan Aluminides Structurization Received by SHS Technology*; TMS: Seattle, WA, USA, 2010; pp. 99–108.

41. Sereda, B.; Kruglyak, I.; Zherebtsov, A.; Belokon', Y. *The Processes Research of Structurization of Titan Aluminides Received by SHS*; Materials Science & Technology: Pittsburg, CA, USA, 2009; pp. 2069–2073.

42. Kurzina, I.A.; Kozlov, E.V.; Sharkeev, Y.P.; Ryabchikov, A.I.; Stepanov, I.B.; Bozhko, I.A.; Kalashnikov, M.P.; Sivin, D.O.; Fortuna, S.V. Influence of ion implantation on nanoscale intermetallic-phase formation in Ti–Al, Ni–Al and Ni–Ti systems. *Surf. Coat. Technol.* **2007**, *201*, 8463. [CrossRef]

43. Imayev, V.M.; Imayev, R.M.; Oleneva, T.I. Current status of γ-TiAl intermetallic alloys investigations and prospects for the technology developments. *Lett. Mater.* **2011**, *1*, 25. [CrossRef]

44. Cordell, T. Titanium Aluminide Intermetallics. Advanced Materials and Processes Technology. *AMPTIAC Newsl.* **2000**, *4*, 9.

45. Raghavan, V. Al-Ti-Y (Aluminum-Titanium-Yttrium). *Phase Equilibria Diffus.* **2005**, *26*, 191. [CrossRef]

46. Zhou, H.; Liu, W.; Yuan, S.; Yan, J. The 500 °C Isothermal Section of the Al-Dy-Ti Ternary System. *Alloy. Compd.* **2002**, *336*, 218–222. [CrossRef]

47. Das, K.; Das, S. A Review of the Ti-Al-Ta (Titanium-Aluminum-Tantalum) System. *Phase Equilibria Diffus.* **2005**, *26*, 322–329.

48. Sridharan, S.; Nowotny, H. Studies in the Ternary System Ti-Ta-Al and in the Quaternary System Ti-Ta-Al-C. *Z. Fuer Met.* **1983**, *74*, 468–472.

49. McCullough, C.; Valencia, J.J.; Levi, C.G.; Mehrabian, R.; Maloney, M.; Hecht, R. Solidification Paths of Ti-Ta-Al Alloys. *Acta Metall. Mater.* **1991**, *39*, 2745–2758. [CrossRef]

 *metals*

Article

Microstructure of In-Situ Friction Stir Processed Al-Cu Transition Zone

Anna Zykova, Andrey Chumaevskii, Anastasia Gusarova, Tatiana Kalashnikova, Sergei Fortuna, Nickolai Savchenko, Evgeny Kolubaev and Sergei Tarasov *

Institute of Strength Physics and Materials Science, Siberian Branch of Russian Academy of Sciences, Tomsk 634055, Russian; zykovaap@ispms.ru (A.Z.); tch7av@gmail.com (A.C.); gusarova@ispms.ru (A.G.); gelombang@ispms.ru (T.K.); s_fortuna@ispms.ru (S.F.); savnick@ispms.ru (N.S.); eak@ispms.ru (E.K.)
* Correspondence: tsy@ispms.ru; Tel.: +7-3822-286-863

Received: 22 May 2020; Accepted: 16 June 2020; Published: 18 June 2020

Abstract: The majority of literature sources dedicated to dissimilar Al-Cu friction stir welding testifies to the formation of intermetallic compounds (IMC) according to diffusion-controlled reactions, i.e., without liquation on the Al/Cu interfaces. Fewer sources report on revealing Al-Cu eutectics, i.e., that IMCs are formed with the presence of the liquid phase. This work is an attempt to fill the gap in the results and find out the reasons behind such a difference. Structural-phase characteristics of an in-situ friction stir processed (FSP) Al-Cu zone were studied. The single-pass FSPed stir zone (SZ) was characterized by the presence of IMCs such as Al_2Cu, Al_2Cu_3, $AlCu_3$, Al_2MgCu, whose distribution in the SZ was extremely inhomogeneous. The advancing side SZ contained large IMC particles as well as Al(Mg,Cu) solid solution (SS) dendrites and Al-Al_2Cu eutectics. The retreating side SZ was composed of Al-Cu solid solution layered structures and smaller IMCs. Such a difference may be explained by different levels of heat input with respect to the SZ sides as well as by using lap FSP instead of the butt one.

Keywords: in-situ friction stir process; aluminum alloys; Al-Cu metallomatrix composite; intermetallic compounds; diffusion-controlled reactions; Al-Cu eutectics

1. Introduction

Metallomatrix composite surface materials modified using friction stir processing (FSP) are state-of-the-art materials, which are intended to combine high strength, wear resistance with high ductility and fatigue resistance of a core metal such as an aluminum alloy [1]. The FSP was originally a process used for surface structural modification, i.e., grain refining, which then was adapted to prepare the metallic matrix composites (MMC) surface coatings by means of introducing various reinforcement particles and admixing them to the matrix metal [1–10]. The FSP utilizes the friction-generated heat for plasticizing the matrix metal, which then is transferred to the rear zone by means of tool rotation and translational motion. The plasticized metal flows along a rather complex trajectory and its adhesion to the tool plays a great role in the metal transfer and stirring. The intense stirring serves to ensure homogeneous distribution of hard particles throughout the stir zone (SZ), and the FSP process parameters such as tool rotation rate and travel speed can be varied to find an optimal degree of mixing as well as temperature conditions.

The hard particles may be introduced into the stirred metal directly [7–9,11–13] or form in-situ inside the metal [14–18] by means of solid-state reactions between the admixed components, between the admixed components and the matrix or between the dissimilar metals processed. The FSP preparation of hybrid composites with the use of in-situ reactions shows up some advantages over those obtained using commercially available reinforcement particles. The first advantage is that in-situ reactions allow us to obtain finer and more homogeneously distributed reinforcing particles [17].

The second advantage is that either coherent or semicoherent boundaries may form between these in-situ prepared particles and the matrix [10,15,19] and therefore more thermodynamically stable and strong particle/matrix interfaces are feasible [20]. The literature sources were analyzed to show that many experimental compositions such as Al7075-Ti-6Al-4V [14]; Al1050-Ni-Ti-C [15]; Al-SiC [15,21]; Al6061-Fly ash [18]; Al1050-Fe_2O_3-Al [17]; Al-1050-Cu [16]; Al-Ni; Al-Nb [22]; Al-graphene [13,21,22] were subjected to FSP in order to prepare the in-situ MMCs.

Commonly, copper is one of the widely used and efficient metals to carry out FSP on aluminum alloys and thus obtain the MMCs reinforced with Al-Cu intermetallic particles [23]. Intermetallic Cu-Al particle reinforced MMCs may be fabricated using friction stir welding (FSW) on dissimilar metals [24–27], multilayer friction stir brazing [28], die-casting, powder metallurgy, etc.

The effect of different tool configurations on friction heat generation, metal flow and formation of intermetallic layers was studied in friction stir spot welding of dissimilar Cu and Al metals [24]. It was shown that thin intermetallic layers were formed from compounds such as $CuAl_2$, CuAl and Al_4Cu_9. Intermetallic compound layers consisting of $CuAl_2$ and Al_4Cu_9 were found at the Cu-Al boundaries along with hot cracking during butt friction stir welding of Cu and Al [25].

AA6061-T6 plates were welded with a copper plate inserted in the butt line between two AA6061 plates and the resulting stir zone structure contained $CuAl_2$ and Al_4Cu_9 intermetallic particles [26]. Both intermetallic particles and interlayers were formed during butt FSW on Cu and Al plates [27]. Ultrasound treatment was used to reduce the thickness of intermetallic layers formed with FSW [29].

Only a few publications were devoted to obtaining the Al-Cu composites using the FSP admixing the Cu powder. Hsu et al. [30] demonstrated that homogeneous Al-Al_2Cu MMC with Young modulus 88 ± 8 GPa, yield stress 450 MPa, ultimate stress 650 MPa and 0.15 plasticity may be obtained using FSP on Al-15 at.% Cu green samples compacted at 225 MPa.

It has been reported [16] that when pure copper powders were FSP admixed to aluminum plates at the tool travel speed 1.66 mm/s and rotation rate 750 rpm, the resulting intermetallic particles were $CuAl_2$ ones. Increasing the FSP pass number and, in particular, the FSP tool rotation rate to 1000 and even to 1500 rpm resulted in precipitation of mainly Al-Cu and Al_4Cu_9 intermetallics. On the contrary, only Al_2Cu precipitates were found in the aluminum irrespective of the FSP pass number [31].

This work was focused on studying the specificity of in-situ synthesis of Al-Cu intermetallic particles by means of lap FSP on a copper and aluminum alloy.

2. Materials and Methods

The hot-rolled AA5056 sheets were cut into 200 mm × 60 mm × 5 mm samples. C11000 copper sheets were cut into 200 mm × 60 mm × 2 mm samples. Chemical compositions of the alloy and copper sheets are shown in Table 1.

Table 1. Chemical composition of A5056 and C11000 plates.

Plates	Chemical Element, wt.%.										
	Al	Mg	Fe	Si	Mn	Cu	Zn	Ti	Ni	Pb	As
A5056	91.9–94.6	4.8–5.8	<0.5	<0.5	0.5–0.8	<0.1	<0.2	<0.02–0.1	-	-	-
C11000	-	-	<0.005	-	-	99.9	<0.004	-	<0.002	<0.005	<0.002

Friction stir processing was carried out with the use of the FSW machine (Sespel, Cheboksary, Russian) at the Institute of Strength Physics and Materials Science of Siberian Branch of Russian Academy of Sciences (Tomsk, Russian) (Figure 1). A truncated cone flute FSW tool with a 2.5 mm height pin and top and bottom diameters of 6 and 4 mm, respectively, was used. The tool shoulder diameter was 12 mm. The FSW tool inclination angle was 3°. The FSW parameters were as follows: rotation rate 500 rpm, travel speed 90 mm/min, plunging force 12,000 N. This set of parameters was found to be optimal as follows from previous experimenting [32]. The FSW tool penetration was 2.5 mm.

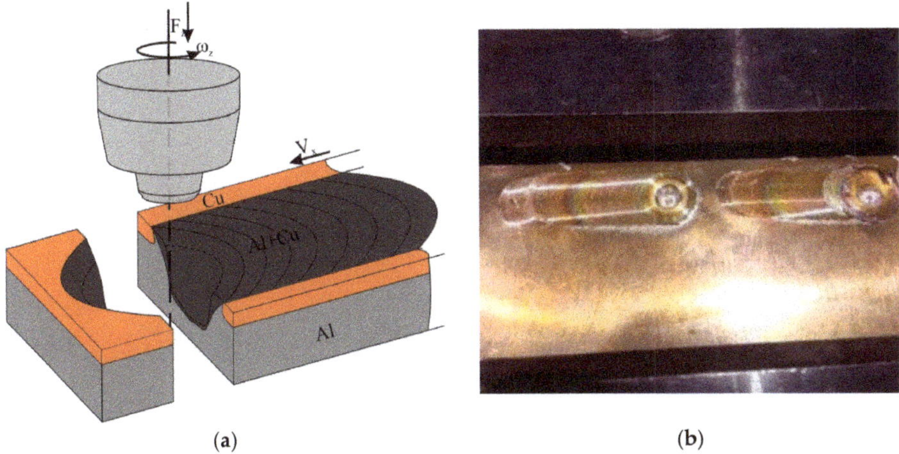

Figure 1. Scheme of FSP on C11000/A5056 sandwich (**a**) and single-pass FSP seams (**b**).

The FSPed samples were cut using electric dischage method (EDM) in planes perpendicular to the joint centerline to obtain specimens for examination and tests (Figure 1a). The microstructural evolution was examined on polished and etched section views prepared according to ASTM standards and visualized using optical microscopes Altami Met 1S (LLC Altami, Sankt Petersburg, Russian) and Olympus LEXT 4100 (Olympus NDT, Inc., Waltham, MA, USA) as well as scanning electron microscopy (SEM) and transmission electron microscopy (TEM) instruments Zeiss LEO EVO 50 (Carl Zeiss, Oberkochen, Germany) and JEOL-2100 (JEOL Ltd., Akishima, Japan), respectively. The chemical composition of precipitates was controlled using an EDS attachment to TEM.

The mean particle sizes were determined using the linear intercept method. Perfect stoichiometric compound component ratios were used to identify intermetallics found in the stir zone and analyzed with EDS. Using an X-ray diffraction (XRD) instrument XRD-7000S (Shimadzu, Kyoto, Japan) operated at 35 kV, 24 mA, irradiation was applied for identifying the Al-Cu phases. Microhardness profiles were obtained using a microhardness tester Duramin 5 (Struers A/S, Ballerup, Denmark) at 100 g load and a dwell time of 10 s.

3. Results

The macrostructure view of the FSW seam cross-sectional area allows us to observe composite structures in both the stirring zone (SZ) and thermomechanically affected zone (TMAZ) (Figure 2a). Both central and bottom areas of the SZ located close to the advancing side of the seam reveal discontinuities which may be shrinkage pores formed during the formation of Cu-Al intermetallics. The stir zone is characterized by alternating Al/Cu layers in its bottom part (Figure 2b) as well as Al-Cu solid solution and intermetallic compound (IMC) layers (Figure 2c). The large IMCs areas are seen on the retreating side of the SZ (Figure 2a).

The XRD pattern in Figure 3 reveals the phases as follows: Al, Cu, Al_2Cu, Al_2Cu_3, $AlCu_3$, Al_2MgCu and thus suggests that in-situ Al-Cu reactions have occurred. Nevertheless, there are large IMC-free areas composed of unreacted Cu and Al. Formation of Al_2Cu IMCs on the Cu/Al interfaces during FSW was noted in the majority of works dedicated to dissimilar FSW [24–27,29,30].

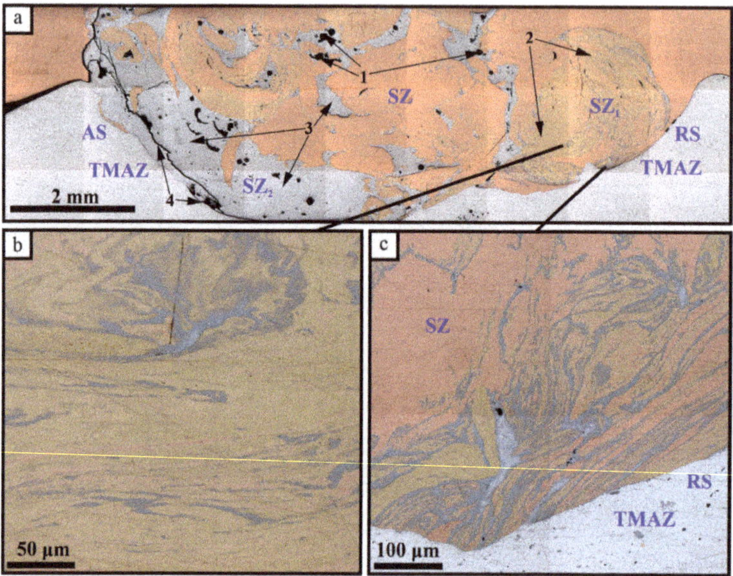

Figure 2. The FSPed SZ cross-sectional zones and Cu-Al intermixed areas: 1—defects; 2—solid solution (SS) areas; 3—IMC areas; 4—cracking at the advancing side; AS—advancing side; RS—retreating side; (**a**) the SZ macrostructure; (**b**) alternating SS and IMC layers; (**c**)—aluminum inflow pattern.

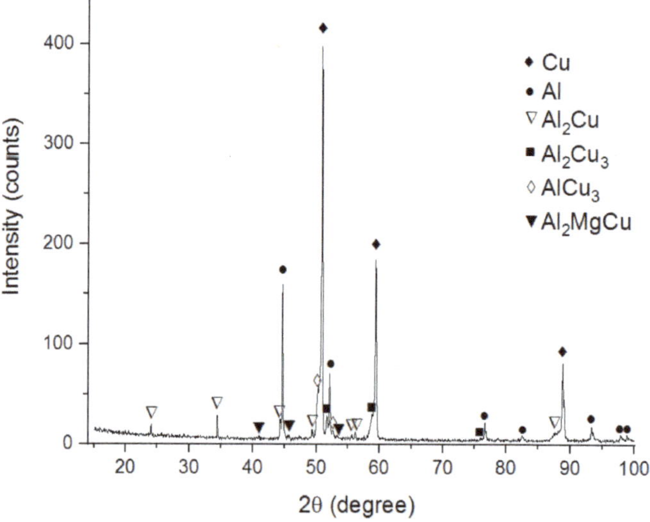

Figure 3. The XRD pattern of the Cu-Al stirring zone metal.

The composite structure of Al-Cu SZ does not look structurally homogeneous since it includes many different phases and microstructures (Figure 4).

The IMC layers formed at the Al/Cu interfaces on the Cu-rich side of SZ are composed of 2 to 5 μm in size Al_2Cu_3 and $AlCu_3$ IMCs (Figure 4a,d). The mean size of the Al_2Cu_3 and $AlCu_3$ particles in these IMCs is about 300 nm. The presence of these phases is confirmed by the results of XRD (Figure 3), TEM (Figure 5) and EDS (Table 2, spectra 4–7).

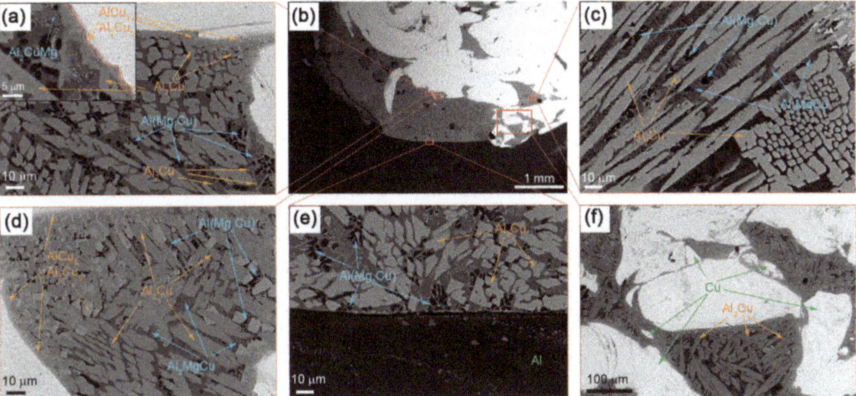

Figure 4. The SEM BSE images of SZ microstructures after a single-pass FSP on the Al-Cu sandwich: (**a**) SS/IMC interface; (**b**) stir zone macrostructures; (**c**) column IMCs; (**d**) IMCs of different compositions; (**e**) substrate/stir zone boundary; (**f**) mixed IMC + SS zone.

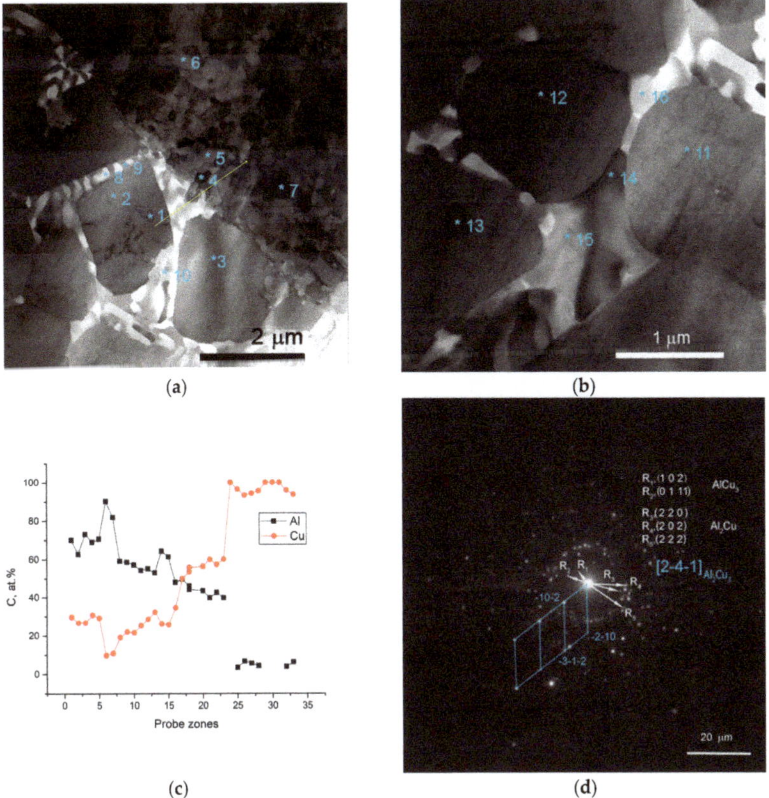

Figure 5. TEM images of SZ areas: Al-Cu interfaces with eutectics (**a**,**b**); EDS profile across the Al$_2$Cu/Cu transition zone (**c**); SAED pattern obtained from the area in Figure 5a and SAED reflection identification (**d**); 1–15 are the EDS probe spots corresponding phases shown in Table 2.

Table 2. Chemical compositions of phases shown in Figure 5.

No.	Element	Content, at.%	Phase Formula	Morphology
1	Al Cu	69.8 30.2	Al_2Cu	Near-spherical
2	Al Cu	71.3 28.7	Al_2Cu	rectangular
3	Al Cu	72.9 27.1	Al_2Cu	plate
4	Al Cu	46.6 53.4	Al_2Cu_3	irregular
5	Al Cu	36.4 63.6	$AlCu_3$	irregular
6	Al Cu	36.8 63.2	$AlCu_3$	irregular
7	Al Cu	39.2 60.8	Al_2Cu_3	irregular
8	Mg Al Cu	9.4 80.0 10.6	AlMg/ Al_2Cu	eutectics
9	Mg Al Cu	26.0 50.6 23.4	AlMg/ Al_2Cu	eutectics
10	Mg Al Cu	14.6 49.2 36.2	AlMg/ Al_2Cu	eutectics
11	Al Cu	68.7 31.3	Al_2Cu	equiaxial
12	Al Cu	68.7 31.3	Al_2Cu	equiaxial
13	Al Cu	63.8 36.2	Al_2Cu	spherical
14	Mg Al Cu	16.6 67.4 16.0	Al_2MgCu	eutectics
15	Mg Al Cu	6.8 83.6 9.6	AlMg/ Al_2Cu	eutectics
16	Mg Al Cu	4.3 88.9 6.8	AlMg/ Al_2Cu	eutectics

The interfaces between fine-crystalline copper and IMCs are shown in Figure 5a as well as Al/Al$_2$Cu eutectics located in between the Al$_2$Cu particles. The selected area electron diffraction (SAED) pattern obtained from the area in Figure 5a shows the reflections, which can be identified as those belonging to AlCu$_3$, Al$_2$Cu and Al$_2$Cu$_3$ (Figure 5d). Analyzing the EDS spectra and taking into account the ideal stoichiometric formulas of the IMCs detected, the theoretical compositions of them were determined and are presented in Tables 2 and 3. It is worthwhile noting that IMCs composed of Al$_2$Cu$_3$ and AlCu$_3$ particles are inherent to all Al/Cu alternating layers (Figure 4a,d–f).

Table 3. Chemical compositions of phases shown in Figure 6.

Spectrum	Element	Content, at.%	Phase Formula	Morphology
1 (Figure 6a)	Al	93.7	Al(Cu,Mg)	dendritic
	Cu	4.0		
	Mg	2.3		
2 (Figure 6a)	Al	66.8	AlMg/ Al_2Cu	eutectics
	Cu	20.4		
	Mg	12.8		
3 (Figure 6a)	Al	62.9	Al_2Cu	rectangular
	Cu	35.7		
	Mg	1.4		
4 (Figure 6a)	Al	61.6	Al_2Cu	angularity
	Cu	37.5		
	Mg	0.9		
1 (Figure 6b)	Al	91.2	Al(Cu,Mg)	dendritic
	Cu	5.7		
	Mg	3.1		
2 (Figure 6b)	Al	71.9	AlMg/ Al_2Cu	eutectics
	Cu	15.8		
	Mg	12.3		
3 (Figure 6b)	Al	63.6	Al_2Cu	angularity
	Cu	35.5		
	Mg	0.9		
4 (Figure 6b)	Al	66.0	Al_2Cu	angularity
	Cu	31.9		
	Mg	2.1		

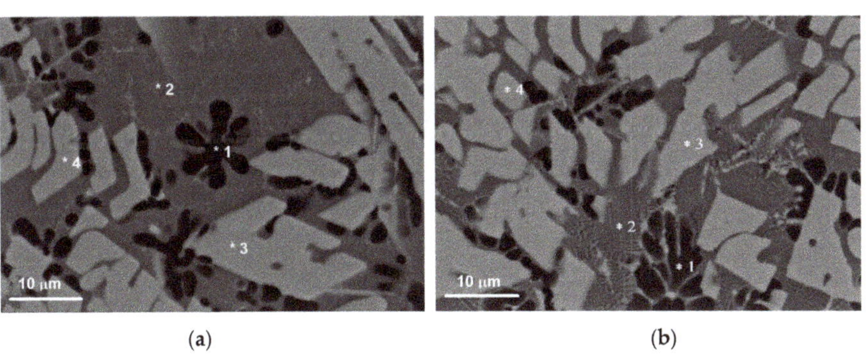

Figure 6. The SEM BSE images of microstructures in the SZ (**a**,**b**), which contain SS Al(Cu,Mg), Al_2Cu and Al_2MgCu particles as detected using the EDS probe on microstructure components denoted 1, 2, 3, 4.

When looking at the Al-rich part of the SZ, more aluminum-rich phases are formed there according to reactions as follows:

$$Al_2Cu_3 + 5Al \rightarrow Al + 3Al_2Cu \tag{1}$$

$$AlCu_3 + 6Al \rightarrow Al + 3Al_2Cu. \tag{2}$$

The microstructure of this zone is composed of SS Al(Cu,Mg) and Al_2Cu particles as confirmed by the EDS and SEM (Figures 4 and 6, Table 3).

The solid solution in this zone contains up to 5.7 at.% Cu and 3.1 at.% Mg in Al, i.e., it can be referred to as an Al(Cu,Mg) phase as shown by the EDS spectra in Figure 6 and Table 3. These Al(Cu,Mg) particles are dendrites and have a mean size of 11.8 μm (Figure 6). Such a dendritic shapes may be evidence in favor of heterogeneous nucleation and growth in a liquid phase.

Two different sorts of $CuAl_2$ particles are formed in the SZ (Figure 4a,c–f; Figure 6) such as thin 7×50 μm^2 area platelets and the fine ones found in eutectics. Figure 5c shows an EDS profile along the line shown in Figure 5a, i.e., a transition zone from IMC to the Al-Al$_2$MgCu eutectics and then to the recrystallized fine-grained copper.

An intermetallic Al-Al$_2$MgCu eutectic phase was also EDS detected in the middle of SZ (Tables 2 and 3) (Figures 3, 4a and 6).

The macroscopic FSP track cross-sectional area in Figure 7a shows lines along which the microhardness number profiles (Figure 7b,c) were obtained. Both Figure 7b and 7c demonstrate that microhardness profiles obtained in two perpendicular directions allow for the differentiation between the matrix and IMCs, and, in fact, reveal the SZ structural inhomogeneity, which means a lack of equality in strength. The microhardness of IMCs is by a factor of 2 to 5 higher than those of the base metals.

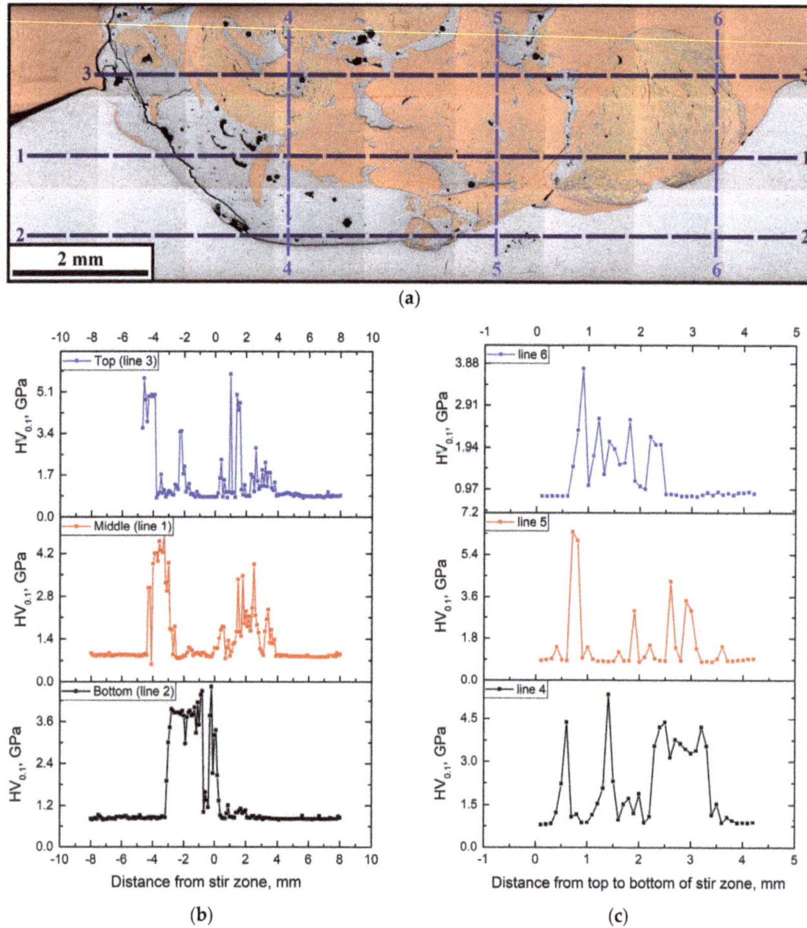

Figure 7. Microhardness profile measurement lines (**a**) and microhardness profiles along horizontal (1, 2, 3) (**b**) and vertical (4, 5, 6) directions (**c**).

4. Discussion

The results of this work clearly show the inhomogeneous structure of the stir zone composed of copper, aluminum, Al-Mg-Cu solid solution and a set of different morphology IMCs. The FSP is a strong nonequilibrium process so that the microstructural evolution of the processed metal is determined by a variety of external factors such as heat generation, mechanical stirring (deformation), heat removal as well as internal process factors such as adhesion-assisted or quasi-viscous transfer of metal portions to the zone behind the tool, dynamic recrystallization, diffusion-controlled precipitation and mechanochemical solid-state reactions. The Cu-Al system is capable of forming intermetallic compounds with a high exothermic effect so that a thin liquid phase layer may form at the Al/Cu interface due to contact melting. Such a phenomenon leads to a fast liquid-phase synthesis of coarse IMC layers and particles, especially when fusion methods are used to obtain the Cu-Al alloys. At the same time, only diffusion-controlled formation of IMCs is possible when the process temperatures are low enough. In FSP on Cu-Al, the temperatures in the stir zone are in the range 400–500 °C [31], i.e., lower than the eutectic temperature T_E = 548.2 °C and no Al-Cu eutectics were detected in this work. All IMCs were formed by means of diffusion-controlled precipitation from a supersaturated solid solution obtained in FSP according to the model suggested by Pretorius et al. [33]. According to such a model, the effective heat of formation ($\Delta H'$) of phases at the binary Al-Me system interfaces can be determined as follows:

$$\Delta H'_i = \Delta H^0_i \cdot \frac{C_e}{C_c} \tag{3}$$

where $\Delta H'_i$ is the effective heat of formation of i-phase, ΔH^0_i is the enthalpy of formation change for phase i; C_e is the effective concentration of the limiting element at the interface; C_c is the concentration of the limiting element in the compound. Taking into account that $\Delta G^0 \approx \Delta H^0$ the effective Gibbs free energy change ($\Delta G'_i$) in case of an i-phase formation is determined as:

$$\Delta G'_i = \Delta G^0_i \cdot \frac{C_e}{C_c}. \tag{4}$$

Table 4 shows the results of calculating the Gibbs free energy changes corresponding to the formation of all Al-Cu binary system phases. It can be noted that only negative $\Delta G'_i$ values were obtained, thus, determining the feasibility of the IMC nucleation and growth. The maximum absolute $\Delta G'_i$ values $\Delta G'_{Al_2Cu_3}$ = −31.28 kJ/mol and $\Delta G'_{AlCu_3}$ = −22.84 kJ/mol were found for Al_2Cu_3 and $AlCu_3$, i.e., these phases were the first ones to form at the Cu/Al interfaces. Therefore, those phases were detected in this work as small particles on the Cu-rich side of the image in Figure 4a. The next phase to form was Al_2Cu with $\Delta G'_{Al_2Cu}$ = −19.54 kJ/mol.

It should be noted that, in general, the diffusion-controlled Al-Cu interaction may lead to the formation of a variety of IMC phases as dependent on Cu-content, sort of source materials (sheet or powders), etc. Since copper and aluminum sheets were used in this work, the local Cu-content may be as high as 40 vol.% due to intense stirring and transfer then three binary phases were formed such as Al_2Cu_3, $AlCu_3$, and Al_2Cu. However, it was reported [16,30,31,34] that only a single Al_2Cu phase was formed when FSP admixing the pure copper powders at a concentration of ≤15 at.% and in-situ synthesizing the Al-Cu composite. Using the Pretorius model [33], it was experimentally established [31] that Al_2Cu had to be the first phase to form since its Gibbs free energy change of formation was the most negative value of all other phases as determined for a Cu at.% concentration corresponding to the lowest temperature eutectic.

Table 4. Effective Gibbs free energy changes at Cu-Al interfaces at 450 °C.

IMC	Composition	Limiting Element	G'_i [35], J/mol	G' at 450 °C, kJ/mol	$G'_{i'}$, kJ/mol
Al$_2$Cu	Al$_{0.67}$Cu$_{0.33}$	Al	$-15{,}826.2 + 2.3T$	-14.16	**-19.54**
AlCu	Al$_{0.50}$Cu$_{0.50}$	Al(Cu)	$-20{,}496.8 + 1.6T$	-19.34	-19.34
Al$_3$Cu$_4$	Al$_{0.43}$Cu$_{0.57}$	Cu	$-20{,}197.4 + 1.9T$	-18.82	-18.82
Al$_2$Cu$_3$	Al$_{0.40}$Cu$_{0.60}$	Al	$-20{,}137.8 + 1.6T$	-18.98	**-31.28**
Al$_4$Cu$_9$	Al$_{0.31}$Cu$_{0.69}$	Cu	$-19{,}707.1 + 1.6T$	-18.55	-18.55
AlCu$_3$	Al$_{0.25}$Cu$_{0.75}$	Al	$-19{,}146.8 + 1.6T$	-17.99	**-22.84**

On the other hand, eutectic structures were observed during friction stir spot welding of AA5083 to copper [36]. The majority of papers devoted to studying the dissimilar Al-Cu butt FSWed SZ did not show any presence of eutectics in distinction to the friction stir spot welding (FSSW) SZ microstructures. The reason is the higher heat removal into a copper sheet in the case of butt FSW as compared to FSSW or lap FSW. Preferential localization of IMCs on the advancing side of the SZ may be related to more intense admixing between aluminum and copper layers (particles) detached off the parent metals and transferred to the stagnation (trailing) zone behind the tool closer to the advancing side where these layers adhere to the already deposited layers; finally, recrystallize and grow the IMCs at rest. Closer to the retreating side these layers are stirred by the tool and deformed so that the strain dissolution of the IMCs prevails over their precipitation, thus, facilitating the formation of the solid solution only. Such a consideration is based on the adhesion-assisted transfer of metal during FSP when a transferred portion (layer) of metal first adheres to the FSW tool, then is transferred to the stagnation zone behind the tool, and adheres back to the previously transferred layers [37].

It is suggested also that more heat is generated on the advancing side as compared to that of the retreating side [38]. This factor may provide higher heating and better conditions for the contact melting between copper and aluminum transferred layers.

The combined effect of severe plastic deformation and heating provided the formation of a row of Al-Cu phases. There are a number of phases whose morphology allows for the suggestion of their origin from a liquid phase. First of all, those phases were Al-Al$_2$Cu eutectics and Al-Mg-Cu solid solution dendrites that formed near the IMC large particles as a result of depleting these zones of copper. Such a scenario is inherent to the advanced side of the SZ zone.

Reaction-diffusion controlled nucleation and growth of IMC smaller particles is inherent to the mechanically alloyed multilayer metal on the retreating side of the SZ. Friction stir processing in this zone resulted in the formation of a Cu-Al mechanical alloy (bronze) which differed also from both parent metals by its color. The smaller IMCs are found in the interlayer spaces whereas the in the main phase there is the Al-Mg-Cu SS.

5. Conclusions

Microstructural evolution and phase composition of stir zone in-situ obtained using friction stir processing on an Al-Cu bimetal workpiece were studied:

1. The single-pass FSP on Cu-Al bimetal plate resulted in the formation of a stir zone with inhomogeneous distribution of intermixed phases identified as unreacted metals, intermetallic phases such as Al$_2$Cu, Al$_2$Cu$_3$, AlCu$_3$, Al$_2$MgCu, Al(Mg,Cu) solid solution and Al-Al$_2$Cu eutectics.
2. Large IMC particles as well as Al-Al$_2$Cu eutectics and Al-Mg-Cu solid solution dendrites were preferentially located on the advancing side of the SZ zone, while the retreating side zone of SZ was characterized by the presence of Al-Cu solid solution layered structures and smaller IMC particles located between the solid solution layers.
3. The microhardness profiles measured across the SZ digitally mirrors the inhomogeneity of the phase distribution there. The microhardness of IMC zones is by a factor of 2–5 higher than that of copper.

4. The IMC areas containing the eutectics and solid solution dendrites, which might originate from Al-Cu liquation, are characterized by large irregular shaped shrinkage pores.

Author Contributions: A.Z., A.C., A.G., T.K., S.F., N.S. performed sample preparation and characterization; A.C., E.K. and S.T. performed project administration, conceptualization, supervision, designed the experiments, and analyzed the data; A.Z., A.C. and S.T. wrote this paper. All authors have read and agreed to the published version of the manuscript.

Funding: Sample preparation and processing were funded by the Government research assignment for ISPMS SB RAS, project No. III.23.2.4. Characterization of microstructures and phases obtained using dissimilar FSP on copper and aluminum alloy was performed with financial support from RSF Grant No. 19-79-00136.

Conflicts of Interest: The authors declare no conflict of interest.

References

1. Sudhakar, M.; Rao, C.H.S.; Saheb, K.M. Production of Surface Composites by Friction Stir Processing-A Review. *Mater. Today Proc.* **2018**, *5*, 929–935. [CrossRef]
2. Padhy, G.K.; Wu, C.S.; Gao, S. Friction stir based welding and processing technologies-processes, parameters, microstructures and applications: A review. *J. Mater. Sci. Technol.* **2018**, *34*, 1–38. [CrossRef]
3. Rao, A.G.; Ravi, K.R.; Ramakrishnarao, B.; Deshmukh, V.P.; Sharma, A.; Prabhu, N.; Kashyap, B.P. Recrystallization Phenomena During Friction Stir Processing of Hypereutectic Aluminum-Silicon Alloy. *Metall. Mater. Trans. A* **2013**, *44*, 1519–1529. [CrossRef]
4. Sun, H.; Yang, S.; Jin, D. Improvement of Microstructure, Mechanical Properties and Corrosion Resistance of Cast Al–12Si Alloy by Friction Stir Processing. *Trans. Indian Inst. Met.* **2018**, *71*, 985–991. [CrossRef]
5. Zhao, H.; Pan, Q.; Qin, Q.; Wu, Y.; Su, X. Effect of the processing parameters of friction stir processing on the microstructure and mechanical properties of 6063 aluminum alloy. *Mater. Sci. Eng. A* **2019**, *751*, 70–79. [CrossRef]
6. Zahmatkesh, B.; Enayati, M.H. A novel approach for development of surface nanocomposite by friction stir processing. *Mater. Sci. Eng. A* **2010**, *527*, 6734–6740. [CrossRef]
7. Bourkhani, R.D.; Eivani, A.R.; Nateghi, H.R. Through-thickness inhomogeneity in microstructure and tensile properties and tribological performance of friction stir processed AA1050-Al2O3 nanocomposite. *Compos. Part B Eng.* **2019**, *174*, 107061. [CrossRef]
8. Abraham, S.J.; Dinaharan, I.; Selvam, J.D.R.; Akinlabi, E.T. Microstructural characterization of vanadium particles reinforced AA6063 aluminum matrix composites via friction stir processing with improved tensile strength and appreciable ductility. *Compos. Commun.* **2019**, *12*, 54–58. [CrossRef]
9. Jain, V.K.S.; Varghese, J.; Muthukumaran, S. Effect of First and Second Passes on Microstructure and Wear Properties of Titanium Dioxide-Reinforced Aluminum Surface Composite via Friction Stir Processing. *Arab. J. Sci. Eng.* **2019**, *44*, 949–957. [CrossRef]
10. Huang, C.W.; Aoh, J.N. Friction stir processing of copper-coated SiC particulate-reinforced aluminum matrix composite. *Materials (Basel)* **2018**, *11*, 599. [CrossRef]
11. Prabhu, M.S.; Perumal, A.E.; Arulvel, S.; Issac, R.F. Friction and wear measurements of friction stir processed aluminium alloy 6082/CaCO3 composite. *Measurement* **2019**, *142*, 10–20. [CrossRef]
12. Barati, M.; Abbasi, M.; Abedini, M. The effects of friction stir processing and friction stir vibration processing on mechanical, wear and corrosion characteristics of Al6061/SiO2 surface composite. *J. Manuf. Process.* **2019**, *45*, 491–497. [CrossRef]
13. Nazari, M.; Eskandari, H.; Khodabakhshi, F. Production and characterization of an advanced AA6061-Graphene-TiB2 hybrid surface nanocomposite by multi-pass friction stir processing. *Surf. Coatings Technol.* **2019**, *377*, 124914. [CrossRef]
14. Adetunla, A.; Akinlabi, E. Fabrication of Aluminum Matrix Composites for Automotive Industry Via Multipass Friction Stir Processing Technique. *Int. J. Automot. Technol.* **2019**, *20*, 1079–1088. [CrossRef]
15. Fotoohi, H.; Lotfi, B.; Sadeghian, Z.; Byeon, J. Microstructural characterization and properties of in situ Al-Al3Ni/TiC hybrid composite fabricated by friction stir processing using reactive powder. *Mater. Charact.* **2019**, *149*, 124–132. [CrossRef]

16. Mahmoud, E.R.I.; Al-qozaim, A.M.A. Fabrication of In-Situ Al-Cu Intermetallics on Aluminum Surface by Friction Stir Processing. *Arab. J. Sci. Eng.* **2016**, *41*, 1757–1769. [CrossRef]

17. Azimi-Roeen, G.; Kashani-Bozorg, S.F.; Nosko, M.; Nagy, Š.; Ma\vTko, I. Correction to: Formation of Al/(Al13Fe4 + Al2O3) Nano-composites via Mechanical Alloying and Friction Stir Processing. *J. Mater. Eng. Perform.* **2018**, *27*, 6800. [CrossRef]

18. Dinaharan, I.; Akinlabi, E.T. Low cost metal matrix composites based on aluminum, magnesium and copper reinforced with fly ash prepared using friction stir processing. *Compos. Commun.* **2018**, *9*, 22–26. [CrossRef]

19. Kumar, P.A.; Madhu, H.C.; Pariyar, A.; Perugu, C.S.; Kailas, S.V.; Garg, U.; Rohatgi, P. Friction stir processing of squeeze cast A356 with surface compacted graphene nanoplatelets (GNPs) for the synthesis of metal matrix composites. *Mater. Sci. Eng. A* **2020**, *769*, 138517. [CrossRef]

20. Zhang, Q.; Xiao, B.L.; Wang, Q.Z.; Ma, Z.Y. In situ Al3Ti and Al2O3 nanoparticles reinforced Al composites produced by friction stir processing in an Al-TiO2 system. *Mater. Lett.* **2011**, *65*, 2070–2072. [CrossRef]

21. Alidokht, S.A.; Abdollah-zadeh, A.; Soleymani, S.; Assadi, H. Microstructure and tribological performance of an aluminium alloy based hybrid composite produced by friction stir processing. *Mater. Des.* **2011**, *32*, 2727–2733. [CrossRef]

22. Zeidabadi, S.R.H.; Daneshmanesh, H. Fabrication and characterization of in-situ Al/Nb metal/intermetallic surface composite by friction stir processing. *Mater. Sci. Eng. A* **2017**, *702*, 189–195. [CrossRef]

23. Ponweiser, N.; Lengauer, C.L.; Richter, K.W. Re-investigation of phase equilibria in the system Al–Cu and structural analysis of the high-temperature phase η1-Al1−δCu. *Intermetallics* **2011**, *19*, 1737–1746. [CrossRef]

24. Zhou, L.; Zhang, R.X.; Li, G.H.; Zhou, W.L.; Huang, Y.X.; Song, X.G. Effect of pin profile on microstructure and mechanical properties of friction stir spot welded Al-Cu dissimilar metals. *J. Manuf. Process.* **2018**, *36*, 1–9. [CrossRef]

25. Shankar, S.; Vilaça, P.; Dash, P.; Chattopadhyaya, S.; Hloch, S. Joint strength evaluation of friction stir welded Al-Cu dissimilar alloys. *Measurement* **2019**, *146*, 892–902. [CrossRef]

26. Khojastehnezhad, V.M.; Pourasl, H.H. Microstructural characterization and mechanical properties of aluminum 6061-T6 plates welded with copper insert plate (Al/Cu/Al) using friction stir welding. *Trans. Nonferrous Met. Soc. China* **2018**, *28*, 415–426. [CrossRef]

27. Xue, P.; Xiao, B.L.; Ni, D.R.; Ma, Z.Y. Enhanced mechanical properties of friction stir welded dissimilar Al–Cu joint by intermetallic compounds. *Mater. Sci. Eng. A* **2010**, *527*, 5723–5727. [CrossRef]

28. Zhang, G.; Yang, X.; Zhu, D.; Zhang, L. Cladding thick Al plate onto strong steel substrate using a novel process of multilayer-friction stir brazing (ML-FSB). *Mater. Des.* **2020**, *185*, 108232. [CrossRef]

29. Muhammad, N.A.; Wu, C.S.; Tian, W. Effect of ultrasonic vibration on the intermetallic compound layer formation in Al/Cu friction stir weld joints. *J. Alloys Compd.* **2019**, *785*, 512–522. [CrossRef]

30. Hsu, C.J.; Kao, P.W.; Ho, N.J. Ultrafine-grained Al–Al2Cu composite produced in situ by friction stir processing. *Scr. Mater.* **2005**, *53*, 341–345. [CrossRef]

31. Huang, G.; Hou, W.; Li, J.; Shen, Y. Development of surface composite based on Al-Cu system by friction stir processing: Evaluation of microstructure, formation mechanism and wear behavior. *Surf. Coatings Technol.* **2018**, *344*, 30–42. [CrossRef]

32. Kalashnikova, T.A.; Gusarova, A.V.; Chumaevskii, A.V.; Knyazhev, E.O.; Shvedov, M.A.; Vasilyev, P.A. Regularities of composite materials formation using additive electron-beam technology, friction stir welding and friction stir processing. *Metall. Work. Mater. Sci.* **2016**, *63*, 2016. (In Russian) [CrossRef]

33. Pretorius, R.; Marais, T.K.; Theron, C.C. Thin film compound phase formation sequence: An effective heat of formation model. *Mater. Sci. Rep.* **1993**, *10*, 1–83. [CrossRef]

34. Azizie, H.M.; Iranparast, D.; Dezfuli, M.A.G.; Balak, Z.; Kim, H.S. Fabrication of Al/Al2Cu in situ nanocomposite via friction stir processing. *Trans. Nonferrous Met. Soc. China* **2017**, *27*, 779–788. [CrossRef]

35. Yang, L.; Mi, B.X.; Lv, L.; Huang, H.J.; Lin, X.P.; Yuan, X.G. Formation Sequence of Interface Intermetallic Phases of Cold Rolling Cu/Al Clad Metal Sheet in Annealing Process. *Mater. Sci. Forum* **2013**, *749*, 600–605. [CrossRef]

36. Shen, J.; Suhuddin, U.F.H.; Cardillo, M.E.B.; dos Santos, J.F. Eutectic structures in friction spot welding joint of aluminum alloy to copper. *Appl. Phys. Lett.* **2014**, *104*, 191901. [CrossRef]

37. Tarasov, S.Y.; Filippov, A.V.; Kolubaev, E.A.; Kalashnikova, T.A. Adhesion transfer in sliding a steel ball against an aluminum alloy. *Tribol. Int.* **2017**, *115*, 191–198. [CrossRef]

38. Hersent, E.; Driver, J.H.; Piot, D.; Desrayaud, C. Integrated modelling of precipitation during friction stir welding of 2024-T3 aluminum alloy. *Mater. Sci. Technol.* **2010**, *26*, 1345–1352. [CrossRef]

Article

Microstructural Analysis of Friction Stir Butt Welded Al-Mg-Sc-Zr Alloy Heavy Gauge Sheets

Tatiana Kalashnikova *, Andrey Chumaevskii, Kirill Kalashnikov, Sergei Fortuna, Evgeny Kolubaev and Sergei Tarasov

Institute of Strength Physics and Materials Science, Siberian Branch of Russian Academy of Sciences, 634055 Tomsk, Russia; tch7av@gmail.com (A.C.); kkn@ispms.tsc.ru (K.K.); s_fortuna@ispms.ru (S.F.); eak@ispms.ru (E.K.); tsy@ispms.ru (S.T.)
* Correspondence: gelombang@ispms.ru; Tel.: +7-3822-286-863

Received: 31 May 2020; Accepted: 15 June 2020; Published: 17 June 2020

Abstract: Friction stir welding (FSW) on a heavy gauge sheet of a hereditary fine-grained Al-Mg-Sc-Zr alloy was carried out to study the specifics of plasticized metal flow and microstructural evolution in different sections and zones of the joint. It was found that the stir zone (SZ) macrostructure may contain either a single or many nugget zones depending on the metal sheet thickness and the seam length. The effect of grain kinking in a thermomechanically affected zone (TMAZ) under pressure from the stir zone metal was discovered. The stir zone metal was fine-grained but had a microhardness lower than that of the base metal, which may be explained by the overaging effect of FSW on the Al_3Sc precipitates. The tensile strength of the joint was almost equal to that of the base metal (BM). The grain size distributions were obtained in different sections below the sheet surface and away from the exit hole, which allowed us to suggest the specific adhesion-assisted layer-by layer metal transfer mechanism in FSW.

Keywords: friction stir welding; aluminum alloy; structure formation; adhesion; metal transfer; mechanical properties

1. Introduction

Al-Mg alloys are well-known structural materials that are widely used in various applications including those where welding is a single option joining method. Improving the Al-Mg alloy strength characteristics is achieved by alloying the metals with scandium and zirconium [1–4], which makes them hereditary fine-grained alloys while retaining their corrosion resistance [2,5].

The alloy fine-grained microstructure is stabilized by the precipitation of nanosized Al_6Mn, Al_3Sc, and $Al_3(Sc,Zr)$ matrix coherent particles, which first precipitate from a solid solution and then have a pinning effect on the grain boundaries so that even annealing at 723 K for 16 h does not increase the grain size [6]. The Al_3Sc precipitates are the most effective precipitates for pinning the grain boundaries [7]. It is known also [8] that the nanosized Al_3Sc precipitates are stable against coalescence so that that their mean diameter increased only from 13.5 to 15.4 nm for 7 days at 350 °C.

Friction stir welding (FSW) is a method used for joining metals in a solid state and is, therefore, often used on metals and alloys that cannot be welded using any fusion welding technique. An Al-Mg-Sc-Zr alloy is an example of such a hardly weldable alloy that and is, therefore, a candidate for joining with friction stir welding. It is common knowledge that this solid state joining method usually forms a fine-grained microstructure in the stirring zone. The presence of the nanosized growth inhibitors in Al-Mg-Sc-Zr is an extra factor that helps retaining the fine-grained structure of the welded metal and thus improving its mechanical characteristics [9]. For example, it was reported [10] that FSW increased the ultimate and yield stresses of the Sc + Zr-alloyed aluminum alloy by 23.8 and 11.9%, respectively. Practically a 100% tensile strength was achieved on the Al-5.4Mg-0.2Sc-0.1Zr alloy FSW seam even after annealing [11]. The mechanical strength advantage of the FSW joint compared to that of the TIG on the Al-Mg-Sc-Zr alloy was thus demonstrated [10].

Another positive factor of the FSW-generated fine-grained structure is its potential to take advantage of superplasticity characteristics [12–16] that allow up to 2000% elongation while retaining microstructural thermal stability at 450 °C [12].

The effect of Al$_3$Sc and Al$_3$(Sc,Zr) nanosized precipitates on pinning the fatigue crack tip was also reported [17]. It was shown [18] that friction stir processing improves the corrosion resistance of Al-Mg alloys in a NaCl aqueous solution.

An important problem in FSW is obtaining a flawless and reliable seam when joining thick sheets. The inhomogeneity and efficiency of metal stirring, enhanced adhesion of metal to the FSW tool, diffusion-controlled FSW tool wear, fast heat removal, and high mechanical loads are factors that aggravate the problems commonly encountered with FSW on aluminum alloy sheets up to 10 mm thickness. Some technologies require welding heavy gauge sheets or pipes, which then are then machined to obtain the final structure. This requirement makes it necessary to study FSW seams on heavy gauge workpieces.

It may be concluded, therefore, that Al-Mg-Sc and Al-Mg-Sc-Zr alloys not only possess high mechanical characteristics and are weldable using FSW but also that their characteristics can be further improved using FSW, which is a promising research field from both practical and scientific points of view.

The severe thermomechanical impact of FSW on the plasticized aluminum alloy in the stirring zone creates the conditions for the strain-induced dissolution of particles, thereby enriching the solid solution with the alloying elements and facilitating the fast precipitation of particles. This is an element of FSW aspect that has not received much attention in the literature, especially when taking into account the dynamic character of dissolution/precipitation, as well as recrystallization, during the process. This issue becomes even more intriguing when applying FSW to an Al-Mg-Sc-Zr alloy, whose microstructure is known to have high temperature stability. The FSW joint structure formation is determined by the specificity of metal transfer and the role played by the adhesion of plasticized metal to the FSW tool.

The objective of this paper is to study the microstructures and mechanical characteristics of the FSW joint zones obtained on heavy gauge Al-Mg-Sc-Zr alloy sheets.

2. Materials and Methods

Hot-rolled 30 and 35 mm thick 300 mm × 300 mm sheets of AA1570 and HSS M2 steel FSW tools were used to carry out friction stir welding using an FSP machine PowerStir 345C (Holroyd Precision Ltd., Milnrow, UK) at the facilities of S.P. Korolev RSC Energia Experimental Machine-building Plant ZEM. The shape and parameters of the tools used for the welding are shown in Figure 1. The process parameters were adjusted experimentally to obtain the FSW joints, as shown in Figure 1. The FSW tool rotation rate, traverse speed, and plunging force were 600 RPM, 10 mm/min, and 40 kN, respectively. The FSW seam can be divided into three zones as shown in Figure 1a, where zones B, S, and H denote the FSW tool plunging zone characterized by insufficient heating plasticization, steady welding, and the FSW tool exit zone, respectively.

The stop-action technique [19] was applied to study the mechanical behavior of the metal in the thermomechanically affected zones (TMAZs) when the FSW tool exit hole was sectioned and then marked as shown in Figure 2. Another series of metallographic views was obtained by EDM cutting the seam by planes normal to the sheet surface and both perpendicular and parallel to the FSW joint line. An EDM cutting machine DK7745 (Suzhou Simos CNC Technology Co., Ltd. Suzhou, China) was used to cut off the samples, which were then ground, polished, and etched to prepare the metallographic section views.

Figure 1. The friction stir welding (FSW) seam zones on a 35 mm thick AA1570 sheet (**a**), the FSW tool used for the 35 mm thick sheet after welding (**b**), the FSW tool used for the 30 mm thick sheet after welding (**c**), and the FSW tool used for the 35 mm thick sheet before welding (**d**). 1–tensile specimens; 2–metallographic section views, d$_{max}$–the larger pin diameter, h–the pin height, α–the taper angle.

The joint strength was tested using a test machine BISS UT-04-0100 (BISS (P) Ltd. Bangalore, India) on samples cut off the seam by a plane perpendicular to the tool travel direction at different distances below the top surface so that the stir zone was located in the middle of the sample's gauge length part (Figure 1). The loading speed was 100 mm/s.

A microstructural examination was performed using optical microscopes Altami MET 1T(Altami Ltd., Saint-Petersburg, Russia), Olympus OLK41102 (Olympus Corp, Tokio, Japan), SEM and TEM instruments Tescan VEGA 3 (TESCAN ORSAY HOLDING, BRNO, Czech Republic), Zeiss LEO EVO 50 (Carl Zeiss AG, Oberkochen Germany), and JEOL JEM-2100 (JEOL Ltd., Akishima, Japan).

Thin foils for TEM were prepared using EDM cutting, mechanical polishing, and ion thinning to represent the stir zone metal in two perpendicular sections with respect to the joint line. A microhardness tester Duramin-5 (Struers A/S, Ballerup, Danemark) was used to obtain microhardness profiles across the FSW seam zones at a 490 mN load.

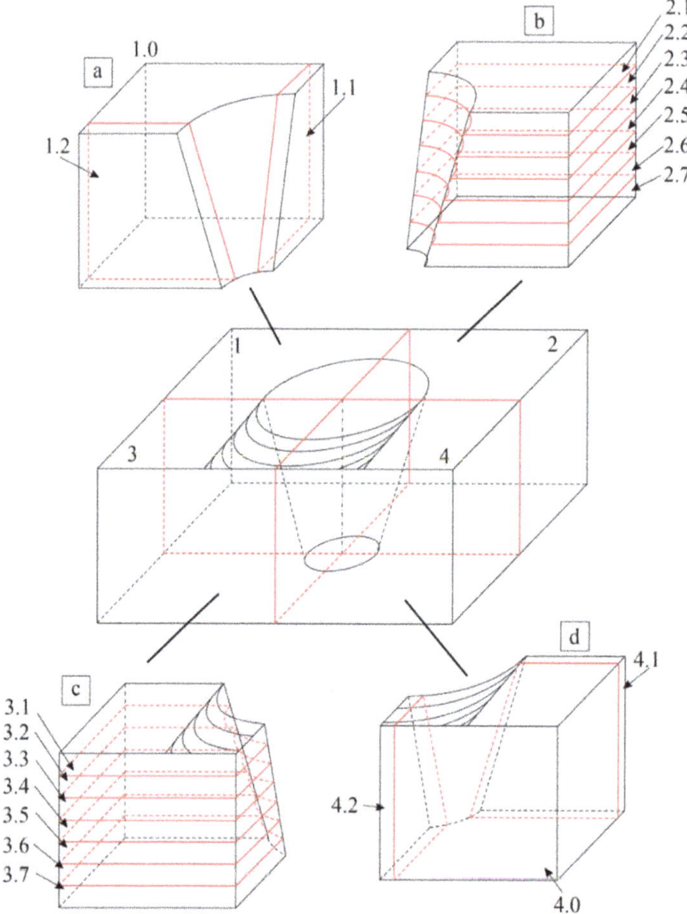

Figure 2. A diagram showing the sample cut-off scheme: (**a**) the leading edge of the exit hole on the retreating side (LE-RS), (**b**) the leading edge of the exit hole on the advancing side (LE-AS), (**c**) the trailing edge of the exit hole on the retreating side (TE-RS), (**d**) the trailing edge of the exit hole on the advancing side (TE-AS).

3. Results

3.1. Stir Zone and Thermomechanically Affected Zone Macrostructures

The transverse cross-sectional view in Figure 3 reveals the inhomogeneous macrostructure of the deep penetration FSW joint stir zone. The onset part of the joint (zone "B" in Figure 1) is characterized by a dense small nugget zone formed in the root part of the seam (Figure 3a,b), while "loose" metal with discontinuities (Figure 3a,b, defect 1) is formed in the top part. The FSW area behind the tool is characterized by the presence of a crack (Figure 3a,b, defect 2).

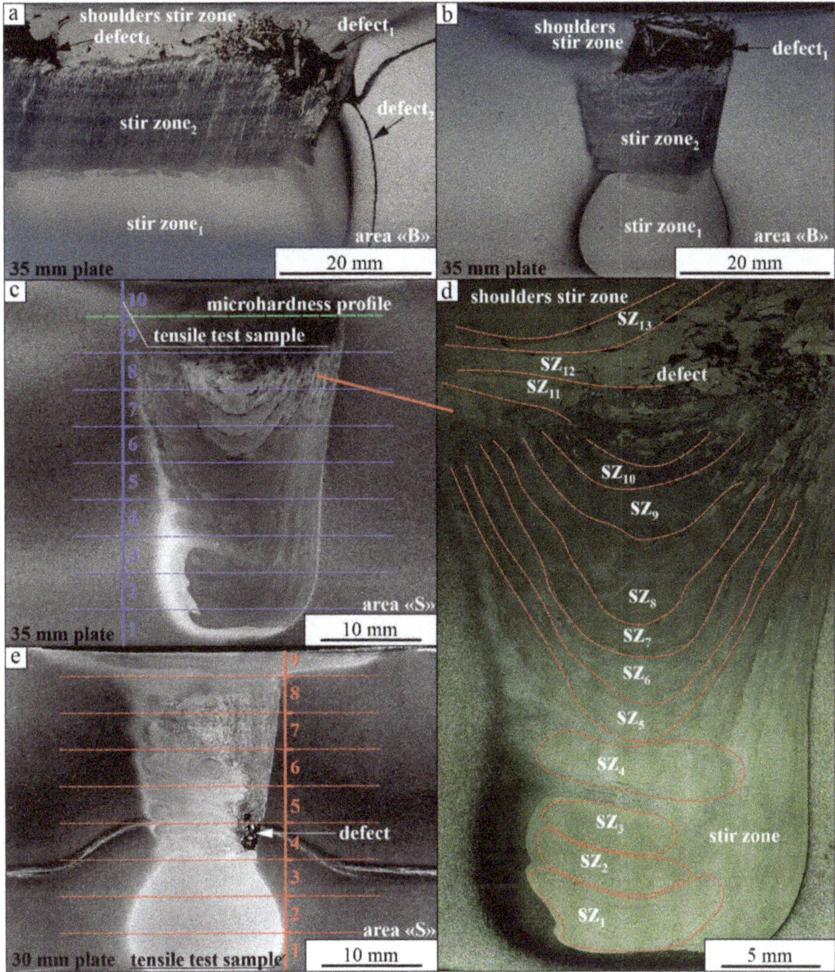

Figure 3. The macrostructures of the FSW joint on 35 mm thickness alloy sheet in section parallel to the joint line: (**a**) the cross section of zone B (**b**); the cross section of zone S (**c**); the multi-nugget macrostructure of the stir zone (**d**); and a cross section view of the joint on the 30 mm thickness metal (**e**).

The steady welding zone "S" is characterized by another type of macrostructure, where numerous nugget structures are formed in the stir zone (SZ) instead of a single nugget (Figure 3c,d). These nugget structures are evidence of change in the metal flow compared to that of zone "B". The top part of the stir zone obtained on the 35 mm thick sheet still contains pores and discontinuities.

This type of macrostructure is also typical for the FSW joint obtained on the 30 mm thick sheet (Figure 3e), with the exception of a wormhole defect formed in the half-height AS zone because this 30 mm thick sheet was made of two thinner ones.

The FSW tool exit zone H revealed the metal flow patterns formed by the tool around the exit hole. Figure 4a shows the microstructure of sample 4.2 cut off the trailing part of the exit hole, as shown in Figure 2. The bottom metal shows a large and dense nugget (Figure 4a, SZ_1), above which a number of metal flow patterns (SZ_2–SZ_7) can be seen. The "loosest" macrostructure is observed on the SZ_8 pattern, which is formed by a shoulder-driven metal flow.

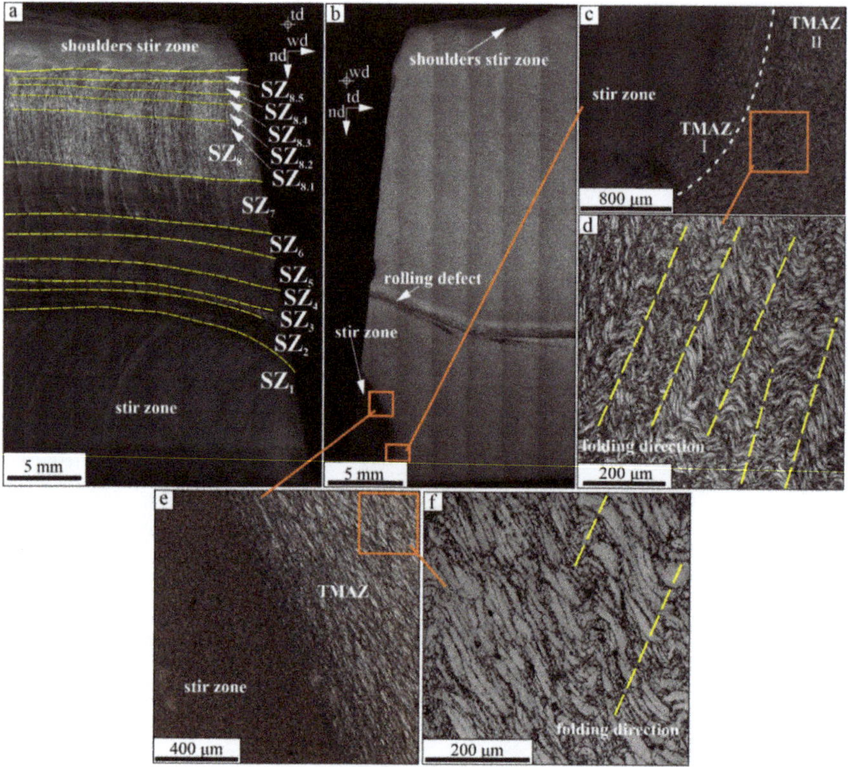

Figure 4. Macrostructures in the FSW tool exit zone as viewed from the plane parallel to the welding direction and perpendicular to the sheet's surface: (**a**) trailing part of the exit hole, sample 4.2; (**b**) the exit hole on the advancing side of the weld under different magnifications (**c**,**e**); kinked TMAZ grains in the top (**f**) and bottom (**d**) parts, respectively.

The macrostructure of the AS exit hole quadrant planar section (Figure 4b) (cut off as shown in in Figure 2, section 4.1) is represented by a difficult-to-see stir zone and a TMAZ that can be divided into the ultrafine grained TMAZ I in the vicinity of the SZ and TMAZ II, which features larger kinked grains located closer to the base metal (BM) (Figure 4c,d). Only a small portion of the SZ can be observed using this view. The rolling welding defect between the two sheets is also clearly seen here (Figure 4b). The TMAZ zone structures show orientation dependence with respect to the different exit hole quadrants. This is also true for such structural characteristics as grain size and grain shapes (Figure 4c,e).

Since the base metal hot-rolled grains were preferentially oriented with their longest axes along the rolling direction and their boundaries were pinned by the nanosized precipitates against continuous recrystallization, mechanical pressure in TMAZ II exerted from the SZ distorted their shapes. Distinct from the TMAZ I superfine grains, whose deformation at the SZ/TMAZ boundary was fully determined by the friction force, the TMAZ II grains oriented with their long axes along the pressure retained their sizes but instead experienced kinking. The alternating kinking band patterns can be delineated as shown in Figure 4d and look similar to those generated with the shear banding mechanism. Those grains whose shorter axes coincide with the pressure direction were more resistant to kinking. Kinking is also more pronounced in the bottom parts of the TMAZ II where the FSW tool-induced metal flow pressure is higher.

The planar section views in Figures 5–7 represent the structures formed at different distances below the metal surface. The structures formed close to the leading edge of the exit hole in sections 2.2–2.6, as well as those formed at the trailing edge of the exit hole in sections 3.2–3.6, are shown in Figures 5 and 6, respectively.

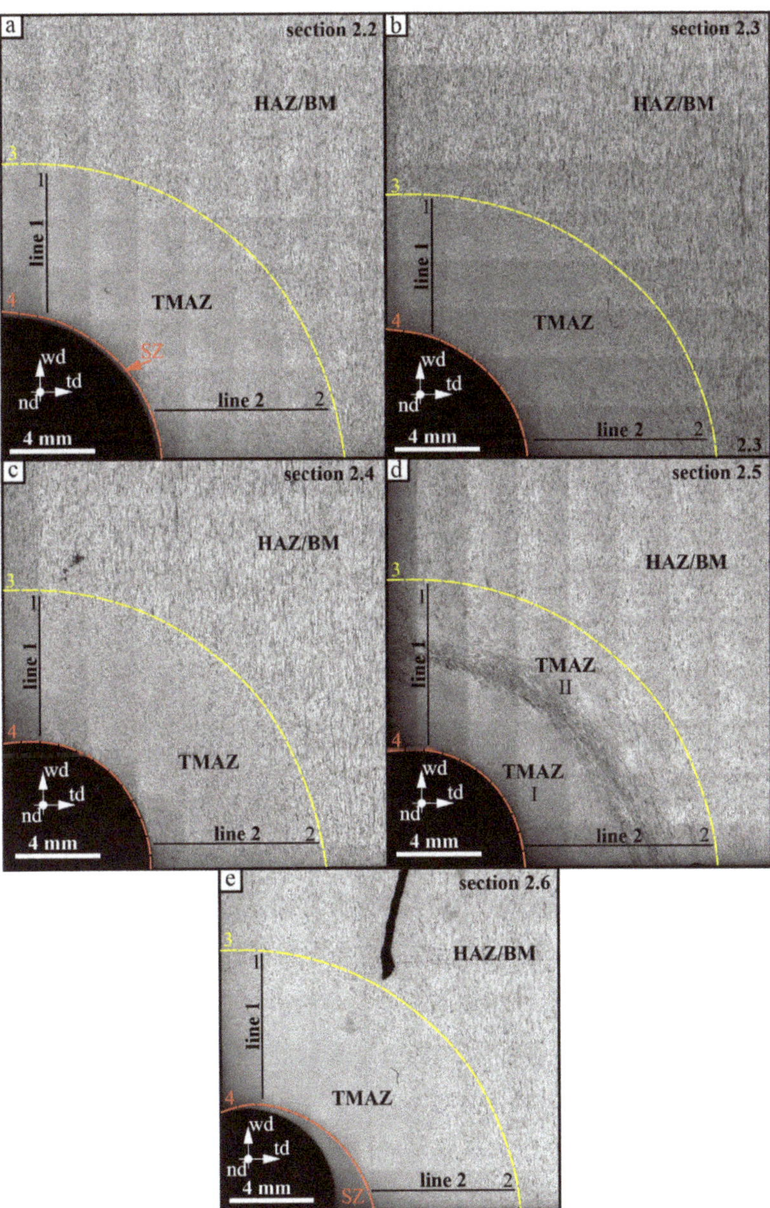

Figure 5. The microstructural evolution on stir zone (SZ), thermomechanically-affected zone (TMAZ), heat-affected zone (HAZ) and base metal (BM)microstructures near the exit hole leading edge on the advancing side (AS), as seen in plane sections 2.2 (**a**), 2.3 (**b**), 2.4 (**c**), 2.5 (**d**), and 2.6 (**e**) (Figure 2).

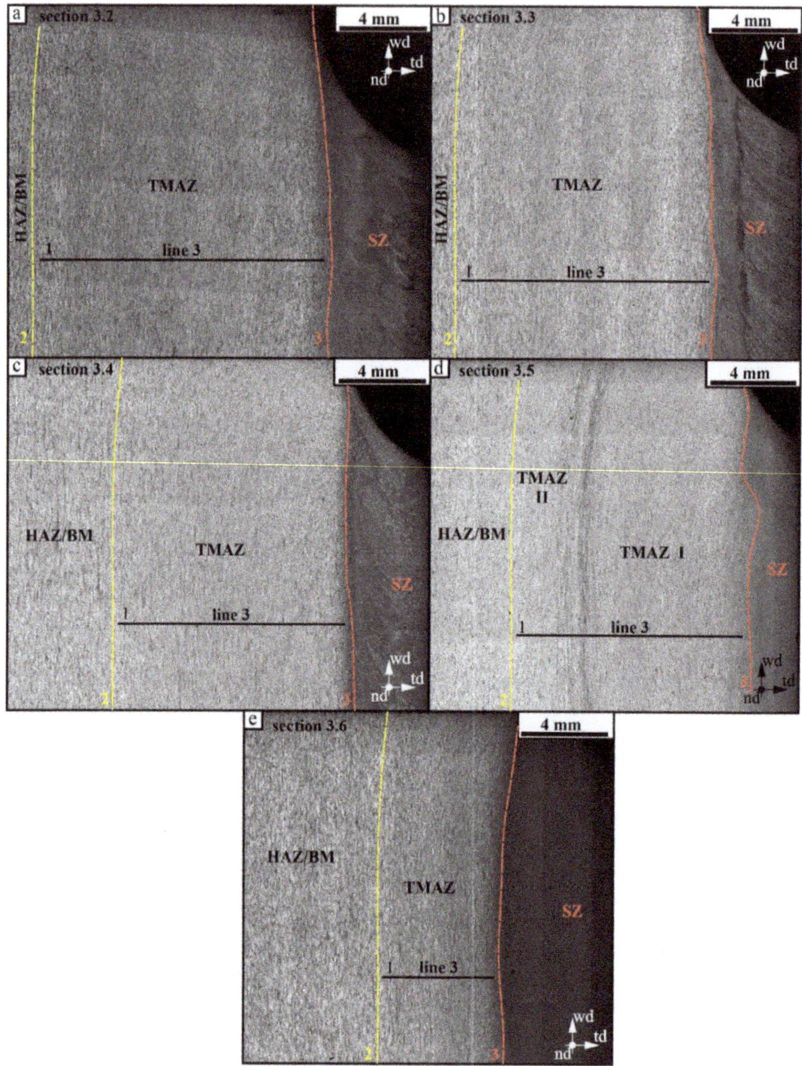

Figure 6. The alloy structures in the exit hole leading edge on the AS side, as viewed in sections 3.2 (**a**), 3.3 (**b**), 3.4 (**c**), 3.5 (**d**) and 3.6 (**e**) (Figure 2).

The leading AS edge planar sections 2.2 to 2.5 (see Figure 2) are represented by a very narrow stir zone (Figure 5a–d) but a wide TMAZ composed of both elongated and almost equiaxed grains, whose origin may be explained by horizontal plane sectioning of the kinked grains. The TMAZs are wide in sections 2.4–2.4 (Figure 5a–c), while those in section 2.5 (Figure 5d) demonstrate a narrow TMAZ that becomes wider again in section 2.6 (Figure 5e).

The grain size in the TMAZ is gradually increased from the SZ/TMAZ to TMAZ/HAZ boundary (Figure 8).

When examining the structures and zones formed on the trailing edge of the hole in sections 3.2–3.6 (Figure 6), it can be noted that the TMAZ is wider in the RS trailing edge (TE-RS) (Figure 6) of the exit hole than the TMAZ in the leading edge AS part (LE-RS) (Figure 5).

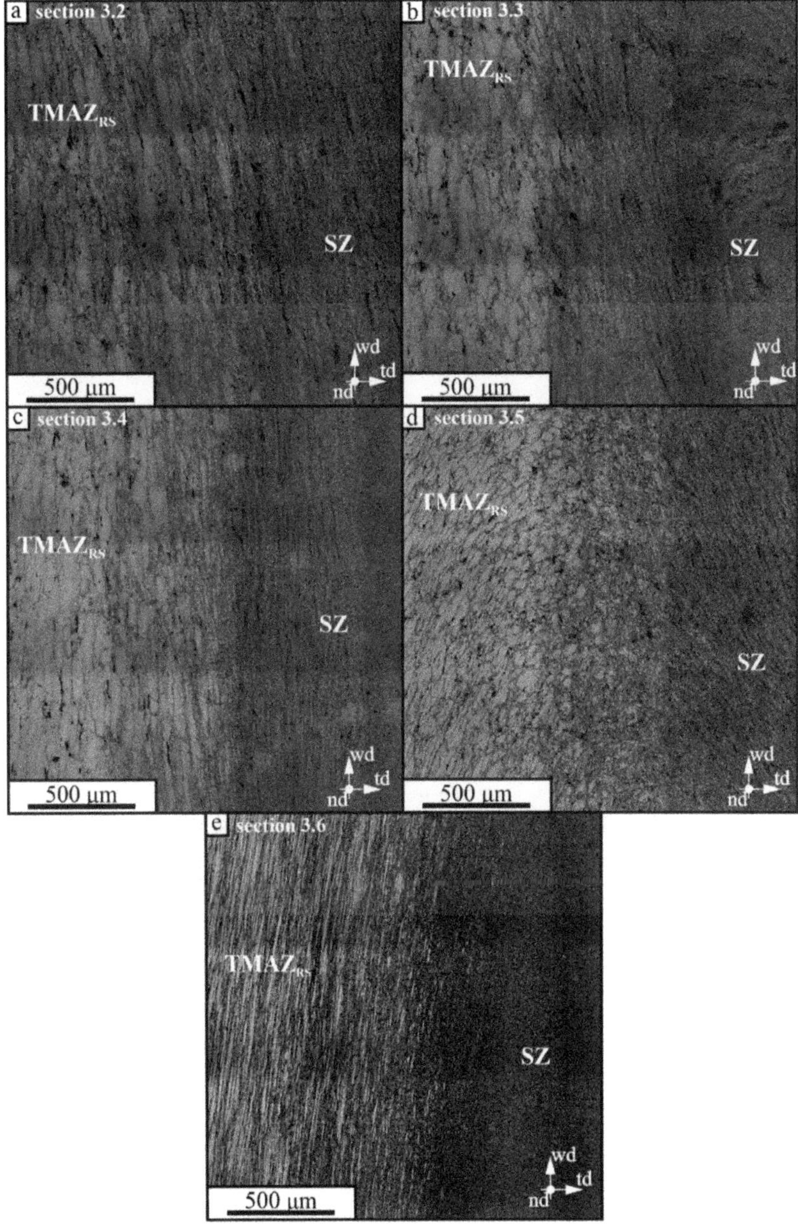

Figure 7. The TMAZ grain structures in planar sections 3.2 (**a**), 3.3 (**b**), 3.4 (**c**), 3.5 (**d**) and 3.6 (**e**), as viewed normal to the sheet plane.

These differences are clearly seen in Figure 7, where the TMAZ grain structures are shown at a higher magnification. sections 3.2 to 3.4 reveal moderately elongated grains whose length is about twice as much their width. sections 3.4 and 3.6, respectively, demonstrate almost equiaxed and extremely elongated thin grains.

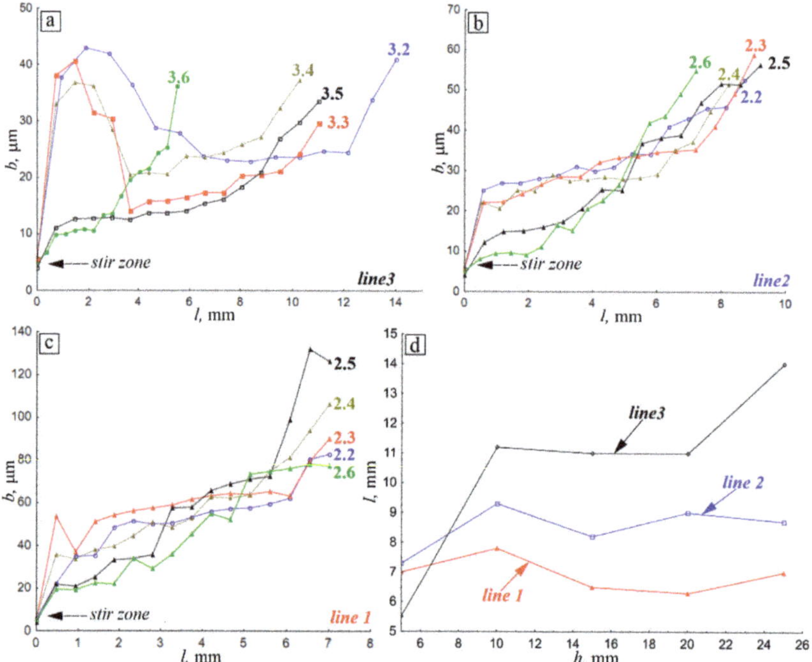

Figure 8. The TMAZ grain size distributions as measured along lines 3 (**a**), 2 (**b**), and 1 (**c**) (in Figures 5 and 6 (**a,c**)) and the dependence of the TMAZ size on the height of the 30 mm thick sample (**d**). The diagram labels are in accordance with those of the diagram in Figure 3.

3.2. Stir Zone and Thermomechanically Affected Zone Microstructures

The specifics of the TMAZ grain structure modification in the exit hole area are shown in Figure 8. Transversely to the SZ RS, along line 3 (shown in Figure 6), the change of the grain size from the stir zone to the heat-affected zone at different depths is determined by two main characteristic features. In samples from the upper part of the welded joint zone (sections 3.2, 3.3, and 3.4 in Figure 2), there is an initial rapid growth of grains with a subsequent drop and a further growth of the grain size, which becomes equal to the grain sizes of the heat-affected zone (HAZ) or the base metal. In samples from the bottom part of the joint (sections 3.5 and 3.6), the grains gradually grow from the stir zone to the heat-affected zone (Figure 8a). There is no grain growth near the stir zone. This feature indicates different deformation and recrystallization processes at different distances from the surface of the FSW-joint. At the leading edge of the tool exit hole on AS along line 2 (shown in Figure 5), the SZ-boundary material in sections 2.2, 2.3, and 2.4 is characterized by larger grains compared to those in sections 2.5 and 2.6 (Figure 8b). At a distance of more than 5 mm from the stir zone, the TMAZ grain size is leveled on all samples.

Along welding line 1 (Figure 5), the increase in the grain size from the stir zone to the heat-affected zone occurs gradually on all samples (Figure 8c). Significant differences are observed only for sections 2.5 and 2.4, far from the stir zone.

Unevenness in the size of the TMAZs can be observed both in different directions (for the welding axis) and at different distances below the weld surface (Figure 8d). It is expected that the smallest size of the TMAZ will be characteristic of the leading edge, where the TMAZ size changes smoothly and identically along line 1 and line 2. In the trailing edge area on the retreating side and across the stir zone along line 3, the size of the stir zone increases from the minimum in the bottom of the joint to the average in the central volume, with the maximum found in the shoulder-affected area. These changes show strain inhomogeneity in different areas of the joint.

The microstructures of the SZ zone in both joints are represented by recrystallized grains and coarse Al_3Sc precipitates (Figure 9a,b). It can be seen from Figure 9c,d that the coarse incoherent particles cannot effectively pin the grain boundaries, so they continue migrating through the precipitates leaving chaotically distributed dislocations and dislocation loops.

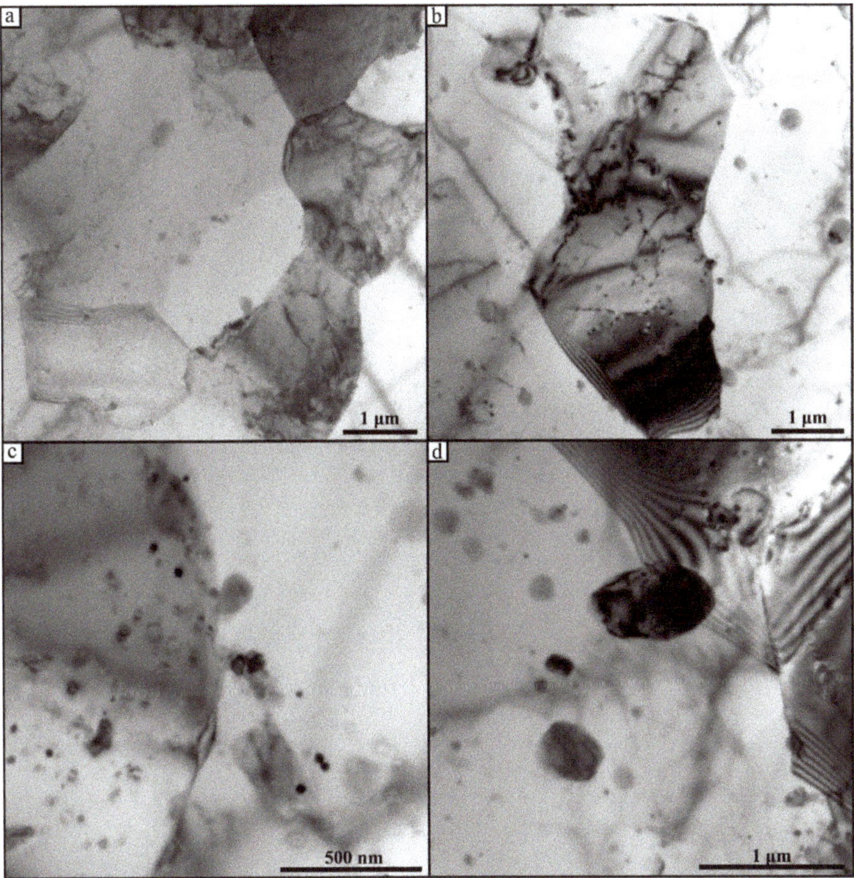

Figure 9. TEM images of the stir zone coarse precipitates, grains, and dislocation loops in 35 mm (**a,c**) and 30 mm (**b,d**) welds.

TEM images of the TMAZ microstructures in both FSW seams are characterized by the presence of numerous extinction contours inside the elongated and bent grains (Figure 10a, shown with arrows). The presence of several bend extinction contours inside an elongated grain is evidence of alternating sign flexural strain (corrugation). These corrugated grains are geometrically subdivided into shorter subgrains to accommodate the bending. Some of these subgrains are separated by high-angle boundaries from the neighboring subgrains (Figure 10b, shown with arrows).

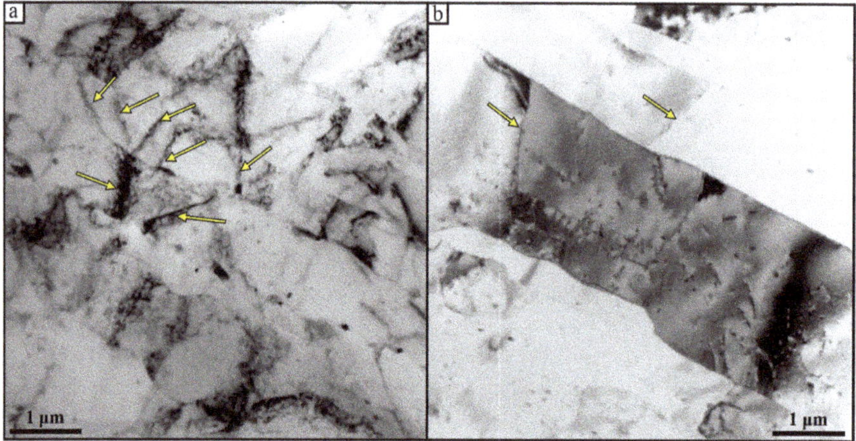

Figure 10. TEM images of the TMAZ structures in the 35 mm (**a**) and 30 mm (**b**) thickness seams.

3.3. Mechanical Strength

The results of the mechanical tests show that the welded metal has a strength close to that of the base metal (Figure 11a). The minimal tensile strength of 235–280 MPa was obtained when testing the samples cut off the top and bottom sections 1.1, 1.2, 1.9, and 1.10 (see Figure 3e). The samples cut off the central sections, with the exception of the section 1.7 sample, had a strength around 355–365 MPa.

Tests in the defect-free zone "S" (Figure 1) areas show that the ultimate tensile strength (UTS) of the specimens is about 350–390 MPa (Figure 11b). The exception is Specimen 2.10, cut from near the facing surface of the sample, whose UTS is about 305 MPa. The dependence of tensile strength on the distance from the weld root is shown in Figure 11c.

Testing of the welded samples shows their high defectiveness due to the formation of large discontinuity in the weld, which is detrimental to the metal's mechanical properties (Figure 11c,d). Samples from the defect-free areas have strength properties at the level of equivalent samples from 35 mm thick welded plates at a level of 400–405 MPa. In the presence of defects, this strength is very low (even close to zero). The difference between zone "S" and zone "B" (Figure 1) is seen in the diagrams. Only one sample cut from the center zone showed a strength close to zero, while about one third of them cut from zone "B" had similar results.

When comparing the mechanical properties in the cross sections of materials of both thicknesses, the defect localized in the structure becomes clearly distinguished (Figure 11e,f). For joints made of 35 mm-thick plates, the poor strength shown on samples from the upper part is typical (Figure 11e). For samples made of 30 mm-thick plates, the strength loss is typical for samples from the central part of the joint (Figure 11f). It should be noted that, generally, samples demonstrate rather low mechanical properties in comparison with that of the base metal, which could be caused by over-ageing of the SZ material during welding. This process is characterized by accelerated precipitation of the particles from the solid solution, making the particles lose their coherence with the matrix and thus impairing the joint mechanical characteristics. At the same time, the mechanical properties of the base metal also have a significant heterogeneity when testing samples cut off along and across the rolling direction (Figure 11g). In this case, the samples cut along the rolling direction have a slightly higher strength and less ductility than those cut across the rolling direction.

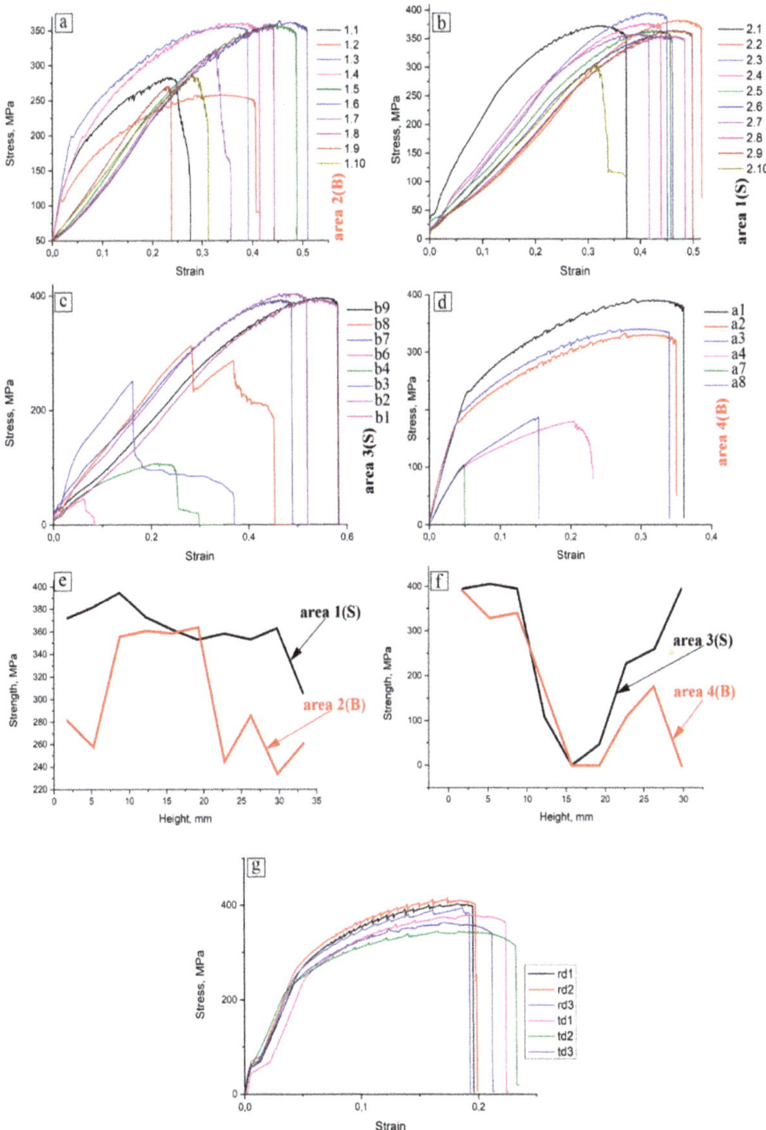

Figure 11. Mechanical properties of specimens cut across the weld at different distances from the weld root for 35 mm-thick (**a**,**b**) and 30 mm-thick (**c**,**d**) plates, the dependence of the mechanical properties on the distance from the weld root for 35 mm-thick (**e**) and 30 mm-thick (**f**) plates in zones "B" and "S" (see Figure 1), and the test diagrams of the base metal tested along and across the rolling direction (**g**).

The discontinuous yielding effect, strongly pronounced during the testing of the base metal specimens (Figure 11g), is eliminated during the testing of the FSW joined samples. During the deformation of specimens along and across the welding axis, the most pronounced manifestation of the Portevin Le Chatelier effect is observed in cases where the tensile axis coincided with the welding axis. This leads to high amplitude serrations under the tensile loading. In samples of the weld material, the phenomenon of discontinuous yielding was observed practically on all specimens cut from the defect-free zones as well as on a small part of the defective ones (Figure 11a–d). Some specimens demonstrated deformation behavior similar to that of the base metal ones tested along the rolling direction, and others showed the same with specimens tested across the rolling direction. For example, curves a1, a2, and a3, shown in Figure 11d, demonstrate similar changes in strain to those in diagrams of the specimens cut along the rolling direction rd1–rd3, while curves 2.1 and 2.3 are similar to those in diagrams of the specimens cut along the transverse direction td1–td3. The main difference between the above groups of specimens is their differently pronounced stress changes during deformation. In this case, the strength of the base metal in the tensile test along the rolling direction is 405 MPa, which coincides with the strength of the 30 mm-thick joint specimens in the transverse direction of 365 MPa, which is close to the strength of the defect-free specimens of the 35 mm-thick joint.

The microhardness profiles obtained from both joints demonstrate the reduced microhardness numbers of the stir zone (Figure 12) compared to those of base metal. Such a finding is typical for heat treatable aluminum alloys because this zone is usually over-aged in FSW such that the precipitates lose their coherence, and dispersion hardening becomes ineffective. No extra hardening was noted in the heat-affected zone, which may be explained by the thermal stability of both grains and precipitates in this zone.

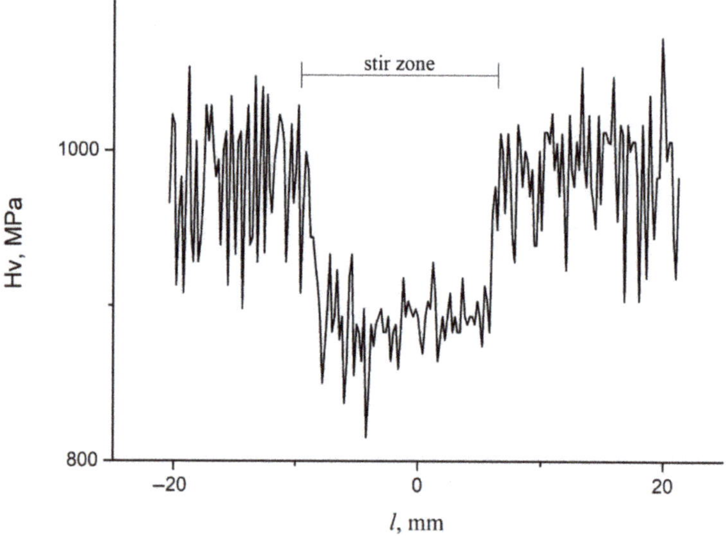

Figure 12. The microhardness profile across the FSW joint at 5 mm below the surface on 35 mm thickness metal.

3.4. Metal Transfer

The macrographs in Figures 13 and 14 show grain refining in the base metal via friction from the rotating FSW tool. It was suggested earlier [20] that metal transfer in FSW may be performed in a layer-by-layer manner when the grain-refined and plasticized metal layer first sticks to the FSW tool surface and is then transferred to the zone behind the tool and deposited there on a previously deposited layer. The stop-action technique allows one to observe plasticized metal layers, which are separated from the bulk of the metal on the advancing side of the hole (Figure 13), transferred to the retreating side (Figure 14), and finally deposited on the previously transferred and bonded layers.

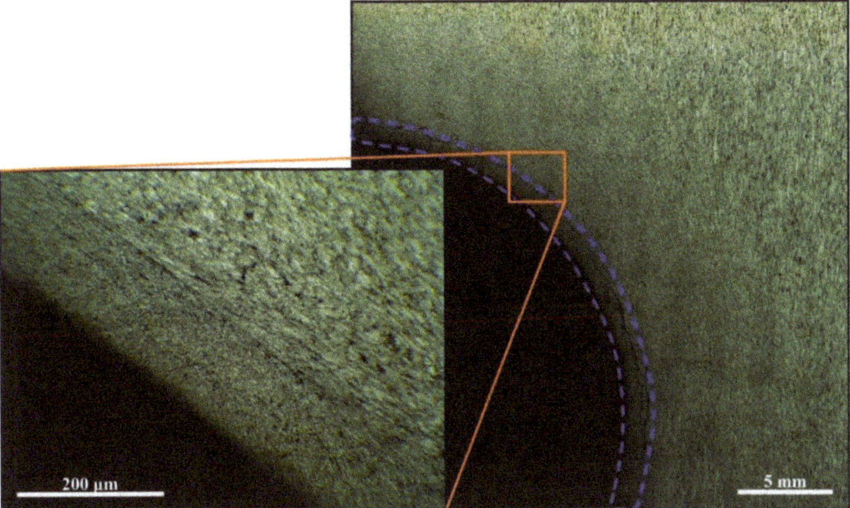

Figure 13. Optical images of the microstructures formed in the FSW tool exit hole metal leading edge metal (section 2.2 in Figure 2). Blue dotted line contour denotes the transfer layer area.

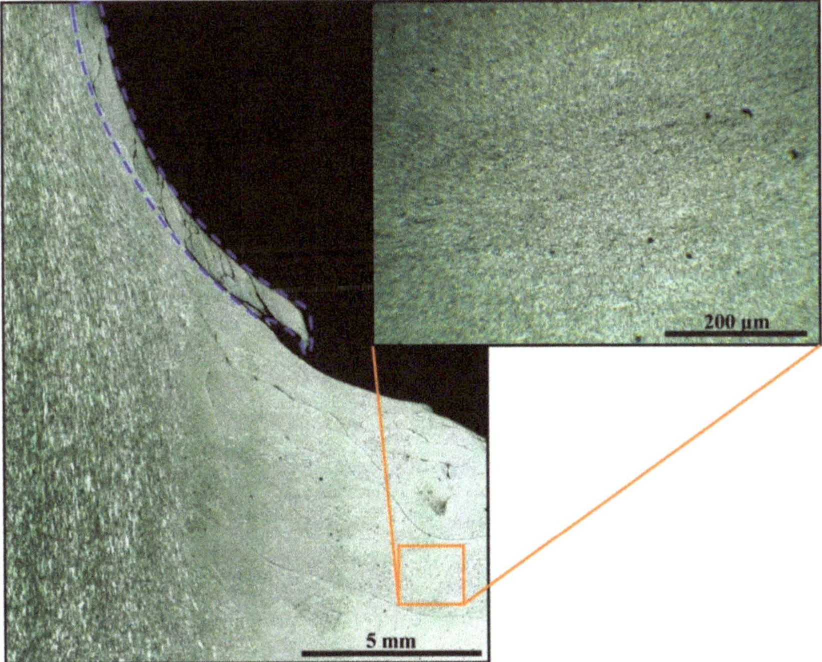

Figure 14. Optical images of the microstructures formed in the FSW tool exit hole metal trailing edge metal (section 3.2 in Figure 2). The blue dotted line contour denotes the transfer layer area.

SEM BSE images allow one to distinguish by contrast the intermetallic precipitates of solid solution grains (Figure 15). The majority of these coarse precipitates look like streaks oriented along the rolling direction (as well as tool travel speed) and are clearly seen in the base metal, while fewer are distributed chaotically.

Figure 15. SEM BSE composite image of the metal and precipitates in the FSW tool exit hole leading edge metal (section 2.2 in Figure 2).

The friction force induced deformation not only distorts the grains but also changes their orientation, which can be observed by the orientation of the precipitate streaks (Figure 15).

Completely recrystallized metal is observed in a thin 0.5–0.6 mm layer, which was stuck to the FSW tool and thereby transferred in a spiral bottom-to-surface trajectory.

4. Discussion

The stir zone formation in high thickness materials may occur in a very complex multiflow manner so that a number of vortex structures with different configurations forms there, partially due to the low pressure in the welding zone. As a result, a welded joint has, in addition to structural irregularities, significant differences in its mechanical properties with respect to the distance below the top surface. Despite the presence of structural irregularities, it is possible to accurately characterize the process of the welded joint formation in the tool pin stir zone (excluding the area of shoulders influence). As described above, the direct material transfer by the tool is predefined by the primary subdivision of the base metal grains through severe plastic deformation with the formation of a wide thermomechanically affected zone (Figure 5). The grain size is expected to increase with distance from the stir zone to the heat-affected zone (Figure 8). The different widths of the TMAZ at different levels of the weld indicates that the process of grain subdivision and, consequently, the formation of the material transfer layer (Figures 14 and 15) occurs with a different intensity depending on the distance below the top surface. This ensures that the stir zone consists of several nuggets and makes the material transfer frequency potentially differ from level to level. This irregularity in the adhesive interaction also likely affects the mechanical characteristics. However, the peculiarities of the formation of structures at each of the studied weld levels demonstrate that the welded joint formation in the tool pin stir zone occurs via the same mechanism.

The results obtained in this work, as well as those obtained earlier, suggest that FSW seam formation occurs according to the scheme shown in Figure 16.

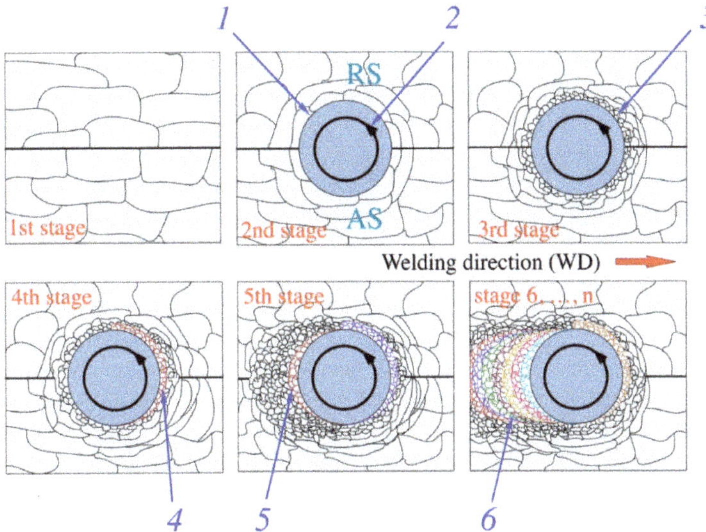

Figure 16. Schematic diagram to illustrate the adhesion-assisted transfer of metal with the FSW pin. 1—tool, 2—rotation direction, 3—primary grain subdivision, 4—formation of transfer layer, 5—reverse transfer (deposition), 6—successive layer-by-layer metal transfer and deposition.

After plunging the FSW tool into a metal, the tool starts travelling along the joint line. The plasticized and grain refined metal layer is transferred by the rotating tool to the trailing zone where it is detached from the tool and sticks back to the hot metal deposited. This cycle is repeated again and again to form the welded seam. The transferred layer thickness is determined by the temperature and shear stress/strain gradients exerted from the tool into the base metal. This model was inferred from numerous experiments on unlubricated and adhesion controlled sliding friction when direct and reversed metal transfer occurs between the contacting bodies [20–22].

This discontinuous transfer mechanism is responsible for the formation of inhomogeneous grain structures, SZ nuggets, and the TMAZ. An overly strong adhesion of the plasticized metal to the FSW tool may have detrimental effects on the stirring and metal flow efficiency. The transferred portion of metal may not detach from the tool surface and thus be involved in the next cycle, thus breaking the regularity of transfer and forming additional metal flows, which can then provide the final multi-nugget pattern shown in this work in Figure 3c,d. This is especially likely in FSW on thick metal sheets with high pressure and tool travel resistance.

On the other hand, overly weak adhesion would not allow the metal to form a welded joint at all. Therefore, the adhesion of the metal to the tool is an important factor in metal transfer, which has never been considered in modeling. It is known, for example, that it is possible to obtain an FSW seam using an absolutely smooth FSW pin surface [23,24] and that it is the adhesion that works here to help transfer the metal.

Our results show that the SZ metal precipitates are not coherent with the Al-Mg matrix; thus, grain structure recrystallization is not stopped and continues by means of grain boundary migration and diffusion-controlled dissolution of the incoherent precipitates, leading to numerous dislocation loops forming in the metal. It is known that the dislocation loops can appear by a vacancy condensation mechanism in diffusion on the grain boundaries. It seems that the coarse precipitates almost have no pinning effect of the grain boundaries so that grains may grow as large as several micrometers. Severe deformation and heating during FSW facilitates strain dissolution of the initial precipitates that, consequently, have to precipitate again. Intense stirring in the SZ dissolves them again and again until stirring is stopped and the metal comes to rest. This is where precipitation overtakes dissolution, and precipitates grow very quickly on the defective structure. That is why the coarsest precipitates are found in the TMAZ (Figure 3), where the strain is low and some particles are hiding there and thus avoid the strain dissolution. In addition, Figure 8a shows that some grain growth can occur in the vicinity of the SZ/TMAZ boundary. This finding can be explained by the same reason, i.e., that the coarse incoherent precipitates are formed in the TMAZ and allow the grains to grow. Nevertheless, even if the precipitates are semicoherent, they may still retain some pinning effects on the grain boundaries.

The TMAZ zone is characterized by kinked grains (Figure 4d), which are the results of deformation under the pressure exerted by the FSW tool. A similar effect was observed under compression deformation of the Al-Mg-Sc alloy (AA5024) at 523 K [25]. TEM mages show that these kinked grains are composed of even smaller subgrains with high-angle boundaries (Figure 13) that are plausibly formed by microshears in the band-like grains, according to the scheme proposed by Belyakov et al. [26]. Another possible mechanism could be geometrical recrystallization [27], which is considered to be a type of continuous recrystallization. It was reported [28] that continuous recrystallization is inherent with the Al-Mg-Sc-Zr alloys.

5. Conclusions

As a result of the study, it was found that the stir zone of FSW joints on heavy gauge Al-Mg-Sc-Zr alloy sheets is a complexly structured zone with the presence of several basic structural elements. At the beginning of the welded joint, there is a nugget with an onion ring structure of a small size, and a highly defective area is located closer to the weld top surface. In the middle and end of the 35 mm-thick joint, there is a stir zone with a complex structure that consists of a large number of vortex or turbulent material flows formed by its movement along the tool contour. In the 30 mm-thick sample, a stir zone with a small nugget was formed over the entire length of the joint.

In samples of both thicknesses, defects can occur throughout the entire weld length and reduce the strength of the joint area to almost zero. The macro-defect formation localization is possible both in the lower third of the thickness in the advancing side of the weld and in the upper part in the stir zone in the shoulder effect area. Alternatively, it can be located on the border of the stir zone in the shoulder-driven area and in the pin-driven area. In the zone of defect formation, the strength properties are about 235–280 MPa for 35 mm-thick plates and for 30 mm-thick plates in the defect areas, the strength decreases until reaching zero, which might be caused by different types of formed defects. The mechanical properties of the defect-free material are thus at a sufficiently high level. Specimens from the defect-free areas of the 30 mm-thick joints have strength properties around 400–405 MPa, while specimens from 35 mm-thick welded joints have a lower strength of about 350–390 MPa. The strength of the base metal during the tensile test along the rolling direction is 405 MPa, and in the transverse direction it is 365 MPa. In terms of strength and discontinuous yielding effect implementation, specimens from the 30 mm-thick plates were closer to the base metal tested along the rolling direction, while specimens from the 35 mm-thick plates were closer to the material tested across the rolling direction.

The structure of welded joints is formed by the successive layer-by-layer transfer of a material by means of adhesive interaction. Due to the high thickness of the plates to be welded, the process of joint formation becomes more complicated, and the layer transferred by the tool is divided into separate sections with the presence of an upward motion component along with the motion around the tool. The high thickness of the Al-Mg-Sc-Zr alloy plates being welded might affect the plastic deformation process during the FSW so that metal flow was rearranged into separate small vortex zones, along with the formation of complex material flows along the tool contour and, as a result of the above, the formation of a multi-nugget stir zone with a complex structure.

Author Contributions: Conceptualization, writing—original draft preparation, methodology, investigation, visualization, T.K. and A.C.; formal analysis, writing—review and editing K.K.; investigation, S.F.; project administration, supervision, E.K.; conceptualization, S.T. All authors have read and agreed to the published version of the manuscript.

Funding: The work was performed according to the government research assignment for ISPMS SB RAS, project No. III.23.2.4.

Conflicts of Interest: The authors declare no conflict of interest.

References

1. Gureeva, M.A.; Grushko, O.E. *Aluminum Alloys in the Modern Transportation Manufacturing and Education*; Moscow State Industrial University (MSIU): Astrakhan, Russia, 2009; Volume 3, pp. 27–41.
2. Filatov, Y.A.; Yelagin, V.I.; Zakharov, V.V. New Al-Mg-Sc alloys. *Mater. Sci. Eng. A* **2000**, *280*, 97–101. [CrossRef]
3. Deng, Y.; Zhang, G.; Yang, Z.; Xu, G. Microstructure characteristics and mechanical properties of new aerospace Al-Mg-Mn alloys with $Al_3(Sc_{1-x}Zr_x)$ or $Al_3(Er_{1-x}Zr_x)$ nanoparticles. *Mater. Charact.* **2019**, *153*, 79–91. [CrossRef]
4. Yang, W.; Yan, D.; Rong, L. The separation of recrystallization and precipitation processes in a cold-rolled Al–Mg–Sc solid solution. *Scr. Mater.* **2013**, *68*, 587–590. [CrossRef]
5. Heinz, A.; Haszler, A.; Keidel, C.; Moldenhauer, S.; Benedictus, R.; Miller, W.S. Recent development in aluminium alloys for aerospace applications. *Mater. Sci. Eng. A* **2000**, *280*, 102–107. [CrossRef]
6. Kumar, N.; Mishra, R.S. Thermal stability of friction stir processed ultrafine grained Al-Mg-Sc alloy. *Mater. Charact.* **2012**, *74*, 1–10. [CrossRef]
7. Nikulin, I.; Kipelova, A.; Malopheyev, S.; Kaibyshev, R. Effect of second phase particles on grain refinement during equal-channel angular pressing of an Al–Mg–Mn alloy. *Acta Mater.* **2012**, *60*, 487–497. [CrossRef]
8. Tuan, N.Q.; Pinto, A.M.P.; Puga, H.; Rocha, L.A.; Barbosa, J. Effects of substituting ytterbium for scandium on the microstructure and age-hardening behaviour of Al–Sc alloy. *Mater. Sci. Eng. A* **2014**, *601*, 70–77. [CrossRef]
9. Meng, Y.; Zhao, Z.; Cui, J. Effect of minor Zr and Sc on microstructures and mechanical properties of Al–Mg–Si–Cu–Cr–V alloys. *T. Nonferr. Metal. Soc.* **2013**, *23*, 1882–1889. [CrossRef]
10. Deng, Y.; Peng, B.; Xu, G.; Pan, Q.; Yin, Z.; Ye, R., Wang, Y.; Lu, L. Effects of Sc and Zr on mechanical property and microstructure of tungsten inert gas and friction stir welded aerospace high strength Al–Zn–Mg alloys. *Mater. Sci. Eng. A* **2015**, *639*, 500–513. [CrossRef]
11. Malopheyev, S.; Kulitskiy, V.; Mironov, S.; Zhemchuzhnikova, D.; Kaibyshev, R. Friction stir welding of an Al–Mg–Sc–Zr alloy in as-fabricated and work-hardened conditions. *Mater. Sci. Eng. A* **2014**, *600*, 159–170. [CrossRef]
12. Liu, F.C.; Ma, Z.Y. Achieving exceptionally high superplasticity at high strain rates in a micrograined Al–Mg–Sc alloy produced by friction stir processing. *Scr. Mater.* **2008**, *59*, 882–885. [CrossRef]
13. Liu, F.C.; Ma, Z.Y. Contribution of grain boundary sliding in low-temperature superplasticity of ultrafine-grained aluminum alloys. *Scr. Mater.* **2010**, *62*, 125–128. [CrossRef]
14. Liu, F.C.; Ma, Z.Y.; Chen, L.Q. Low-temperature superplasticity of Al–Mg–Sc alloy produced by friction stir processing. *Scr. Mater.* **2009**, *60*, 968–971. [CrossRef]
15. Liu, F.C.; Xue, P.; Ma, Z.Y. Microstructural evolution in recrystallized and unrecrystallized Al–Mg–Sc alloys during superplastic deformation. *Mater. Sci. Eng. A* **2012**, *547*, 55–63. [CrossRef]
16. Malopheyev, S.; Mironov, S.; Vysotskiy, I.; Kaibyshev, R. Superplasticity of friction-stir welded Al–Mg–Sc sheets with ultrafine-grained microstructure. *Mater. Sci. Eng. A* **2016**, *649*, 85–92. [CrossRef]
17. Li, M.; Pan, Q.; Wang, Y.; Shi, Y. Fatigue crack growth behavior of Al–Mg–Sc alloy. *Mater. Sci. Eng. A* **2014**, *598*, 350–354. [CrossRef]
18. Argade, G.R.; Kumar, N.; Mishra, R.S. Stress corrosion cracking susceptibility of ultrafine grained Al–Mg–Sc alloy. *Mater. Sci. Eng. A* **2013**, *565*, 80–89. [CrossRef]

19. Prangnell, P.B.; Heason, C.P. Grain structure formation during friction stir welding observed by the 'stop action technique'. *Acta Mater.* **2005**, *53*, 3179–3192. [CrossRef]

20. Tarasov, S.Y.; Filippov, A.V.; Kolubaev, E.A.; Kalashnikova, T.A. Adhesion transfer in sliding a steel ball against an aluminum alloy. *Tribol. Int.* **2017**, *115*, 191–198. [CrossRef]

21. Das, S.; Morales, A.T.; Alpas, A.T. Microstructural evolution during high temperature sliding wear of Mg–3% Al–1% Zn (AZ31) alloy. *Wear* **2010**, *268*, 94–103. [CrossRef]

22. Johnson, K.L.; Kauzlarich, J.J. Transfer of material between rolling and sliding surfaces. *Int. J. Mech. Sci.* **2004**, *46*, 343–357. [CrossRef]

23. Tongne, A.; Jahazi, M.; Feulvarch, E.; Desrayaud, C. Banded structures in friction stir welded Al alloys. *J. Mater. Process. Tech.* **2015**, *221*, 269–278. [CrossRef]

24. Schmidt, H.N.B.; Dickerson, T.L.; Hattel, J.H. Material flow in butt friction stir welds in AA2024-T3. *Acta Mater.* **2006**, *54*, 1199–1209. [CrossRef]

25. Chhangani, S.; Masa, S.K.; Mathew, R.T.; Prasad, M.J.N.V.; Sujata, M. Microstructural evolution in Al–Mg–Sc alloy (AA5024): Effect of thermal treatment, compression deformation and friction stir welding. *Mater. Sci. Eng. A* **2020**, *772*, 138790. [CrossRef]

26. Belyakov, A.; Kimura, Y.; Tsuzaki, K. Microstructure evolution in dual-phase stainless steel during severe deformation. *Acta Mater.* **2006**, *54*, 2521–2532. [CrossRef]

27. De Pari, L., Jr.; Misiolek, W.Z. Theoretical predictions and experimental verification of surface grain structure evolution for AA6061 during hot rolling. *Acta Mater.* **2008**, *56*, 6174–6185.

28. Masyukov, S.A. The Effect of Rolling on the Texture, Inhomogeneity of Precipitation of Dispersed Phases and Recrystallization AmG6, 1420 and 1570 Alloys. Ph.D. Thesis, MATI RGTU, Moscow, Russia, 2004.

Article

Wear of ZhS6U Nickel Superalloy Tool in Friction Stir Processing on Commercially Pure Titanium

Alihan Amirov *, Alexander Eliseev *, Evgeny Kolubaev *, Andrey Filippov * and Valery Rubtsov *

Institute of Strength Physics and Materials Science of Siberian Branch Russian Academy of Sciences 2/4, pr. Akademicheskii, 634055 Tomsk, Russia

* Correspondence: amirov@ispms.tsc.ru (A.A.); alan@ispms.ru (A.E.); eak@ispms.ru (E.K.); avf@ispms.ru (A.F.); rvy@ispms.ru (V.R.)

Received: 29 May 2020; Accepted: 15 June 2020; Published: 16 June 2020

Abstract: The use of electric arc or gas welding in the manufacture of titanium components often results in low quality welded joints due to large residual stresses and strains. A successful solution to this problem can be found in the application of friction stir welding. However, friction stir welding (FSW) of titanium alloys is complicated by rapid tool wear under high loads and temperatures achieved in the process. This paper studies the durability of a tool made of ZhS6U Ni-based superalloy used for friction stir processing of commercially pure titanium and the effect of the tool wear on the weld quality. The total length of the titanium weld formed by the tool without failure comprised 2755 mm. The highest wear of the tool is observed at the base of the pin, which brings about the formation of macrodefects in the processed material. The tool overheating causes an increase in the dendrite element size of ZhS6U alloy. The transfer layer contains chemical elements of this alloy, indicating that the tool wear occurs by diffusion and adhesion. As a result of processing, the tensile strength of commercially pure titanium increased by 25%.

Keywords: friction stir welding; grade 2 titanium alloy; ZhS6U Ni-based superalloy; microstructure; welding tool; tool wear

---

1. Introduction

Titanium alloys have found wide applications in the aerospace, transport, and chemical industries, as well as in medicine due to their high specific strength, heat resistance, and corrosion resistance [1]. These alloys considered to be readily weldable [1], but the use of conventional fusion welding techniques on them may lead to some undesirable effects, e.g., the formation of a coarse cast structure, porosity, strains, and residual stresses. Friction stir welding (FSW), which emerged in the early 1990s, is a solid-state process [2] and is successfully used for joining various types of aluminum [3,4], magnesium [5], and copper alloys [6,7], as it has many advantages including low residual stresses. Therefore, FSW may be more suitable for joining titanium alloys than conventional fusion welding methods. In addition, the FSW is widely used for joining metallic structures including those made of dissimilar metals and are not weldable by means of fusion methods. The use of FSW for joining aluminum and titanium alloys may allow avoiding unreliable riveted joints and applying modern light materials. Furthermore, friction stir processing (FSP) may be used for surface hardening metallic materials as well as for producing bulk processed materials including composites.

A problem of the FSW and FSP techniques is the rapid tool wear under high heat and load conditions [8,9]. In view of this fact, welding tools for FSW of titanium alloys are usually made of refractory materials, including alloys based on tungsten [10–18], cobalt [19–22], and molybdenum [23–25], as well as polycrystalline cubic boron nitride [26,27]. As reported [27], the use of a polycrystalline cubic boron nitride tool for FSW of titanium alloys led to the presence of tool wear products in the weld, such as titanium boride, nitrogen, and oxygen. In addition, fabricating the cubic boron FSW tool is both a costly and tedious process. Commercially pure tungsten has good strength characteristics at elevated temperatures, but it has low fracture toughness and therefore is worn out rapidly in FSW of hard materials (steel, titanium alloys) [28]. The heat and wear resistance of a tungsten tool can be improved by alloying it with such a rare earth and scarce metal as rhenium, which makes the tool's cost even higher.

A very popular tool material for FSW of titanium alloys is cemented tungsten carbide. This material has high 1650 HV and it is slightly sensitive to abrupt temperature changes [28]. There is evidence, however, that an adhesive titanium carbide layer is formed on the surface of the FSW joints due to high affinity of titanium to carbon [29]. In addition, carbon monoxide (CO) can be formed during FSW due to high temperatures and high pressure [28]. Although, this remark is not relevant for FSW performed with argon shielding.

Some FSW tools are made of tungsten with the addition of lanthanum. It was found that the given tool material diffuses to the titanium alloy surface [15]. This fact negatively affects both the weld quality and the tool wear.

Interesting results were obtained for a cobalt-based alloy tool. It was shown that the tool wear during FSW is caused by repeated titanium alloy adhesion to and pulling out the fragments from the tool [29]. However, the durability of cobalt tools was not studied in detail. Other rarer materials were also used for FSW tools, but they have not become widespread, e.g., a molybdenum-based alloy tool [23–25]. Unfortunately, the cited authors do not describe the tool wear process and provide no data on how the use of this type of tool affects the weld quality. Another example is a zirconium diboride tool with silicon carbide additives [30], the use of which proved to be undesirable due to rapid adhesive failure. Despite the large number of studies, the available materials for the manufacture of tools are either not durable enough, do not ensure reliable weld quality, or are technologically unfeasible due to complicated processing and high cost.

The use of a polycrystalline nickel-based superalloy as a tool material for FSW of steel [31] allowed yielding positive results with respect to the tool wear despite the presence of defects in the welded joint. An iridium-containing Ni-based superalloy used for stainless steel welding showed slight wear and deformation over the welding the 3 m length seam [32]. In practice, this superalloy has been successfully applied since the 1960s for the manufacture of turbine blades operating under high temperatures and mechanical loads [33]. A recent study [34] has shown the possibility of using the Ni-based superalloy as a material for making a tool for FSW on titanium alloys. However, the tool life has not yet been considered. All the above problems also apply to friction stir processing. FSP is similar to FSW in its effects and working principle, but it is slightly simpler technologically. The objective of this study is to investigate the durability of a ZhS6U Ni-based superalloy tool and the effect of its wear on a friction stir processed commercially pure titanium for the first time.

2. Experimental Procedure

2.1. Materials and Experimental Set-Up

The superalloy tool life was investigated in friction stir processing (FSP) of titanium alloy workpieces under constant conditions. This process is similar to friction stir welding and differs only in the absence of a joint line between two welded parts. For convenience of description, the processed material will be called a weld. In order to avoid oxidation of titanium alloy, the processing was carried out with argon shielding. Argon was fed through a special nozzle into the processing zone. An AISI 304 stainless steel plate was used as the substrate (Figure 1). The processed titanium alloy workpieces were fixed on the welding machine table using the clamps. A rotating tool was plunged into the workpiece with an axial force and then moved forward along a straight line. Friction between the tool and the workpiece resulted in heating, plasticization and stirring of the material. The FSP was carried out at constant axial load, which is a preset parameter to maintain the metal flow both at the tool plunging stage and during processing. The tool travel speed was not a parameter to control. The FSP tool travel started immediately after the shoulder contacted the metal surface. The tool inclination angle is 1.5°.

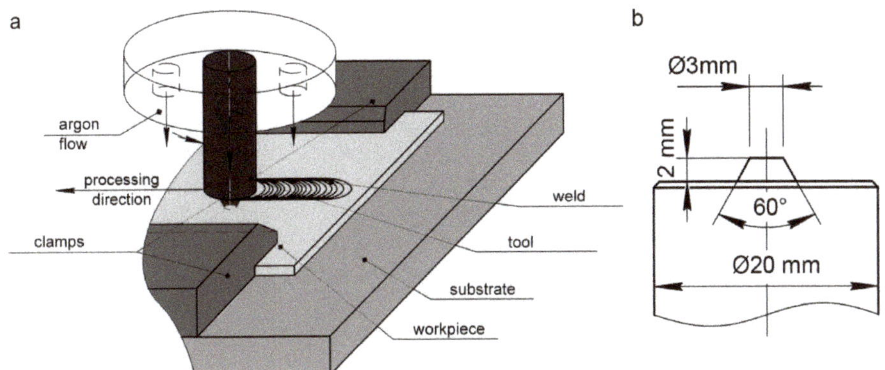

Figure 1. Scheme of friction stir processing (**a**) and tool schematic (**b**).

The superalloy tool had a 2 mm length pin and a 20 mm diameter shoulder (Figure 1b). It was made of as-cast ZhS6U Ni-based superalloy of composition presented in Table 1. This alloy is an analogue of the Mar-M247 alloy. The tool was made using the lathe turning from a cylindrical billet.

Table 1. Element composition of ZhS6U alloy (wt%).

Fe	Nb	Ti	Cr	Co	W	Ni	Al	Mo	S
≤1	0.8–1.2	2–2.9	8–9.5	9–10.5	9.5–11	54.3–62.7	5.1–6	1.2–2.4	≤0.01
Ce	**Si**	**Mn**	**P**	**C**	**Zr**	**Bi**	**B**	**Pb**	**Y**
≤0.02	≤0.4	≤0.4	≤0.015	0.13–0.2	≤0.04	≤0.0005	≤0.035	≤0.01	≤0.01

The workpieces were made of 2.5 mm thick hot-rolled sheets of grade 2 titanium alloy. The initial composition of the alloy is presented in Table 2. The FSP was carried out in the direction parallel to the rolling direction. The processing was conducted using the process parameters at which the highest weld strength was achieved as reported elsewhere [34]. The axial force on the tool during both plunging (F_{pl}) and processing (F_{pr}) was maintained at 7848 N, the traverse speed (V) was 180 mm/min, and the tool rotation rate (ω) was 950 rpm. The total weld length was more than 2.5 m. Since the laboratory FSW machine was not attached with a liquid cooling system of, unlike industrial machines, the processing was carried out in successive 100 mm length passes to avoid the tool overheating. The process was stopped after each pass to cool the tool in air. The FSP tool was not cleaned in between the passes not to interfere with the conditions created in the usual FSW process.

Table 2. Element composition of titanium alloy (wt%).

Fe	C	Si	N	Ti	O	H	Impurity
≤0.25	≤0.07	≤0.1	≤0.04	99.24–99.7	≤0.2	≤0.01	the rest 0.3

2.2. Investigative Techniques

Metallographic examinations were performed on both the tool and the weld materials. Samples of the processed metal were cut in a plane perpendicular to the weld. The tool material structure was examined in the cross section views. The views were obtained using polishing on abrasive papers and diamond pastes. Then, the processed titanium samples were etched in a 2% hydrofluoric acid solution for two minutes to reveal the microstructure and then washed in a 40% nitric acid solution. The tool microstructure was revealed with a reagent composed of 8 g $CuSO_4$, 40 mL HCl, and 40 mL H_2O.

Qualitative as well as quantitative analysis of the weld and tool microstructure was carried out using an Altami MET-1C metallographic microscope(Altami, St Petersburg, Russia) and a Microtrac SM 3000 scanning electron microscope (Nikkiso Co. Ltd., Tokyo, Japan) attached with an energy-dispersive X-ray microanalyzer (IXRF Systems). The worn surfaces of the tool were examined using an Olympus LEXT OLS4000 confocal microscope (Olympus, Tokyo, Japan). The tool was also photographed in the intervals between the passes to document its shape change.

Quasi-static tensile tests of the weld material were conducted in a UTC 110M-100 testing machine (Testsystems, Ivanovo, Russia) at room temperature. Flat specimens for mechanical tests were cut from stirring zone with their tensile axis coincided with the seam centerline (Figure 2). Such a test scheme was used to study the processed metal strength and thus evaluate the FSP efficiency. The base metal tensile test samples were used as reference ones. The loading rate was 1 mm/s according to the standard [35]. Microhardness numbers were obtained to evaluate the local hardness in the stir zone and in the base metal. Microhardness tester Duramin5 (Struers, Ballerup, Denmark) at 100 g load and dwell time 10 s was used.

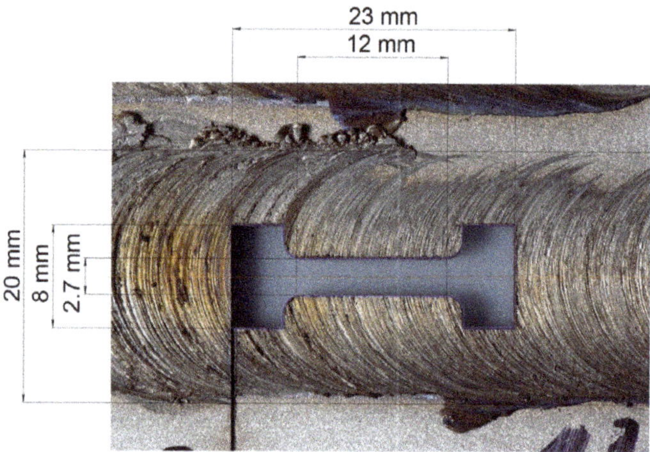

Figure 2. Scheme of a tensile test specimen cut from the weld zone.

3. Results and Discussion

3.1. Tool Wear

The general views of the tool at different stages of the experiment are shown in optical images in Figure 3. One can see, that the tool worn surfaces are covered with an adherent layer of oxidized titanium. The edges of the pin and the shoulder are the most severely worn places during the processing. After processing the 1105 mm length, the pin shape is still conical and the depression is not observed (Figure 3b). Further processing results in gradual changing the pin shape from conical to nearly cylindrical one. A circular shape depression is gradually formed on the shoulder around the base of the pin in the course of processing.

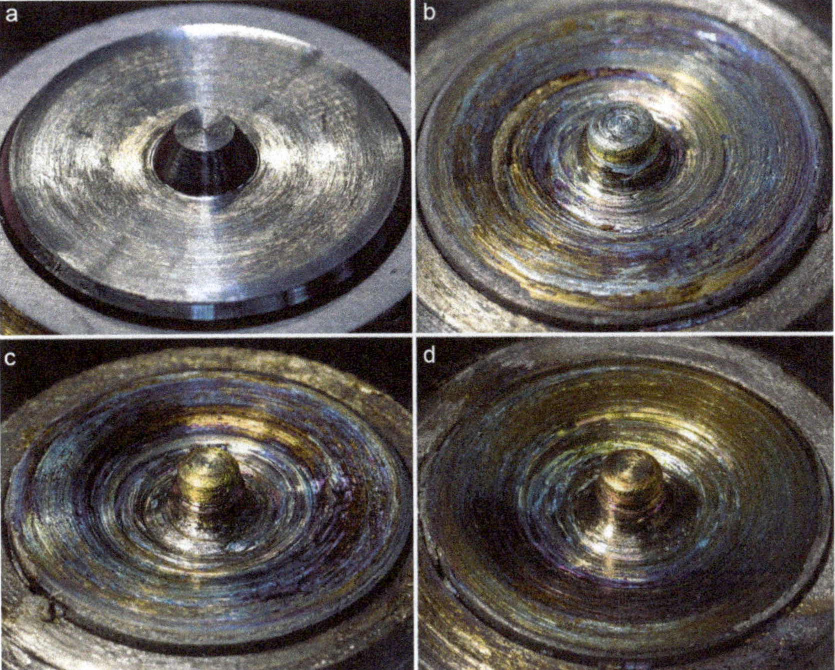

Figure 3. General view of the tool before friction stir processing (FSP) (**a**), after traversing 1105 mm (**b**), 2335 mm (**c**), and 2755 mm (**d**).

Figure 4 shows typical 3D images of the tool worn surfaces before processing and after traversing 2755 mm. The surface morphology changed significantly. The adherent layers are represented by ridges generally oriented along the direction of rotation, with sparse pits formed due to adhesion and detachment of the transfer layer during processing. Despite the fact that the maximum wear of the shoulder is observed closer to the pin, there are no regular morphological features other than the fact that the shoulder roughness is higher in this area. In particular, the roughness of the pin tip surface after FSP was Ra = 15.6 μm, the average shoulder roughness was Ra = 13.6 μm, and the shoulder roughness around the pin was Ra = 26.2 μm (the pin and shoulder roughness before processing was Ra = 1.55 and Ra = 1.15 μm, respectively).

Figure 5 shows an SEM backscattered-electron (BSE) image of the tool worn surface after traversing 2755 mm and energy-dispersive X-ray spectroscopy (EDX) maps of Ti and W. Lighter band areas in the SEM BSE image correspond to the metal enriched with the tool material elements. It is suggested that the transfer layer was oxidized, embrittled, and detached from the tool surface when retracted from the titanium and no more protected by the argon. Therefore, this layer's fragments may be easily detached during the next FSP pass. Tungsten is also observed in deposited titanium layers, indicating the diffusion of the tool elements into the transfer layer.

The axial cross-sectional view of the tool after traversing 2755 mm is shown in Figure 6 with superimposed drawing of the initial tool geometry. The highest wear is observed in the shoulder region at the base of the pin. The initial profile height loss in this zone is as high as 1 mm. The shoulder region closer to the edge is less worn so that its height loss was 0.25 mm. The pin height remained unchanged. In our opinion, such a non-uniform wear type may be of consequences as follows. Firstly, the reduced diameter at the base of the pin can lead to insufficient heat production because of lower friction velocity and hence to poor stirring. Secondly, the pin height does not change with the shoulder wear and therefore the tool may plunge deeper into the material. As a result, the pin can come into contact with the substrate and stick to it, or the substrate material can be introduced into the stirring zone.

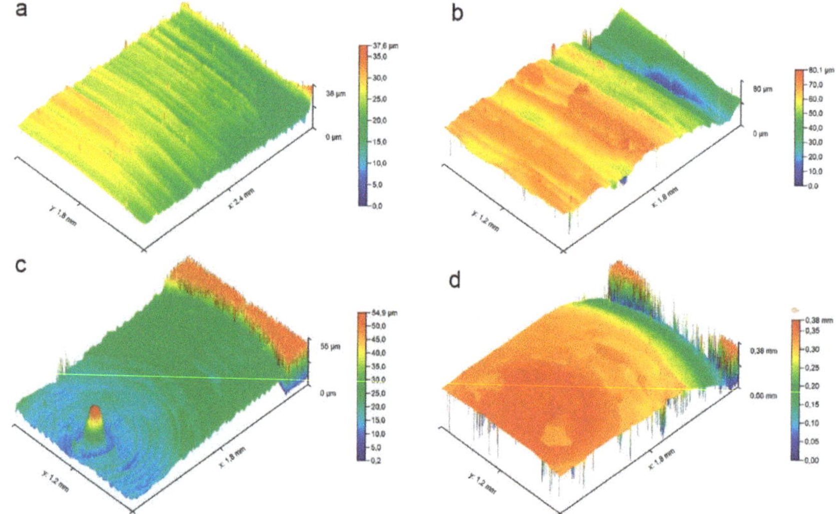

Figure 4. Typical 3D images of the surface morphology of the tool shoulder (**a**) and pin (**c**) before processing and the shoulder (**b**) and pin (**d**) after traversing 2755 mm.

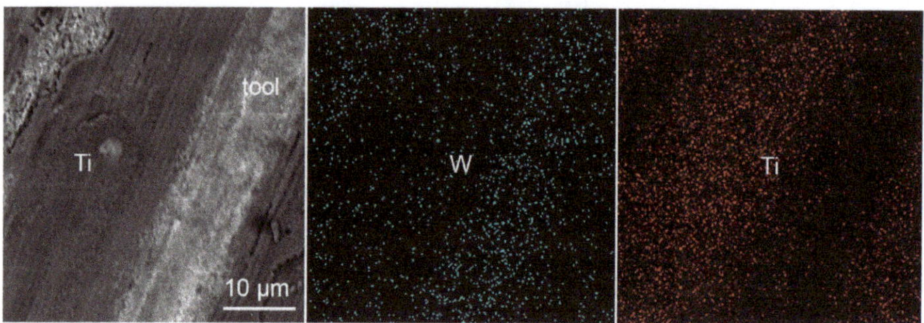

Figure 5. SEM BSE image of the tool friction surface after traversing 2755 mm and EDX maps of Ti and W.

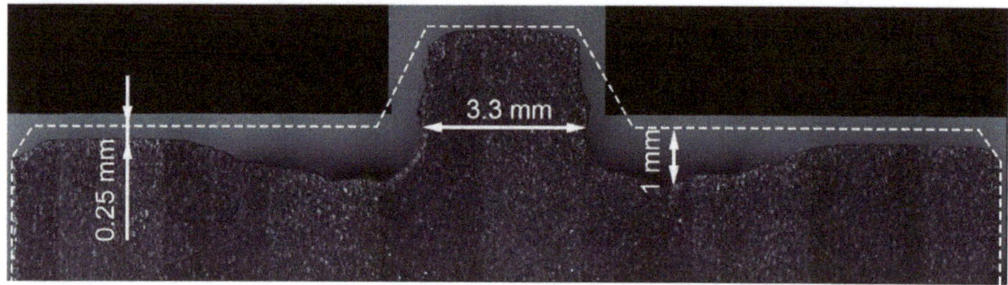

Figure 6. Axial cross-sectional view of the tool after traversing 2755 mm with superimposed drawing of the initial tool geometry.

Costa et al. [31], despite a different pin geometry, observed a similar wear pattern of a tool from a similar Mar-M247 alloy. The authors explained the wear behavior by material overheating that causes recrystallization of the alloy, carbide formation, and oxygen diffusion, which led to cracking of the surface layers and adhesive wear. The tool damage was avoided, as reported, using the argon shielding. However, there is no explanation why the maximum wear occurs closer to the tool axis. Formation of the annular depression around the pin was explained by Liu et al. [36] using the contact melting theory when they observed a similar concavity formed by friction in the center of a coating rod after deposition. Despite the fact that the friction velocity decreases closer to the center, heat transfer is impeded in this region, as the thermal energy mainly flows from the outer radius inwards and outwards. Therefore, the rod center becomes the point where the energy is concentrated and a quasi-liquid layer is formed. That is why the wear in the center is maximum. Such an explanation seems to hold good for the case of tool wear in FSP. However, as shown [31], the formation of a depression was avoided simply by using argon shielding, which could not greatly affect the heat transfer conditions. Nevertheless, contact melting might explain the fact that the surface layer material was recrystallized.

Generally, the heat-resistant superalloy tool wear is very specific in the case of welding on titanium. For instance, in FSW with steel tool on aluminum alloys the most severe wear is experienced by the pin's end. It was shown [37] even that the pin end experienced plastic deformation. The similar wear pattern was observed on the H13 tool after welding on CuCrZr alloy [38]. FSW on Ti-6Al-4V carried out using similar configuration tools made of other materials also demonstrated different type of wear [16]. In particular, a W-La tool experienced severe plastic deformation while those made of WC-Co demonstrated brittle fracture.

An indication of the adhesive wear mechanism can be the diffusion of the tool elements into the transfer layer that remained on the friction surface after processing. Figure 7 shows an SEM image of the tool axial section and EDX maps of Ti and tool elements. The maps demonstrate that the main elements of the tool diffuse into the adherent titanium layer. The diffusion of Al is especially high. However, there is almost no titanium in the tool material. This is probably due to the fact that titanium rich layers of the tool are immediately removed during processing. Yet, the observed diffusion may be evidence of adhesive wear.

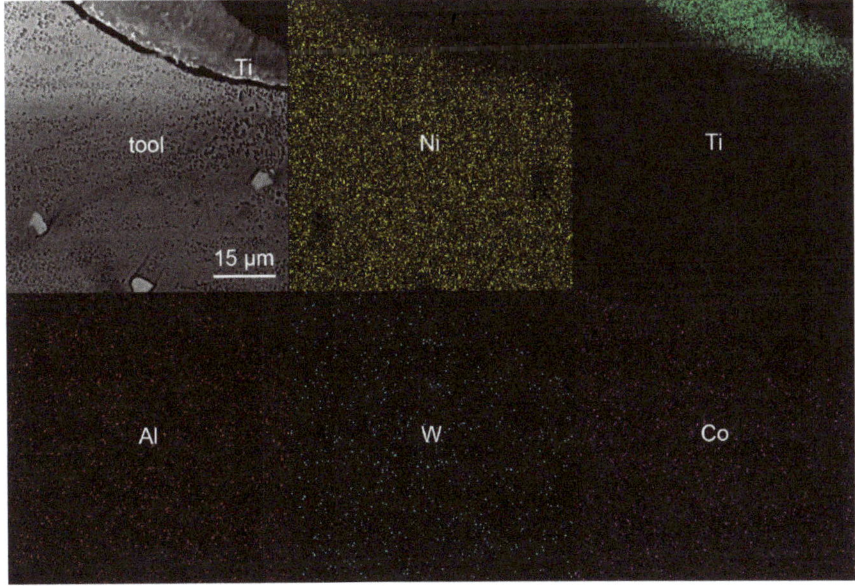

Figure 7. SEM image of the tool axial section and EDX maps of Ti and tool elements.

3.2. Structure of the Tool

According to metallographic images of the tool and the base ZhS6U alloy (Figure 8), the structure at this scale is represented by arrays of γ-phase dendrites. The high-contrast bright objects at the array boundaries are carbides characteristic of the alloy. The shape of the arrays generally tends to be equiaxed after FSP both in the base material and in the tool. As a result of the thermal effect, the structural elements grew considerably. The size of dendritic arrays, the primary and secondary dendrite arm spacings increased by a factor of 3. The measurement results are given in Table 3. It is characteristic that the structure changed uniformly throughout the tool. The array sizes and the dendrite arm spacings did not change as well as the amount of carbides did not increase in the subsurface region. The thickness of the adherent titanium layer on the tool is in the range of 30–80 μm, which is consistent with the surface morphology measurements. The change in contrast near the worn surface is caused by over-etching due to surface tension at the edge of the specimen when it was removed from the reagent.

Table 3. Characteristics of microstructural elements.

Characteristics	Base Material	Tool
Dendritic array size, mm	0.27 ± 0.09	0.8 ± 0.3
Primary dendrite arm spacing, μm	66 ± 5	191 ± 31
Secondary dendrite arm spacing, μm	14 ± 1	44 ± 6
Volume fraction of carbides, %	1.28 ± 0.07	1.14 ± 0.04
Average carbide size, μm	1.8 ± 1.5	3 ± 3
Average γ′ cube size	0.6 ± 0.2	0.6 ± 0.2

Figure 8. Metallographic images of the tool axial section after traversing 2755 mm (**a,c**) and the base ZhS6U alloy (**b**).

Figure 9 shows SEM BSE images of the base ZhS6U alloy structure and the tool structure. The images highlight the dendritic array boundaries decorated with eutectics and carbides of heavy metals, which mainly include Ti, W, Mo, and Nb [39]. In addition, irregular carbides and eutectics are often located in the interdendritic spaces [40]. At a finer level, the alloy structure is represented by γ'-phase cubes [41]. The measurement results for these structural elements show that the carbide volume fraction decreased slightly after FSP (Table 3), while the average carbide size increased by a factor of more than 1.5. The carbide size measurements have low accuracy because of the large size scatter in the range of 0.7–5 µm. The size change of carbides at their almost constant volume fraction is related to the coagulation of particles that occurs as dendrites grow during recrystallization. The size of γ' cubes remained unchanged as a result of the thermal effect. The material below the tool friction surface as well as below the adherent titanium layer also revealed no specific feature. Figure 9b was obtained in the same region as the optical image of Figure 8c.

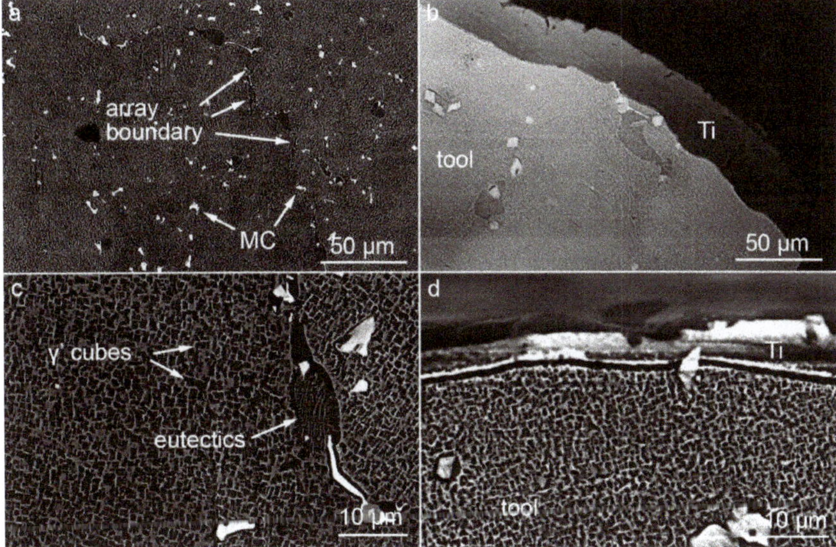

Figure 9. Cross-sectional SEM images of the base ZhS6U alloy (**a,c**) and the tool material after traversing 2755 mm (**b,d**). Figure 9b corresponds to Figure 8c.

3.3. Quality of Friction Stir Processing

The processing quality here refers to the structure of the weld and its mechanical properties. Figure 10 shows the cross-sectional metallographic images of FSPed material in different regions: at distances 5, 560, and 2755 mm from the starting point of the weld. According to metallographic analysis, the general view of the weld and the characteristic view of structural zones in all processing regions are the same for all samples. The weld structure exhibits two typical zones: stir zone (SZ) consisting of small recrystallized grains of titanium alloy, and thermomechanically affected zone (TMAZ). The rest is the base material. The heat affected zone, which is usually located between the base metal and TMAZ, is often absent or not detected in titanium alloys. The TMAZ in FSPed titanium alloy is narrow as compared to that of obtained on materials, with the average thickness of about 0.25 mm. The geometry of the structural zones is generally symmetrical about the weld axis. Their shapes in different processing regions differ slightly. The SZ area gradually decreases with the tool wear: from 11.11 mm^2 in the first sample to 8.03 mm^2 in the last one. This occurs because the pin diameter is reduced as the tool is worn.

Samples for metallographic analysis were cut approximately every 250 mm along the weld length. The sample cut at a distance of 5 mm from the starting point has no discontinuities and appears typical of the FSPed metals. The first discontinuities are observed in the sample cut at 560 mm distance, at the SZ boundary on the retreating side of the weld. These discontinuities grew with the processing time as well as new ones formed, including those on the advancing side. A sharp increase in the number of discontinuities is observed at ≈2300 mm distance. There are also oxides in the SZ, despite the fact that the processing was conducted in argon. As in the case with discontinuities, the first samples exhibit almost no traces of oxides, which become more pronounced with distance. The formation of discontinuities is associated specifically with the tool wear, because the tool geometry change induces a change in the mass transfer trajectory. At constant processing parameters, this leads to a change in heat production and deformation and, as a consequence, to non-optimal stirring conditions. The gradual emergence of oxides in the weld is presumably due to oxide accumulation on the tool surface. When processing is stopped, the tool is retracted from the material and the argon shield is removed. The extracted FSP tool is always covered with a layer of the processed metal. In our case, the adherent transfer layer was actively oxidized at high temperature in air and could be introduced into the material during the next pass. Oxides are gradually stirred into the material during processing together with the already present defects. The oxidized metal is located at the edges of the transfer layers, due to which the cohesive forces between the layers are reduced and delamination occurs. Ultimately, the oxide regions become the sites of discontinuity formation. The total area of all discontinuities in the cross section was 0.07 mm^2 at 5 mm distance, 0.15 mm^2 at 560 mm, and 0.79 mm^2 at 2755 mm.

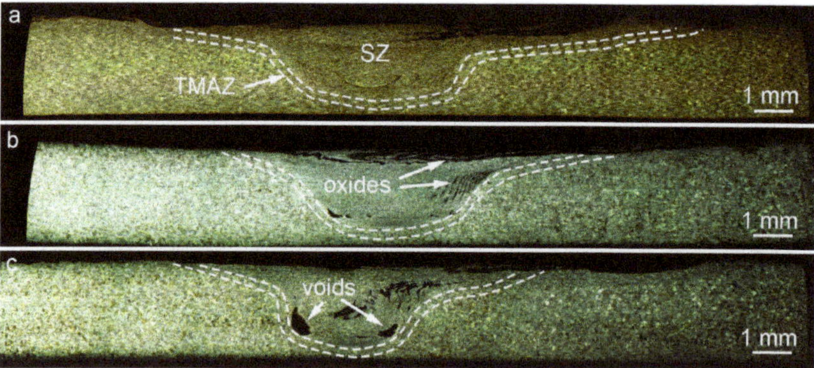

Figure 10. Cross-sectional metallographic images of FSPed material at 5 mm (**a**), 560 mm (**b**), and 2755 mm (**c**) from the weld start.

Friction stir processing resulted in a recrystallized and considerably refined alloy structure, which is typical of FSP [42]. In particular, the SZ exhibits equiaxed solid solution grains of size 8 ± 2 μm (Figure 11a), while the base metal consisted of 54 ± 11 μm grains. The grain size in the SZ did not change with tool wear. The solid solution grains also exhibit banded structures similar to twins. Twinning is not typical of commercially pure titanium under ordinary conditions, but it occurs at low temperatures, e.g., in cold rolling [43], and/or at high strain rates. The twining was earlier observed in FSWed on commercially pure titanium [44] and explained by the lack of slip systems at a high strain rate.

The admixed oxides in the SZ are surrounded by martensite (Figure 11b), which is not characteristic of the studied alloy. Commercially pure titanium belongs to α-phase materials. As shown [45], heating above the phase transition temperature and rapid cooling (>100 °C/s) can cause the formation of α′ martensite from the melt under high thermal stresses. However, such temperatures and cooling rates are not achieved during FSP, while high stresses are present. Therefore, the occurrence of martensite may be explained by alloying with the tool metal. The oxidized martensitic regions contain tool particles stirred during wear together with transfer layers into the processed metal. As a result of diffusion, these particles dope the surrounding material with W, Al, Ni, and other atoms and induce the formation of martensite.

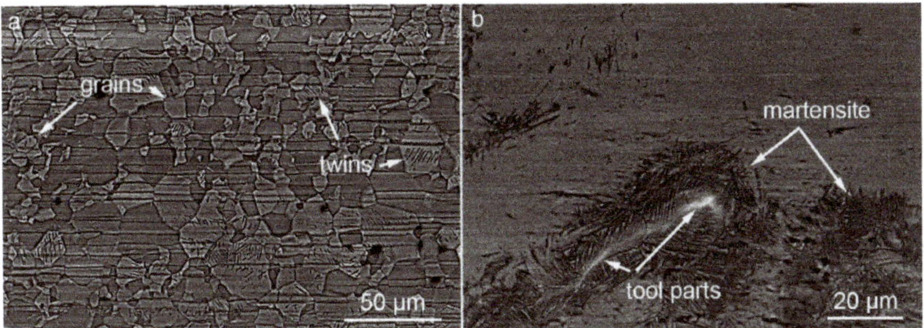

Figure 11. SEM images of the grain structure in the stir zone (SZ) (**a**) and oxidized regions (**b**).

Figure 12 shows the results of quasi-static tensile tests for the SZ of specimens. The ultimate tensile strength of the processed material first increased by ≈25% as compared to the base metal strength of 350 MPa. With tool wear, the tensile strength of the specimens decreased and reached 375 MPa after traversing 2700 mm, which is by ≈7% higher than the base metal strength. A sharp decrease in strength is observed at 500 mm distance corresponding to the formation of first macrodefects. The second sharp drop at 2300 mm is also related to a drastic increase in the number of discontinuities. Therefore, the strength reduction is associated primarily with the growth of defects during tool wear, which leads to insufficient heating and poor stirring of the material. Despite the weld strength exceeds the base alloy strength, this result is not satisfactory, because the presence of macrodefects makes the weld performance unreliable under fatigue loading. The maximum elongation at fracture diminished by a factor of 2. The processed material showed the plasticity level to be independent of the FSW tool wear. The same was true for the stir zone microhardness numbers, which were 1.5 ± 0.1 GPa, 1.4 ± 0.1 GPa, and 1.6 ± 0.1 GPa for 20 mm, 500 mm, and 2700 mm processed length, respectively. The base metal microhardness was 1.3 ± 0.1 GPa. Such a moderate hardening was the result of severe plastic deformation during FSP.

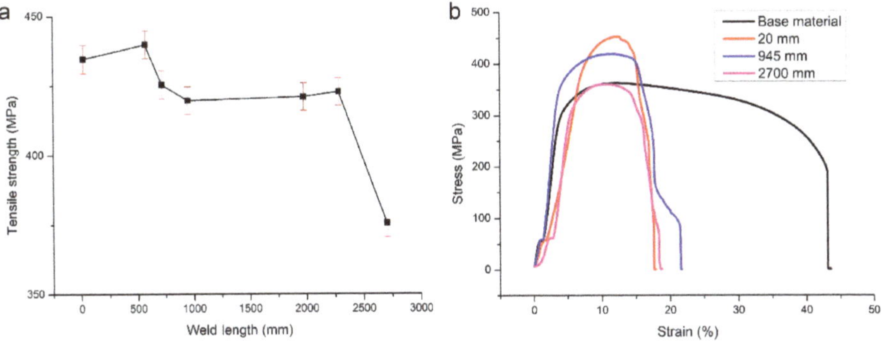

Figure 12. Dependence of the weld strength on the tool wear (**a**) and stress–strain curves of friction stir processed welds at different weld lengths (**b**).

The formation of defects during tool wear can be avoided by changing the processing parameters, e.g., the axial load. An increase in the load will lead to greater heat production, enhanced adhesion, and therefore better stirring. Although this method is complicated, it may be a temporary solution. The tool life can be extended by using liquid cooling on the back side of the tool to minimize the contact melting and recrystallization effects. One more cause of softening is an increase in the amount of oxides introduced into the SZ from the tool. This problem can be solved by blowing the tool with argon after retracting it from titanium. The given solution requires a large consumption of argon, but it can help avoiding the oxidation of titanium transferred on the friction surface of the tool.

4. Conclusions

This study explored the durability of a ZhS6U Ni-based superalloy tool in FSP of commercially pure titanium and the effect of the tool wear on the weld quality. The total length of the weld produced by the tool without failure comprised 2755 mm. The highest wear was observed at the base of the pin which lost its conical shape during processing and a 1 mm deep circular depression formed in the shoulder near the pin. Analysis of the adherent transfer layer on the tool retracted from titanium revealed the presence of ZhS6U alloy constituents, which is indicative of diffusion during FSP. Thus, the wear occurs by an adhesive mechanism. It was also found that the tool overheating during FSP leads to the recrystallization of ZhS6U alloy. In particular, the size of dendritic arrays as well as the primary and secondary dendrite arm spacings increased by a factor of 3. The size of heavy metal carbides increased almost twice, while their volume fraction did not change. The size of γ' cubes also remained unchanged.

The tool wear resulted in macrostructural changes in the processed material. The SZ area was reduced, while the number and size of macrodefects increased with the tool wear. The defects included discontinuities caused by insufficient heating as well as oxide particles admixed. The oxides formed on the tool after it was retracted from titanium and then were stirred into the workpiece during the next pass. As a result of FSP, commercially pure titanium recrystallized. The solid solution grain size decreased by a factor of $\approx$6.5, inducing significant hardening. The first FSP samples had tensile strength of $\approx$435 MPa, which was 25% higher than that of the base metal (350 MPa). After traversing 2700 mm the tensile strength decreased to 375 MPa due to the increased number and size of macrodefects.

In general, ZhS6U alloy proved to be suitable for making the FSP tool for titanium. However, its use requires liquid cooling of the tool during FSP to avoid overheating. Argon shielding is also needed to avoid oxidation of the adherent transfer layer on the extracted tool while it is cooled.

Author Contributions: Conceptualization, A.E. and V.R.; investigation, A.A., A.E., and A.F.; resources, E.K. and V.R.; writing—original draft preparation, A.A. and A.E.; writing—review and editing, E.K. All authors have read and agreed to the published version of the manuscript.

Funding: The reported study was funded by RFBR, Project No. 19-33-90187. SEM investigations was performed under the government statement of work for ISPMS SB RAS Project No. III.23.2.4.

Conflicts of Interest: The authors declare no conflict of interest.

References

1. Peters, M.; Hemptenmacher, J.; Kumpfert, J.; Leyens, C. Structure and Properties of Titanium and Titanium Alloys. In *Titanium and Titanium Alloys*; Leyens, C., Peters, M., Eds.; Wiley-VCH: Weinheim, Germany, 2005.

2. Sizova, O.; Shlyakhova, G.; Kolubaev, A.; Kolubaev, E.A.; Psakhie, S.G.; Rudenskii, G.; Chernyavsky, A.G.; Lopota, V. Microstructure Features of Aluminum Alloys Welded Joint Obtained by Friction Stir Welding. *Adv. Mater. Res.* **2013**, *872*, 174–179. [CrossRef]

3. Eliseev, A.A.; Fortuna, S.V.; Kalashnikova, T.A.; Chumaevskii, A.V.; Kolubaev, E.A. Structural Phase Evolution in Ultrasonic-Assisted Friction Stir Welded 2195 Aluminum Alloy Joints. *Russ. Phys. J.* **2017**, *60*, 1022–1026. [CrossRef]

4. Tarasov, S.Y.; Rubtsov, V.E.; Kolubaev, E.A. A proposed diffusion-controlled wear mechanism of alloy steel friction stir welding (FSW) tools used on an aluminum alloy. *Wear* **2014**, *318*, 130–134. [CrossRef]

5. Sato, Y.S.; Park, S.H.C.; Matsunaga, A.; Honda, A.; Kokawa, H. Novel production for highly formable Mg alloy plate. *J. Mater. Sci.* **2005**, *3*, 637–642. [CrossRef]

6. Kalashnikova, T.A.; Kalashnikov, K.N.; Shvedov, M.A.; Vasilyev, P.A. Structure and properties of copper compensator joints obtained by hybrid friction stir welding technology. *Met. Work. Mater. Sci.* **2019**, *21*, 85–93. [CrossRef]

7. Kalashnikova, T.A.; Gusarova, A.V.; Chumaevskii, A.V.; Knyazhev, E.O.; Shvedov, M.A.; Vasilyev, P.A. Regularities of composite materials formation using additive electron-beam technology, friction stir welding and friction stir processing. *Met. Work. Mater. Sci.* **2019**, *21*, 94–112. [CrossRef]

8. Mishra, R.S.; Ma, Z.Y. Friction stir welding and processing. *Mater. Sci. Eng. R* **2005**, *50*, 1–78. [CrossRef]

9. Nandan, R.; Debroy, T.; Bhadeshia, H. Recent advances in friction-stir welding—Process, weldment structure and properties. *Prog. Mater. Sci.* **2008**, *53*, 980–1023. [CrossRef]

10. Farias, A.; Batalha, G.F.; Prados, E.F.; Magnabosco, R.; Delijaicov, S. Tool wear evaluations in friction stir processing of commercial titanium Ti–6Al–4V. *Wear* **2013**, *302*, 1327–1333. [CrossRef]

11. Liu, H.J.; Zhou, L.; Liu, Q.W. Microstructural evolution mechanism of hydrogenated Ti–6Al–4V in the friction stir welding and post-weld dehydrogenation process. *Scr. Mater.* **2009**, *61*, 1008–1011. [CrossRef]

12. Zhou, L.; Liu, H.J. Effect of 0.3 wt.% hydrogen addition on microstructural evolution of Ti–6Al–4V alloy in the friction stir welding and post-weld dehydrogenation process. *Mater. Charact.* **2011**, *62*, 1036–1041. [CrossRef]

13. Wu, L.H.; Xue, P.; Xiao, B.L.; Ma, Z.Y. Achieving superior low-temperature superplasticity for lamellar microstructure in nugget of a friction stir welded Ti-6Al-4V joint. *Scripta Mater.* **2016**, *122*, 26–30. [CrossRef]

14. Li, B.; Shen, Y.; Hu, W.; Luo, L. Surface modification of Ti–6Al–4V alloy via friction-stir processing: Microstructure evolution and dry sliding wear performance. *Surf. Coat. Technol.* **2014**, *239*, 160–170. [CrossRef]

15. Pilchak, A.L.; Tang, W.; Sahiner, H.; Reynolds, A.P.; Williams, J.C. Microstructure Evolution during Friction Stir Welding of Mill-Annealed Ti-6Al-4V. *Metall. Mater. Trans. A* **2010**, *42*, 745–762. [CrossRef]

16. Wang, J.; Su, J.; Mishra, R.S.; Xu, R.; Baumann, J.A. A Preliminary Study of Deformation Behavior of Friction Stir Welded Ti-6Al-4V. *J. Mater. Eng. Perform.* **2014**, *23*, 3027–3033. [CrossRef]

17. Lippold, J.C.; Livingston, J.J. Microstructure Evolution During Friction Stir Processing and Hot Torsion Simulation of Ti-6Al-4V. *Metall. Mater. Trans. A* **2013**, *44*, 3815–3825. [CrossRef]

18. Fall, A.; Fesharaki, M.; Khodabandeh, A.; Jahazi, M. Tool Wear Characteristics and Effect on Microstructure in Ti-6Al-4V Friction Stir Welded Joints. *Metals* **2016**, *6*, 275. [CrossRef]

19. Edwards, P.D.; Ramulu, M. Comparative study of fatigue and fracture in friction stir and electron beam welds of 24mm thick titanium alloy Ti-6Al-4V. *Fatigue Fract. Eng Mater. Struct.* **2016**, *39*, 1226–1240. [CrossRef]

20. Muzvidziwa, M.; Okazaki, M.; Suzuki, K.; Hirano, S. Role of microstructure on the fatigue crack propagation behavior of a friction stir welded Ti–6Al–4V. *Mater. Sci. Eng. A* **2016**, *652*, 59–68. [CrossRef]

21. Yoon, S.; Ueji, R.; Fujii, H. Effect of initial microstructure on Ti–6Al–4V joint by friction stir welding. *Mater. Des.* **2015**, *88*, 1269–1276. [CrossRef]

22. Sato, Y.S.; Susukida, S.; Kokawa, H.; Omori, T.; Ishida, K.; Imano, S.; Park, S.H.C.; Sugimoto, I.; Hirano, S. Wear of cobalt-based alloy tool during friction stir welding of Ti-6Al-4V alloy. In Proceedings of the 11th International Symposium on FrictionStir Welding, Cambridge, UK, 17–19 May 2016.

23. Mironov, S.; Zhang, Y.; Sato, Y.S.; Kokawa, H. Crystallography of transformed b microstructure in friction stir welded Ti–6Al–4V alloy. *Scr. Mater.* **2008**, *59*, 511–514. [CrossRef]

24. Mironov, S.; Zhang, Y.; Sato, Y.S.; Kokawa, H. Development of grain structure in b-phase field during friction stir welding of Ti–6Al–4V alloy. *Scr. Mater.* **2008**, *59*, 27–30. [CrossRef]

25. Zhang, Y.; Sato, Y.S.; Kokawa, H.; Park, S.H.C.; Hirano, S. Microstructural characteristics and mechanical properties of Ti–6Al–4V friction stir welds. *Mater. Sci. Eng. A* **2008**, *485*, 448–455. [CrossRef]

26. Wu, L.H.; Wang, D.; Xiao, B.L.; Ma, Z.Y. Tool wear and its effect on microstructure and properties of friction stir processed Ti–6Al–4V. *Mater. Chem. Phys.* **2014**, *146*, 512–522. [CrossRef]

27. Zhang, Y.; Sato, Y.S.; Kokawa, H.; Park, S.H.C.; Hirano, S. Stir zone microstructure of commercial purity titanium friction stir welded using pcBN tool. *Mater. Sci. Eng. A* **2008**, *488*, 25–30. [CrossRef]

28. Rai, R.; De, A.; Bhadeshia, H.K.D.H.; DebRoy, T. Review: Friction stir welding tools. *Sci. Technol. Weld. Join.* **2011**, *16*, 325–342. [CrossRef]

29. Mironov, S.; Sato, Y.S.; Kokawa, H. Friction-stir welding and processing of Ti-6Al-4V titanium alloy: A review. *J. Mater. Sci. Technol.* **2018**, *34*, 58–72. [CrossRef]

30. Amirov, A.I.; Utyaganova, V.R.; Beloborodov, V.A.; Eliseev, A.A. Formation features of a welding joint of alloy Grade2 by the friction stir welding using temperature resistant tools. *Met. Work. Mater. Sci.* **2019**, *21*, 72–82. [CrossRef]

31. Costa, A.M.S.; Oliveira, J.P.; Pereira, V.F.; Nunes, C.A.; Ramirez, A.J.; Tschiptschin, A.P. Ni-based Mar-M247 superalloy as a friction stir processing tool. *J. Mater. Process. Tech.* **2018**, *262*, 605–614. [CrossRef]

32. Nakazawa, T.; Sato, Y.S.; Kokawa, H.; Ishida, K.; Omori, T.; Tanaka, K.; Sakairi, K. Friction Stir Welding of Steels Using a Tool Made of Iridium-Containing Nickel Base Superalloy. In *Friction Stir Welding and Processing VIII*; Mishra, R.S., Mahoney, M.W., Sato, Y., Hovanski, Y., Eds.; Springer: Cham, Switzerland, 2015.

33. Kablov, E.N.; Petrushin, N.V.; Parfenovich, P.I. Design of castable refractory nickel alloys with polycrystalline structure. *Met. Sci. Heat Treat.* **2018**, *60*, 106–114. [CrossRef]

34. Amirov, A.I.; Eliseev, A.A.; Rubtsov, V.E.; Utyaganova, V.R. Butt friction stir welding of commercially pure titanium by the tool from a heat-resistant nickel alloy. *AIP Conf. Proc.* **2019**, *2167*, 020016.

35. ISO 6892-1-2016 Metallic Materials. *Tensile Testing. Part 1: Method of Test at Room Temperature*; CENELEC: Brussel, Belgium, 2016.

36. Liu, X.M.; Zou, Z.D.; Zhang, Y.H.; Qu, S.Y.; Wang, X.H. Transferring mechanism of the coating rod in friction surfacing. *Surf. Coat. Technol.* **2008**, *202*, 1889–1894. [CrossRef]

37. Bevilacqua, M.; Ciarapica, F.; Forcellese, A.; Simoncini, M. Comparison among the environmental impact of solid state and fusion welding processes in joining an aluminium alloy. *Proc. Inst. Mech. Eng. Part B J. Eng. Manuf.* **2020**, *234*, 140–156. [CrossRef]

38. Sahlot, P.; Jha, K.; Dey, G.K.; Arora, A. Quantitative wear analysis of H13 steel tool during friction stir welding of Cu-0.8%Cr-0.1%Zr alloy. *Wear* **2017**, *378–379*, 82–89. [CrossRef]

39. Zhou, T.; Ding, H.; Ma, X.; Feng, W.; Zhao, H.; Li, A.; Meng, Y.; Zhang, H. A comparison study of the thermal fatigue properties of three Ni-based cast superalloys cycled from 20 to 1100 °C. *Adv. Eng. Mater.* **2019**, *21*, 1900054. [CrossRef]

40. Wang, L.N.; Sun, X.F.; Guan, H.R. Effect of melt heat treatment on MC carbide formation in nickel-based superalloy K465. *Results Phys.* **2017**, *7*, 2111–2117. [CrossRef]

41. Jonšta, P.; Vlčková, I.; Jonšta, Z.; Podhorná, B. Materialographic Analysis of MAR M-247 Superalloy. *Key Eng. Mater.* **2015**, *647*, 66–71. [CrossRef]

42. Kalashnikov, K.N.; Kalashnikova, T.A.; Chumaevskii, A.V.; Ivanov, A.N. Production of materials with ultrafine-grained structure of aluminum alloy by friction stir processing. *J. Phys. Conf. Ser.* **2018**, *1115*, 042048. [CrossRef]

43. Zherebtsov, S.V.; Dyakonov, G.S.; Salishchev, G.A.; Salem, A.A.; Semiatin, S.L. The influence of grain size on twinning and microstructure refinement during cold rolling of commercial-purity titanium. *Metall. Mater. Trans. A* **2016**, *47*, 5101–5113. [CrossRef]

44. Lee, W.-B.; Lee, C.-Y.; Chang, W.-S.; Yeon, Y.-M.; Jung, S.-B. Microstructural investigation of friction stir welded pure titanium. *Mater. Lett.* **2005**, *59*, 3315–3318. [CrossRef]

45. Zhang, X.D.; Hao, S.Z.; Li, X.N.; Dong, C.; Grosdidier, T. Surface modification of pure titanium by pulsed electron beam. *Appl. Surf. Sci.* **2011**, *257*, 5899–5902. [CrossRef]

Article

Structure and Properties of Al–0.6 wt.%Zr Wire Alloy Manufactured by Direct Drawing of Electromagnetically Cast Wire Rod

Nikolay Belov [1,2,*], Maxim Murashkin [3,4], Natalia Korotkova [1], Torgom Akopyan [1,2] and Victor Timofeev [5]

[1]   Department of Metal Forming, National University of Science and Technology MISiS, Leninsky prospekt 4, Moscow 119049, Russia; kruglova.natalie@gmail.com (N.K.); nemiroffandtor@yandex.ru (T.A.)
[2]   Laboratory for Gradient Materials, Nosov Magnitogorsk State Technical University, Lenin pr., 38, Magnitogorsk 455000, Russia
[3]   Institute of Physics of Advanced Materials, Ufa State Aviation Technical University, K. Marx 12, Ufa 450008, Russia; m.murashkin.70@gmail.com
[4]   Laboratory for Mechanics of Bulk Nanostructured Materials, St. Petersburg State University, Universitetskaya nab. 7/9, St. Petersburg 199034, Russia
[5]   Department of Electrical Engineering, Siberian Federal University, 79 Svobodnyy Prospekt, Krasnoyarsk 660041, Russia; viktortim0807@mail.ru
*   Correspondence: nikolay-belov@yandex.ru; Tel.: +79-154-145-945

Received: 5 May 2020; Accepted: 6 June 2020; Published: 9 June 2020

Abstract: The method of electromagnetic casting (EMC) was used to produce the long-length rod billet (with a diameter 12 mm) of aluminum alloy containing 0.6 wt.% Zr, 0.4%Fe, and 0.4%Si. The combination of high cooling rate ($\approx 10^4$ K/s) during alloy solidification and high temperature before casting (≈ 830 °C) caused zirconium to dissolve almost completely in the aluminum solid solution (Al). Additions of iron and silicon were completed in the uniformly distributed eutectic Al_8Fe_2Si phase particles with an average size of less than 1 µm. Such fine microstructure of the experimental alloy in as-cast state provides excellent deformability during wire production using direct cold drawing of EMC rod (94% reduction). TEM study of structure evolution in the as-drawn 3 mm wire revealed the onset of Al_3Zr ($L1_2$) nanoparticle formation at 300 °C and almost-complete decomposition of (Al) at 400 °C. The distribution of Zr-containing nanoparticles is quite homogeneous, with their average size not exceeding 10 nm. Experimental wire alloy had the ultimate tensile strength (UTS) and electrical conductivity (EC) (234 MPa and 55.6 IACS, respectively) meeting the AT2 type specification. At the same time, the maximum heating temperature was much higher (400 versus 230 °C) and meets the AT4 type specification.

Keywords: aluminum-zirconium wire alloys; electromagnetic casting; drawing; electrical conductivity; phase composition; nanoparticles

1. Introduction

Aluminum alloys containing the addition of zirconium (0.2–0.4 wt.%) are widely used for manufacturing of heat resistant wires. Such alloys possess the improved combination of electrical conductivity, mechanical properties, and, particularly, thermal stability [1–3]. The most common commercial technology for wire rods production (including Al–Zr alloys) is the technology of continuous casting and rolling (CCR) realized in Properzi and Southwire rolling mills [4,5]. To meet the American Society for Testing and Materials (ASTM) specification (AT1-AT4 types) for the heat resistance [1], the microstructure should have a sufficient amount of Zr-containing precipitates (of metastable phase Al_3Zr having $L1_2$ crystal lattice) [6]. These precipitates should have a size within 10 nm and uniform distribution. To obtain such nanoparticles, zirconium should be fully dissolved in the aluminum solid solution (hereinafter (Al)) during solidification. Nanoparticles of Al_3Zr phase form during decomposition of the supersaturated aluminum solid solution (hereinafter (Al)) [7–11]. At the same time, to meet the ASTM specification for electrical conductivity, the concentration of Zr in (Al) should be as minimal as possible. The microstructure of Al–Zr wire alloy depends considerably on the processing root, particularly melting and casting temperatures and parameters of annealing [6,12,13]. As a rule, the CCR wire rods are subjected to annealing at temperatures above 350 °C during a long time of holding. Moreover, complete (Al) decomposition and formation of a sufficient number of $L1_2$ nanoparticles requires holding CCR wire rods for hundreds of hours.

As shown earlier [14], electromagnetic casting (EMC) provides sufficiently high casting temperatures and ultra-high solidification rates (10^3–10^4 K/s) and allows the production of long rod billets with a diameter of 8–12 mm. This rapid solidification in the EMC method corresponds to the solidification of granules, which is realized in the RS/PM (rapid solidification/powder metallurgy) method [15,16]. This method has been most successfully applied to obtain alloys with a high content of transition metals, including zirconium. Recently, it was shown that for the experimental Al–0.6Zr–0.4Fe–0.4Si (wt.%) alloy in the form of a long 8 mm diameter ingot that the EMC method was very suitable for wire alloys with increased content of zirconium [17]. The technological route used in this work included operations of rolling and intermediate annealing. Based on these results, the aim of this work was to study the structure and properties of the aluminum wire alloy containing 0.6%Zr, obtained using direct cold drawing from an as-cast 12 mm diameter EMC rod.

2. Experimental Methods

The experimental alloy in form of the long rod billets with a diameter of 12 mm and ≈20 m in length was obtained by EMC using equipment of the Research and Production Centre of Magnetic Hydrodynamics (Krasnoyarsk, Russia) [18]. This equipment consisted of an induction furnace, magnetic hydrodynamic (MHD)-stirrer, dosing trough, and electromagnetic crystallizer. Coolant from a cooling reservoir was supplied to the workpiece directly, which provides high cooling speed. The alloy was smelted in a graphite-chased crucible from commercial aluminum (99.5 wt.%) and master alloys (containing zirconium, iron, and silicon) at ≈850 °C. The temperature of melt before casting (≈830 °C) was higher than the liquidus temperature (822 °C).

The wire (3 mm diameter) was manufactured from the as-cast EMC rod on a dragging mill (the reduction ratio was 94%). This wire was studied in as-drawn state and after annealing at 300–400 °C (Figure 1). To estimate the effect of deformation on the decomposition of (Al) during annealing, a 2 mm strip was made additionally from the as-cast EMC billet by cold rolling. This strip was annealed together with the EMC rod at 300–600 °C. The annealing modes used are summarized in Table 1. It should be noted that the annealing modes used were previously substantiated in [6] for the Al alloys, which hardened due to the nanoparticles of the $L1_2$ phase. Values for the Vickers hardness and electrical resistivity (ER) were measured at each regime.

The microstructure was examined by means of scanning electron microscopy (SEM, TESCAN VEGA 3, Tescan Orsay Holding, Brno, Czech Republic), electron microprobe analysis (EMPA, OXFORD Aztec, Oxford Instruments, Oxfordshire, UK), and transmission electron microscopy (TEM, JEM 2100, JEOL, Tokyo, Japan), at an accelerating voltage of 200 kV. To study the microstructure, thin foils were used, produced by jet polishing on a Tenupol-5 machine (Struers, Ballerupcity, Denmark) with the chemical solution consisting of 20% nitric acid and 80% methanol at a temperature of −25 °C and a voltage of 15 V. The SEM specimens were electrolytically polished at 12 V in an electrolyte containing six parts C_2H_5OH, one part $HClO_4$, and one part glycerin.

The Vickers hardness (HV) was measured using a DUROLINE MH-6 universal tester (METKON Instruments Inc., Bursa, Turkey), with a load of 1 kg and a dwell time of 10 s. The hardness was measured at least five times at each point. The specific electrical conductivity (EC) of the EMC rod and the cold rolled strip was determined using the eddy current method with a VE-26NP eddy structures cope (CJSC Research institute of introscopy SPEKTR, Moscow, Russia). The electrical resistivity (ER) of the EMC rod and the strip was calculated from the EC data. The electrical resistivity of the cold drawn wire was measured in accordance with IEC 60468:1974 standard [19] for straightened samples of at least 1 m in length in the rectified part.

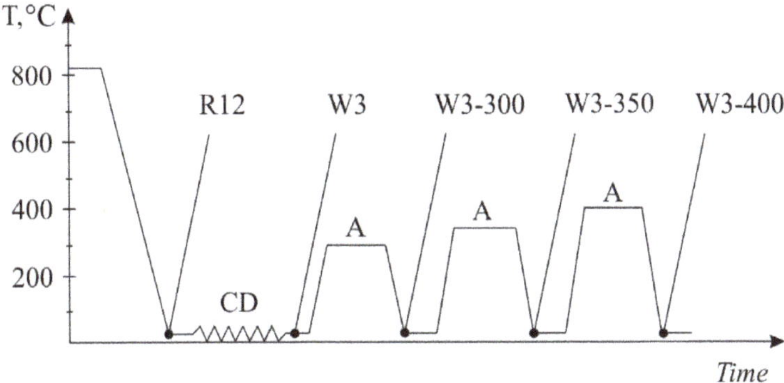

Figure 1. Processing root for experimental wire alloy (see Table 1). CD—cold drawing, A—annealing, R12—as-cast electromagnetic casting (EMC) rod, and W3—wire.

Table 1. Annealing regimes for EMC rod, strip and wire.

Designation	Regime Treatment
R12-EMC casting rod (diameter 12 mm)/S2-cold rolled strip (thickness 2 mm)	
R12/S2	As-cast/as-cold rolled
R12-300/S2-300	Annealing at 300 °C, 3 h
R12-350/S2-350	R12-300/S2-300 + annealing at 350 °C, 3 h
R12-400/S2-400	R12-350/S2-350 + annealing at 400 °C, 3 h
R12-450/S2-450	R12-400/S2-400 + annealing at 450 °C, 3 h
R12-500/S2-500	R12-450/S2-450 + annealing at 500 °C, 3 h
R12-550/S2-550	R12-500/S2-500+ annealing at 550 °C, 3 h
R12-600/S2-600	R12-550/S2-550 + annealing at 600 °C, 3 h
W–Wire (diameter 3 mm) manufactured by cold drawing of as-cast EMC rod	
W3	As-drawn
W3-300	Annealing at 300 °C, 3 h
W3-350	W300 + annealing at 350 °C, 3 h
W3-400	W350 + annealing at 400 °C, 3 h

The as-processed wire specimens were tensile tested at room temperature on a Zwick Z250 (ZwickRoell AG, Ulm, Germany) universal testing machine at a loading rate of 10 mm/min. Yield strength (YS), ultimate tensile strength (UTS), and elongation to failure (El.) were determined. To obtain consistent results, five specimens were tested.

The size of the dendritic cells was experimentally determined using metallography from high-contrast microstructural images processed with appropriate software, Sizer (National University of Science and Technology MISiS, Moscow, Russia). The Horizontal Lines option was used for implementing the stereological method of measuring the relative length of the phase regions. To obtain reliable data, we analyzed at least 10 fields in the microstructure to define the content of each structural component. The experimental dendritic cell size data were used for evaluating the cooling rate in the alloy crystallization temperature range using a well-known function [20].

3. Results and Discussion

3.1. Characterization of as-Cast EMC Rod

Due to the high solidification rate of the as-cast ingot, the microstructure of the experimental alloy has a fine structure. Only a small quantity of primary Al_3Zr phase crystals were revealed (Figure 2a). According to EMPA, the concentration of zirconium in (Al) was very close to its total content in the alloy (0.58 versus 0.62%). The measured average size of the dendritic cells was 4.2 ± 1.6 μm (Figure 2b), which, according coefficients given in [20] for the technical grade aluminum 1050, corresponded to a cooling rate of about 2.5×10^4 K/s. Iron-bearing particles in the form of thin veins were located along the boundaries of the dendritic cells. As shown earlier [17], these veins correspond to the Al_8Fe_2Si phase. It should be noted that the microstructure was analyzed over the entire cross section of the ingot, and no significant differences were found between the structural parameters (as well as hardness values measured in different sections of the ingot). It can be assumed that a relatively small cross section (12 mm in diameter) and a high cooling rate during solidification provided the formation of a relatively uniform structure.

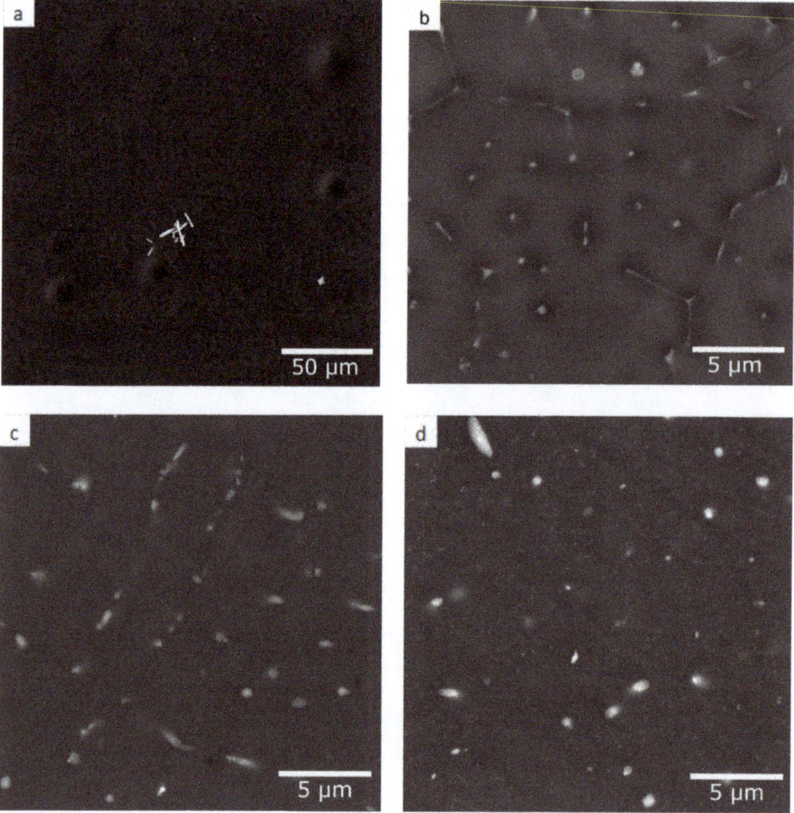

Figure 2. Effect of annealing temperature on microstructures of the EMC rod, SEM: (**a**,**b**) as-cast (R0), (**c**) 500 °C (R500), and (**d**) 600 °C (R600).

Annealing of the as-cast rod at up to 400 °C made no visible changes to its microstructure. The morphology of the Al_8Fe_2Si phase changed at higher temperatures. Annealing at 450 °C produced spherical particles of this phase, which is preferable to most of the other iron bearing phases (particularly,

Al$_3$Fe and Al$_5$FeSi) that have a needle-shaped morphology [21–23]. Most of the veins transformed to globular particles upon annealing at 500 °C (Figure 2c). Increasing the annealing temperature leads to particle enlargement, and the maximum particle size in the R12-600 state (Table 1) reaches 1 μm (Figure 2d). In this state, the alloy also contains precipitates attributable to a stable Zr-bearing phase [8,24], i.e., Al$_3$Zr (D0$_{23}$). Calculations show [17] that these two phases should be in equilibrium with (Al) in this state.

3.2. Effect of Annealing on Hardness and Electrical Conductivity

Alloys containing at least 0.4%Zr are known to undergo substantial hardening during the precipitation of these particles [5]. The EMC rod hardness vs. final annealing stage temperature curve (Figure 3a) showed that the hardening was significant at 350 °C, reaching the highest level at 400 °C (the R12-400 state). Increasing the annealing temperature leads to a significant decrease in HV, which can be accounted for, primarily, by coarsening of the Al$_3$Zr precipitates [7–11]. The hardness of the cold rolled strip was higher compared to that of the EMC rod (650 vs. 400 MPa). Deformation hardening was retained upon strip annealing to 400 °C. Annealing at higher temperatures leads to a significant softening which can be attributed to the formation of recrystallized grains. In a range of 450–500 °C, the hardness of the EMC rod was much higher than that of the strip, but the difference decreased noticeably with an increase in the annealing temperature. After annealing at 600 °C, they had approximately the same hardness (67–70 HV), mainly because of the coarsening of the Al$_3$Zr precipitates (at this temperature they should transform completely to the equilibrium D0$_{23}$ phase) [25–28].

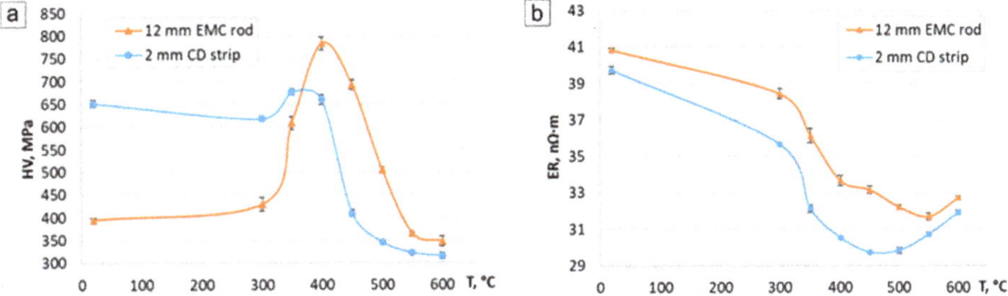

Figure 3. Hardness (HV) (**a**) and electrical resistivity (ER) (**b**) vs. temperature of annealing curves for EMC rod and cold rolled strip.

The formation of L1$_2$ (Al$_3$Zr) nanoparticles leads to a decrease in the Zr concentration in (Al) according Al–Zr phase diagram [6]. This process promotes the decrease of electrical resistivity [29,30], as shown in Figure 3b. The ER decreased from 40.8 to 31.2 nΩm for the EMC rod, and from 39.7 to 29.8 nΩm for the cold rolled strip. The decomposition rate of (Al) in the cold rolled strip is higher than that in the EMC rod. Indeed, in the former case, the minimum ER was reached after annealing at 450 °C, while in the latter case it was reached at 550 °C. In both cases, these temperatures are higher than the maximum hardening point (Figure 3a). The increase of ER at higher temperatures can be accounted for by an increase in the solubility of Zr in (Al) [6,27]. The non-recrystallized structure was preserved in strip after annealing at up to 400 °C (states S2-400). Since the best combination of hardness and ER in the cold rolled strip was reached after annealing at this temperature, we did not consider annealing the wire at higher temperatures.

3.3. Structure and Properties of Wire

We supposed the manufacturability of the as-cast EMC rod during drawing was very good. This was favored as its fine microstructure did not contain coarse intermetallic particles or inhomogeneities. Cold drawing led to the formation of elongated grains and deformation hardening. Drawing also crushed initial eutectic Al$_8$Fe$_2$Si phase veins into submicron globular particles (Figure 4a,b).

It was seen that high-temperature annealing (W400 mod) leads to coarsening of the particles (Figure 4b). Since the as-drawn state (W3 in Table 1) did not imply heating, (Al) retained all the zirconium, which dissolved in it during crystallization. Therefore, the ER for all the initial states (R12, S2 and W3) proved to be close (about 40 nΩm). Figure 5 shows that the ER of the wire and the strip are very close after annealing.

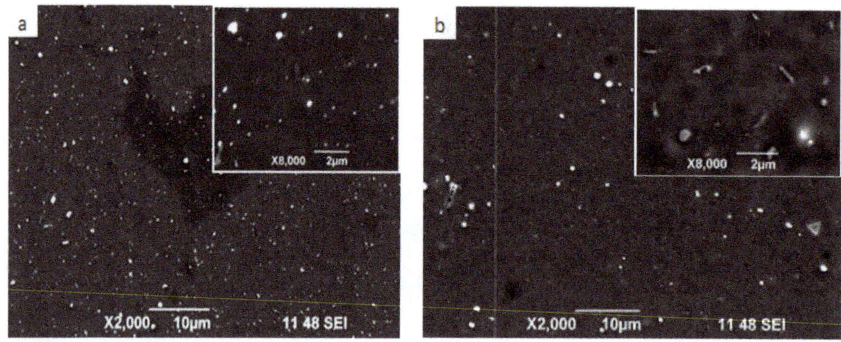

Figure 4. SEM structure of the 3 mm wire after cold drawing (from as-cast EMC rod) and annealing at 300 °C (W300, see in Table 1) (**a**) and 300 + 350 + 400 °C (W400 see in Table 1) (**b**).

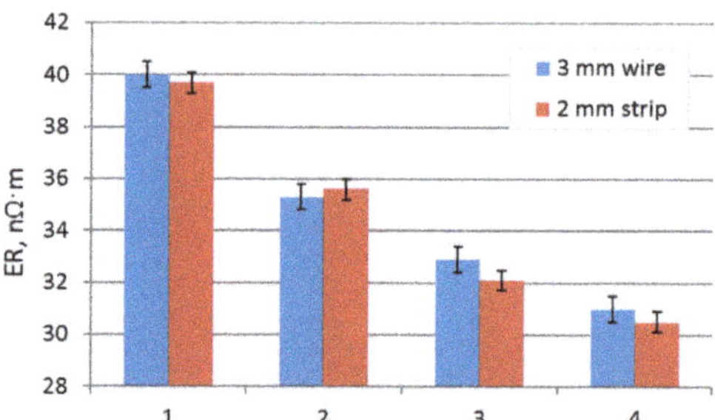

Figure 5. Comparison of electrical resistivity (ER) of 3mm wire and 2 mm strip for Al–0.6%Zr–0.4%Fe–0.4%Si alloy: 1—as-cast, 2—300 °C, 3—300 + 350 °C, 4—300 + 350 + 400 °C.

The structural changes caused by annealing were analyzed in detail using TEM. The structure annealed at 300 °C (state W300 in Table 1) contained (Al) grains and subgrains (Figure 6a), and showed the onset of Al_3Zr ($L1_2$) nanoparticles formation (Figure 6b,c). However, relatively large particles reaching ≈50 nm in size also precipitated at subgrain boundaries (Figure 6d). The presence of Fe- and Zr-bearing particles were confirmed by the direct energy-dispersive X-ray spectroscopy (EDX) analysis of their composition. According to ER data (see Figure 4), the decomposition of (Al) after annealing at 300 °C should be uncompleted. Thus, part of Zr should remain in (Al).

Annealing at 400 °C (state W400 in Table 1) largely changes the TEM structure because of the formation of multiple $L1_2$ nanoparticles inside subgrains (Figure 7b). The distribution of these nanoparticles were quite homogeneous, with their average size being within 10 nm. The reflections in the selected-area electron diffraction (SAED) patterns are much brighter (Figure 7c), as compared with those for the W300 state (Figure 6c). Taking into account the decrease of ER (Figure 3), one can conclude that (Al) decomposition after annealing at 400 °C was almost complete. This structure provides for the optimal combination of strength, electrical resistivity, and, particularly, thermal stability [31].

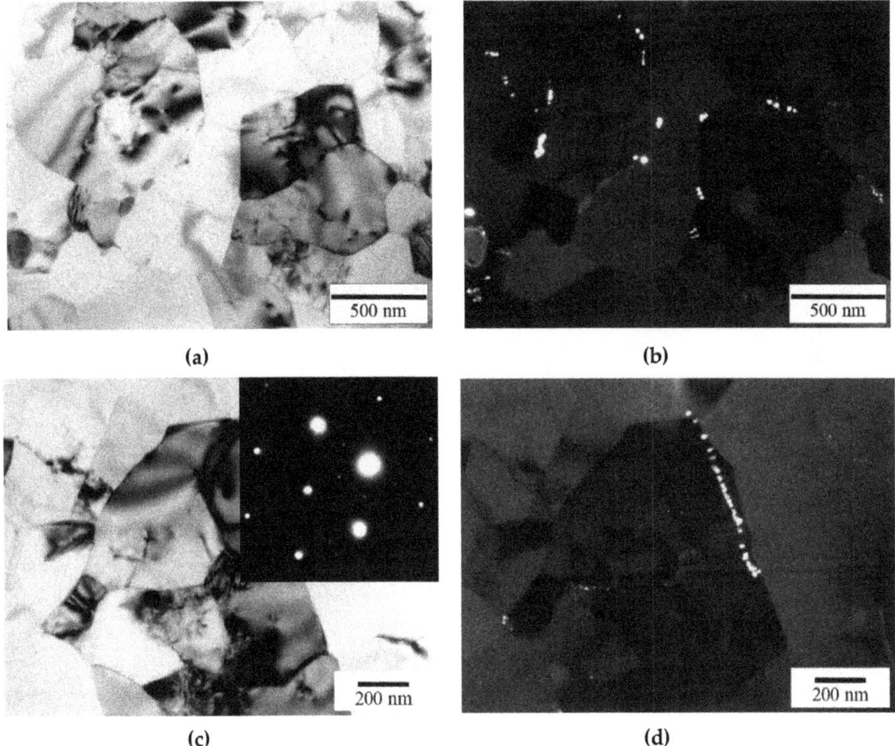

(a)

(b)

(c)

(d)

Figure 6. TEM image in the cross section of the experimental alloy after cold drawing and annealing at 300 °C (W300, see in Table 1): (**a**,**c**) bright field (BF); (**b**,**d**) dark field (DF); (**c**) BF TEM image and corresponding SAED patterns (the crystal is close to the [111] zone axis orientation) and (**d**) DF TEM image showing Al_3Zr phase nanoparticles present in the Al matrix.

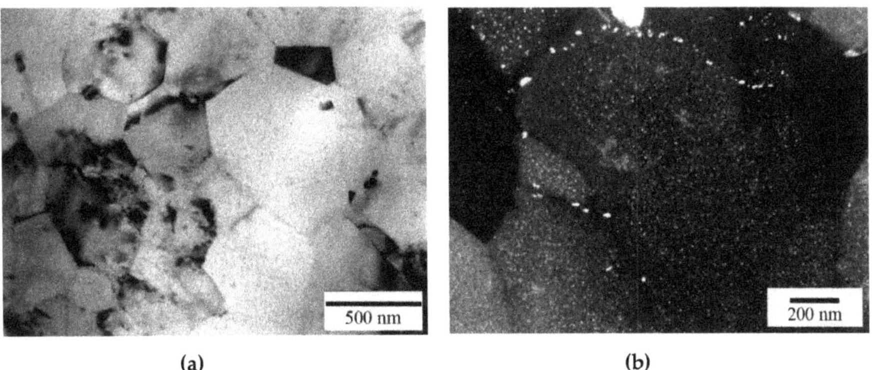

(a)

(b)

Figure 7. *Cont.*

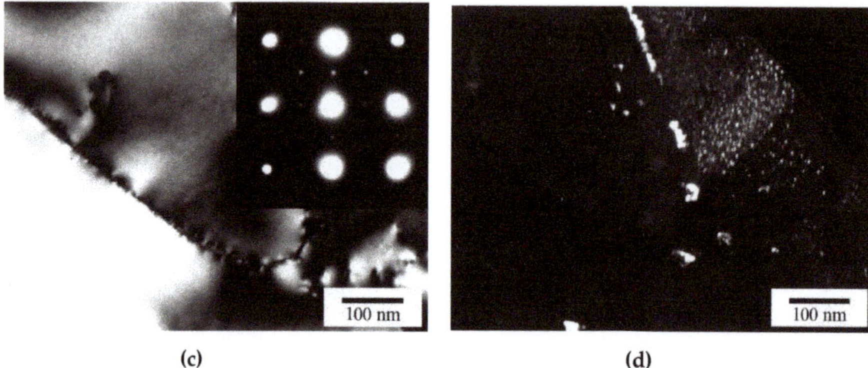

Figure 7. TEM image in the cross section of the Al alloy after cold drawing and annealing at 300 + 350 + 400 °C (W400 see in Table 1): (**a,c**) bright field (BF); (**b,d**) dark field (DF); (**c**) BF TEM image and corresponding SAED patterns (the crystal is close to the [001] zone axis orientation) and (**d**) DF TEM image showing Al_3Zr phase nanoparticles present in the Al matrix.

At the same time, the precipitates located at the boundaries of the grains and subgrains were much larger (Figure 7d). They probably formed during the first step of annealing, i.e., at 300 °C (Figure 6d). The fraction of large precipitates were small enough, compared with nanoparticles located inside subgrains. It should be noted that the size of subgrains do not change significantly, compared to the W300 state (see Figures 6a and 7a). Thus, Zr-containing nanoparticles stabilize the structure upon heating up to 400 °C, which is extremely important for heat resistant conductive alloys [32,33].

As can be seen from Table 2, the experimental alloy has a good combination of strength and electrical conductivity (the UTS and EC are 234 MPa and 55.6 IACS, respectively), meeting the AT2 type specification [1,32]. At the same time, the maximum heating temperature (standard heat resistance test) was much higher, and mechanical properties obtained meet the AT4 type specification. As can be seen from Table 1, the total annealing time was less than 10 h and probably can be further decreased (if to eliminate intermediate steps, see Figure 1). In any case, it is much less than the typical annealing time for CCR wire rods (hundreds of hours).

Table 2. Mechanical and electrical properties of Al–Zr wire alloys.

Alloy	Maximal Temperature of Heating, °C	UTS, MPa	YS, MPa	El, %	ER, nΩm	Conductivity, %IACS
EMC Al–0.6%Zr(Fe,Si)	400	234 ± 5	207 ± 7	6.8 ± 0.7	31.0 ± 0.1	55.6
AT2/KTAL (High-Strength Thermal Resistant Aluminum Alloy)	230	225–248	-	1.5–2.0	31.347	55.0
AT4/XTAl (Extra Thermal Resistant Aluminum Alloy)	400	159–169	-	1.5–2.0	29.726	58.0

The fracture surface of the experimental wire alloy after a tensile test has a homogeneous ductile fine-dimpled pattern (Figure 8a). The size of the dimples is significantly larger than that of iron-bearing particles inside the dimples (Figure 8b). No oxides or nonmetallic inclusions were found. Thus, the EMC technology provides for melt refining, despite high process temperature.

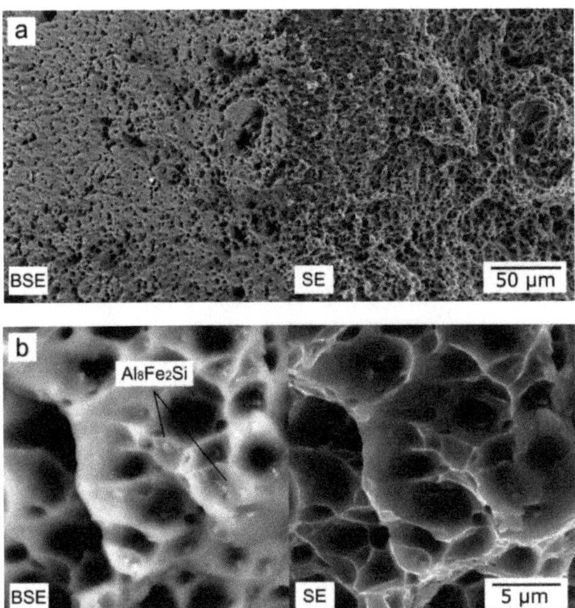

Figure 8. Fracture surfaces of experimental wire alloy (W400 see in Table 1), SEM, **left**—back scattered electron image, **right**—secondary electron image, (**a**) small magnification, (**b**) high magnification.

4. Summary

1. The experimental aluminum alloy, containing 0.6%Zr, 0.4%Fe and 0.4%Si (wt.%), was manufactured by the method of electromagnetic casting (EMC) in the form of long-length rod 12 mm in diameter. The as-cast EMC rod has high ductility when cold drawing a wire with a high degree of deformation (94%). High deformability of as-cast rods can be explained by favorable microstructure, e.g., small size of the dendritic cells (about 4 μm), submicron eutectic particles of Al_8Fe_2Si phase, and almost full dissolving of Zr in Al solid solution.

2. The effect of annealing temperature (up to 600 °C) on the hardness and electrical resistivity (ER) of EMC rod, cold rolled strip, and cold drawn wire were studied. It was shown that the temperature dependences of ER for the cold deformed strip and the wire were very close. The best combination of hardness and ER in the cold rolled strip was reached after annealing at 400 °C.

3. TEM study of structure evolution in the as-drawn wire revealed the onset of Al_3Zr (L1$_2$) nanoparticle formation at 300 °C, and almost complete decomposition of (Al) at 400 °C. The distribution of the nanoparticles was quite homogeneous, with their average size not exceeding 10 nm. At the same time, the precipitates at subgrain boundaries were much larger. Zr-containing nanoparticles allow one to stabilize the structure upon heating up to 400 °C, which is extremely important for heat resistant conductive alloys.

4. The experimental wire alloy has UTS and EC (234 MPa and 55.6 IACS, respectively) meeting the AT2 type specification. At the same time, the maximum heating temperature was much higher (400 versus 230 °C) and mechanical properties obtained meet the AT4 type specification. The possibility of electromagnetic casting of wire rods suitable for direct cold drawing would be a substantial economic advantage.

Author Contributions: Conceptualization, N.B.; investigation, N.K. and M.M.; methodology, V.T. and T.A.; supervision, N.B.; writing—original draft, N.B.; Writing—review & editing, M.M., T.A., and N.K. All authors have read and agreed to the published version of the manuscript.

Funding: The study was carried out within the framework of the implementation of the Resolution of the Government of the Russian Federation of April 9, 2010 No. 220 (Contract No. 074-02-2018-329 from May 16, 2018).

Acknowledgments: The results were obtained by using the equipment of RPC Magnetic hydrodynamics LLC, Krasnoyarsk, Russia; Institute of Physics of Advanced Materials, Ufa State Aviation Technical University, Ufa, Russia; and Department of metal forming, National University of Science and Technology MISiS, Moscow, Russia.

Conflicts of Interest: The authors declare no conflict of interest.

References

1. ASTM B941-16. *Standard Specification for Heat Resistant Aluminum-Zirconium Alloy Wire for Electrical Purposes;* ASTM International: West Conshohocken, PA, USA, 2016; pp. 1–4.

2. Brubak, J.P.; Eftestol, B.; Ladiszlaidesz, F. Aluminium Alloy, a Method of Making it and an Application of the Alloy. IFI CLAIMS Patent Services. U.S. Patent 5,067,994, 26 November 1991. Available online: https://patents.google.com/patent/US5067994A/en?oq=5067994 (accessed on 20 February 2019).

3. Knych, T.; Jablonsky, M.; Smyrak, B. New aluminium alloys for electrical wires of fine diameter for automotive industry. *Arch. Metall. Mater.* **2009**, *54*, 671–676. Available online: https://www.researchgate.net/publication/263734063_New_aluminium_alloys_for_electrical_wires_of_fine_diameter_for_automotive_industry (accessed on 20 May 2009).

4. Properzi, I. Machine for the Continuous Casting of Metal Rod. IFI CLAIMS Patent Services. U.S. Patent 2,659,948A, 24 November 1953. Available online: https://patents.google.com/patent/US2659949A/en?oq=2659949 (accessed on 25 February 2019).

5. Southwire Company. Available online: https://www.southwire.com (accessed on 28 February 2019).

6. Belov, N.A.; Alabin, A.N.; Matveeva, I.A.; Eskin, D.G. Effect of Zr additions and annealing temperature on electrical conductivity and hardness of hot rolled Al sheets. *Trans. Nonferrous Met. Soc. China* **2015**, *25*, 2817–2826. [CrossRef]

7. Knipling, K.E.; Karnesky, R.A.; Lee, C.P.; Dunand, D.C.; Seidman, D.N. Precipitation evolution in Al–0.1Sc, Al–0.1Zr and Al–0.1Sc–0.1Zr (at.%) alloys during isochronal aging. *Acta Mater.* **2010**, *58*, 5184–5195. [CrossRef]

8. Deschamp, A.; Guyo, P. In situ small-angle scattering study of the precipitation kinetics in an Al–Zr–Sc alloy. *Acta Mater.* **2007**, *55*, 2775–2783. [CrossRef]

9. Lefebvre, W.; Danoix, F.; Hallem, H.; Forbord, B.; Bostel, A.; Marthinsen, K. Precipitation kinetic of Al$_3$(Sc,Zr) dispersoids in aluminium. *J. Alloys Compd.* **2009**, *470*, 107–110. [CrossRef]

10. Forbord, B.; Lefebvre, W.; Danoix, F.; Hallem, H.; Marthinsen, K. Three dimensional atom probe investigation on the formation of Al$_3$(Sc,Zr)-dispersoids in aluminium alloys. *Scrip. Mater.* **2004**, *51*, 333–337. [CrossRef]

11. Clouet, E.; Barbu, A.; Lae, L.; Martin, G. Precipitation kinetics of Al$_3$Zr and Al$_3$Sc in aluminum alloys modeled with cluster dynamics. *Acta Mater.* **2005**, *53*, 2313–2325. [CrossRef]

12. Çadırl, E.; Tecer, H.; Sahin, M.; Yılmaz, E.; Kırındı, T.; Gündüz, M. Effect of heat treatments on the microhardness and tensile strength of Al–0.25 wt.% Zr alloy. *J. Alloy Compd.* **2015**, *632*, 229–237. [CrossRef]

13. Robson, J.D.; Prangnell, P.B. Dispersoid precipitation and process modelling in zirconium containing commercial aluminium alloys. *Acta Mater.* **2001**, *49*, 599–613. [CrossRef]

14. Avdulov, A.A.; Usynina, G.P.; Sergeev, N.V.; Gudkov, I.S. Otlichitel'nyye osobennosti struktury i svoystv dlinnomernykh slitkov malogo secheniya iz alyuminiyevykh splavov, otlitykh v elektromagnitnyy kristallizator. (Distinctive features of the structure and properties of long ingots of small cross section from aluminum alloys cast in an electromagnetic mold). *Tsvet. Met.* **2017**, *7*, 73–77. [CrossRef]

15. Dobatkin, V.I.; Elagin, V.I.; Fedorov, V.M. *Bystrozakristallizovannyealyuminievyesplavy (Rapidly Solidified Aluminum Alloys);* VILS: Moscow, Russia, 1995; pp. 43–59.

16. Polmear, I.J. *Light Alloys. From Traditional Alloys to Nanocrystals,* 5th ed.; Butterworth-Heinemann: Oxford, UK, 2006; pp. 129–130.

17. Belov, N.A.; Korotkova, N.O.; Akopyan, T.K.; Timofeev, V.N. Structure and properties of Al–0.6%Zr–0.4%Fe–0.4%Si (wt%) wire alloy manufactured by electromagnetic casting. *JOM* **2020**, *72*, 1561–1570. [CrossRef]

18. RPC Magnetic Hydrodynamics, LLC. Available online: http://www.npcmgd.com (accessed on 25 April 2020).

19. IEC 60468:1974. *Method of Measurement of Resistivity of Metallic Materials;* IEC: Geneva, Switzerland, 1974.

20. Bäckerud, L.; Chai, G.; Tamminen, J. *Solidification Characteristics of Aluminum Alloys. Vol. 1: Foundry Alloys*; Skanaluminium: Oslo, Norway, 1986; pp. 9–26.
21. Glazoff, M.V.; Khvan, A.V.; Zolotorevsky, V.S.; Belov, N.A.; Dinsdale, A.T. *Casting Aluminum Alloys. Their Physical and Mechanical Metallurgy*; Elsevier: Oxford, UK, 2019; pp. 180–186.
22. Belov, N.A.; Aksenov, A.A.; Eskin, D.G. *Iron in Aluminum Alloys: Impurity and Alloying Element*; Fransis and Tailor: London, UK, 2002; pp. 43–68.
23. Belov, N.A.; Eskin, D.G.; Aksenov, A.A. *Multicomponent Phase Diagrams: Applications for Commercial Aluminum Alloys*; Elsevier: Amsterdam, The Netherlands, 2005; pp. 19–31.
24. Gao, T.; Ceguerra, A.; Breen, A.; Liu, X.; Wu, Y.; Ringer, S. Precipitation behaviors of cubic and tetragonal Zr–rich phase in Al–(Si–)Zr alloys. *J. Alloys Compd.* **2016**, *674*, 125–130. [CrossRef]
25. Yea, J.; Guana, R.; Zhaoa, H.; Yinc, A. Effect of Zr content on the precipitation and dynamic softening behavior in Al–Fe–Zr alloys. *Mater. Charact.* **2020**, *162*, 110–181. [CrossRef]
26. Vlach, M.; Stulíková, I.; Smola, B.; Žaludová, N.; Černá, J. Phase transformations in isochronally annealed mould-cast and cold-rolled Al–Sc–Zr-based alloy. *J. Alloy Compd.* **2010**, *492*, 143–148. [CrossRef]
27. Belov, N.A.; Korotkova, N.O.; Alabin, A.N.; Mishurov, S.S. Influence of a silicon additive on resistivity and hardness of the Al–1% Fe–0.3% Zr alloy. *Russ. J. Non-Ferrous Metals* **2018**, *59*, 276–283. [CrossRef]
28. Jiang, J.; Jiang, F.; Zhang, M.; Tang, Z.; Tonga, M. Recrystallization behavior of Al-Mg-Mn-Sc-Zr alloy based on two different deformation ways. *Mater. Lett.* **2020**, *265*, 127455. [CrossRef]
29. Fu, J.; Yang, Z.; Deng, Y.; Wu, Y.; Lu, J. Influence of Zr addition on precipitation evolution and performance of AlMg-Si alloy conductor. *Mater. Charact.* **2020**, *159*, 110021. [CrossRef]
30. Zhang, J.; Wang, H.; Yia, D.; Wanga, B.; Wang, H. Comparative study of Sc and Er addition on microstructure, mechanical properties, and electrical conductivity of Al-0.2Zr-based alloy cables. *Mater. Charact.* **2018**, *145*, 126–134. [CrossRef]
31. Vlach, M.; Stulikova, I.; Smola, B.; Piesova, J.; Cisarova, H.; Danis, S.; Plasek, J.; Gemma, R.; Tanprayoon, D.; Neuber, V. Effect of cold rolling on precipitation processes in Al–Mn–Sc–Zr alloy. *Mat. Sci. Eng. A* **2012**, *548*, 27–32. [CrossRef]
32. Knych, T.; Piwowarska, M.; Uliasz, P. Studies on the process of heat treatment of conductive AlZr alloys obtained in various productive processes. *Arch. Metal. Mater.* **2011**, *56*, 687–692. [CrossRef]
33. Orlova, T.S.; Mavlyutov, A.M.; Latynina, T.A.; Ubyivovk, E.V.; Murashkin, M.Y.; Schneider, R.; Gerthsen, D.; Valiev, R.Z. Influence of severe plastic deformation on microstructure strength and electrical conductivity of aged Al-0.4Zr (wt.%) alloy. *Rev. Adv. Mater. Sci.* **2018**, *55*, 92–101. [CrossRef]

Article

Microstructure and Hardness Evolution of Al8Zn7Ni3Mg Alloy after Casting at very Different Cooling Rates

Pavel Shurkin [1,*], Torgom Akopyan [1,2], Nataliya Korotkova [1], Alexey Prosviryakov [3], Andrey Bazlov [3], Alexander Komissarov [4] and Dmitry Moskovskikh [5]

[1] Department of Metal Forming, National University of Science and Technology MISiS, Leninsky Ave. 4, 119049 Moscow, Russia; nemiroffandtor@yandex.ru (T.A.); darkhopex@mail.ru (N.K.)

[2] Baikov Institute of Metallurgy and Materials Science, Russian Academy of Sciences, Leninsky Ave. 49, 119991 Moscow, Russia

[3] Department of Physical Metallurgy of Non-Ferrous Metals, National University of Science and Technology MISiS, Leninsky Ave. 4, 119049 Moscow, Russia; pro.alex@mail.ru (A.P.); bazlov@misis.ru (A.B.)

[4] Laboratory of Hybrid Nanostructured Materials, National University of Science and Technology MISiS, Leninsky Ave. 4, 119049 Moscow, Russia; komissarov.alex@gmail.com

[5] Centre of Functional Nanoceramics, National University of Science and Technology MISiS, Leninsky Ave. 4, 119049 Moscow, Russia; mos@misis.ru

* Correspondence: pa.shurkin@gmail.com

Received: 13 May 2020; Accepted: 5 June 2020; Published: 7 June 2020

Abstract: In this study, we combined both a high strength Al-8%Zn-3%Mg aluminum matrix and a reinforcing contribution of Al_3Ni intermetallics in Al8Zn7Ni3Mg hypereutectic alloy with a tuned microstructure via a variation of cooling rates from 0.1 K/s to 2.3×10^5 K/s. Using the Thermo-Calc software, we analyzed the effect of nickel content on the phase equilibria during solidification and found out that 7%Ni provides a formation of equal fractions of primary (6.5 vol.%) and eutectic (6.3 vol.%) crystals of the Al_3Ni phase. Using microstructural analysis, a refinement of intermetallics with an increase in cooling rate was observed. It is remarkable that the structure after solidification at ~10^3 K/s across 1 mm flake casting consists of a quasi-eutectic with 1.5 μm Al_3Ni fibers, while an increase in the cooling rate to ~10^5 K/s after melt spinning leads to the formation of 50 nm equiaxed Al_3Ni particles. Under these conditions, the alloy showed an aging response at 200 °C, resulting in hardness of 200 HV and 220 HV, respectively. After 470 °C annealing, the fibers in the 1 mm sample evolved to needles. However, in melt-spun ribbons, the particles were kept globular and small-sized. Overall, the results may greatly contribute to the development of new eutectic type composites for rapid solidification methods.

Keywords: composite materials; hypereutectic aluminum alloys; Al-Zn-Mg alloys; rapid solidification; eutectic; CALPHAD; microstructure; intermetallics; precipitation hardening

---

1. Introduction

Due to the growing demand in strong lightweight materials, numerous studies have been dedicated to increase aluminum alloys strength. Since the Al-Zn-Mg-(Cu) alloys (7xxx) are the class of Al-based materials with the highest strength, their application has made advances for years in aircraft and space components [1,2]. The most common approaches toward the strengthening of 7xxx alloys are precipitation hardening and work hardening [3–5]. According to [6], conventional 7075 alloy may have 600 MPa tensile strength (T6), while by using the complex alloying and processing techniques the properties can be further enhanced. For example, according to [7], the von Mises yield rule of the dislocation theory, based on the yield prediction of the isotropic material under complex loading condition, says that they can reach more than 1148 GPa yield strength. This value was partially confirmed in some studies [8–10], where about 700–900 MPa tensile strength was achieved as a result of a raise in the overall content (Zn + Mg + Cu), addition of precipitation inoculants like Ag or Sc, and complex thermomechanical treatment, including cold working or severe plastic deformation. However, the increase in strength can also be achieved by creating a composite material with a high fraction of in-situ reinforcing particles [11,12]. The accelerated solidification of the alloys, such as in the case of melt spinning or selective laser melting (hereafter referred to as SLM), can minimize the decrease in ductility caused by reinforcing particles, while the strength can be significantly increased via the Orowan looping mechanism [13] or a load transfer [14]. Therefore, we find it worth considering to strengthen Al-Zn-Mg alloy using an in-situ particulate reinforcing approach which has been widely implemented for various Al-based materials.

Aluminum matrix composites (AMCs) have become popular during the last decades, and nowadays the amount of their demand grows along with the requirements for properties of products and the technology of their fabrication [15]. Most AMCs are post-produced by the introduction of ceramic particles (e.g., Al_2O_3, SiC, AlN, etc.) via casting methods (stir/squeeze casting), powder and granule metallurgy, combustion synthesis, etc. [15–20]. Nevertheless, their fabrication faces challenges like difficulty in producing a smaller particle size [15] and poor wettability between ceramics and molten aluminum [20]. These disadvantages cause strong restrictions on their applications, especially at a high volume of the reinforcement.

Aluminum-based alloying systems offer opportunities to fabricate AMCs reinforced by particles formed after natural (in-situ) crystallization which provide their homogenous distribution. For example, the most common commercial eutectic-based Al-Si alloys share many properties of ceramic reinforced composites [21]. Since the mechanical properties of the AMCs are tuned by controlling the type, size, morphology, and volume fraction of the filler, the prevalent number of the Al-Si in-situ composites are hypereutectic (>12%Si) and fabricated using special techniques providing rapid solidification (hereafter referred to as RS methods) in order to avoid a coarse primary silicon phase and achieve a quasi-eutectic microstructure phenomena [22,23]. In this case, the strengthening is provided by the formation of a homogeneously distributed eutectic mixture.

However, the tensile properties of Al–Si composites are low (e.g., UTS < 200 MPa for consolidated powder product [24]). Moreover, when it comes to 7xxx alloys, it is recognized that silicon is a harmful impurity since it causes Mg_2Si phase formation which leads to a reduction of precipitation strengthening [25]. This statement is supported by official data presented in the Aluminium Association standard, e.g., commercial 7075 and 7068 alloys contain up to 0.4 wt.% and 0.12% Si, respectively [26].

Meanwhile, another eutectic-forming element, nickel, does not interact with zinc and magnesium. The Al_3Ni compound has a tensile strength of about 2160 MPa and an acceptable Young modulus of 116–152 GPa that makes it reasonable for reinforcing of aluminum [27]. Moreover, its orthorhombic lattice provides incoherency toward aluminum [28]. Some near eutectic (<4 wt.%Ni) Al–Zn–Mg–Ni based on (Al) + Al_3Ni eutectic alloys are recognized to be promising as structural materials fabricated by casting and metal forming technologies [29,30], while Al–Zn–Mg/Al_3Ni composites with a high volume of reinforcement (e.g., at hypereutectic concentrations of nickel) have not been described in depth. The enhancement of the aging rate along with Al_3Ni phase volume fraction was reported in [31]. The 7005/(0–10.4 vol.%) Al_3Ni composite was prepared by metal mold gravity casting and there were particles of more than 50 μm in size. Thus, their contribution into reinforcing is reasonable to be low. The approach toward refining of Al_3Ni primary crystals up to 5 μm by rolling was suggested in the study on 7050/(5–10 wt.%) Ni composite [7]. It is alleged that the yield strength calculated by the Orowan equation can get 630 MPa, while the brittle fracture surface was shown. Both these papers do not consider the RS methods to obtain a quasi-eutectic structure. In an earlier research [32], the formation of the fully eutectic structure in Al-7%Ni hypereutectic alloy after unidirectional solidification is reported. In Martínez-Villalobos et al.'s study [33], the same result on Al-8%Ni after melt spinning was achieved and an effectiveness of reinforcement was shown to be due to ultrafine Al_3Ni particles formation.

As for matrix, the experiments on melt-spun ribbons of Al-Zn-Mg alloys showed an opportunity of obtaining a supersaturated solid solution without prior annealing for quenching [34]. The same result was achieved after rapid solidification released during SLM of 7068 alloy along with providing precipitation hardening via the aging of as-built parts [35]. In this study, we chose the copper-free matrix system Al-8%Zn-3%Mg as basis related to the strongest commercial 7xxx alloys (7001, 7090, 7055, etc. [26]). We do not consider copper addition because it causes unreasonable complication of phase composition, as it is clearly shown in a pivotal research [36] on the optimization of Al-Zn-Mg-Cu-Ni alloys.

As for reinforcement, the key requirements are to possess a high enough volume of aluminides particles and an absence of the primary crystals in the structure. Thus, it is mainly desirable to obtain a fully eutectic structure on the supersaturated matrix field. To ensure a high concentration of reinforcement particles, the study is dedicated on nickel content of 7% that is substantiated in the computational section. Since the morphology of the Al_3Ni phase can be tuned on the cooling rate basis, we followed the investigation path via the study of the phase diagrams and experimental simulation of cooling rate enhancement.

From the above, the current paper aims to investigate the evolution of the solidification path, structure, and hardness depending on cooling rate variation from slow to rapid solidification of the Al8Zn7Ni3Mg alloy by thermodynamic calculation and experimental study.

2. Materials and Methods

Initially, we investigated details of Al-8%Zn-3%Mg alloys with various nickel contents, in particular, solidification path and phase composition using the Thermo-Calc software (Version 3.1, TCAl4 Al-based alloy database, Thermo-Calc Software AB, Stockholm, Sweden) [37]. Single point equilibrium, phase diagram, property diagram, and Scheil–Goulliver solidification simulation options were used.

For the experimental section, the Al8Zn7Ni3Mg alloy was the main test material. The samples were prepared by melting high-purity aluminum (99.99%Al), zinc (99.97%Zn), magnesium (99.9%Mg), and Al20%Ni master alloy in a graphite–chamotte crucible using a Nabertherm K 1/13 (Nabertherm GmbH, Lilienthal, Germany) resistance furnace in an air atmosphere. The melt temperature was kept at 850 °C and the total time of the melting process was about 90 min. The Al20%Ni master alloy was mixed into the molten aluminum using a graphite stick. Before casting, the melt was purified by dry C_2Cl_6 powder. The chemical composition as determined by spectral analysis is presented in Table 1.

Table 1. Chemical composition of the experimental alloy.

Designation	Concentrations, wt. %			
	Zn	**Mg**	**Ni**	**Al**
Al8Zn7Ni3Mg	7.79	3.13	7.16	Balance

In order to obtain a variety of cooling rates, we provided different solidification conditions. A portion of molten metal ~50 g was solidified in the furnace, and thus the lowest cooling rate was achieved (FC sample). Three cooling conditions were provided via casting. We obtained cylindrical samples of 30 mm and 5 mm in diameter, and a thin flake of less than 1 mm via pouring onto a cold steel heat sink. The highest cooling rate was induced using a melt spinning (MS sample) of experimental alloy ingot via pouring a molten metal onto a rotating copper wheel of DVX-II apparatus (Dexing Magnet Tech. CO. Ltd., Xiamen, China) in an argon gas atmosphere. The linear rotation speed of the copper wheel was 30 m/s. A general view of the samples is demonstrated in Figure 1.

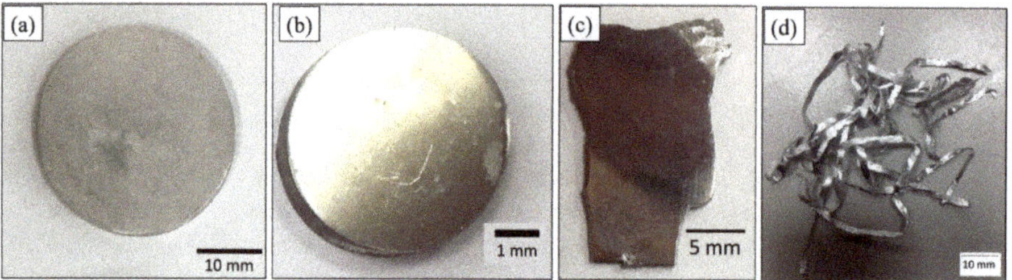

Figure 1. Experimental cast samples and melt-spun ribbons: (**a**) 30 mm cast sample; (**b**) 5 mm cast sample; (**c**) 1 mm cast sample; (**d**) melt-spun ribbon.

The microstructure was examined by optical microscopy (OM, Axio Observer MAT, Carl Zeiss Microscopy GmbH, Oberkochen, Germany), scanning electron microscopy (SEM, TESCAN VEGA3, Tescan Orsay Holding, Brno, Czech Republic) with an electron microprobe analysis system (EMPA, Oxford Instruments plc, Abingdon, UK), and the Aztec software (Version 3.0, Oxford Instruments plc, Abingdon, UK). The metallographic samples were ground with SiC abrasive paper and polished with 1 μm diamond suspension. A total of 1% hydrogen fluoride (HF) water solution was used for etching. To investigate the structure of the melt-spun ribbons, we used transmission electron microscopy (TEM, JEM-2100, JEOL Ltd., Tokyo, Japan).

The size of the dendritic cells (dendritic parameter, d), as well as of the intermetallics, was experimentally determined using metallography from high-contrast microstructural images processed with the appropriate software, ImageJ (National Institutes of Health, Bethesda, MD, USA). The Horizontal Lines option was used for implementing the stereological method of measuring the relative length of the phase regions. To obtain reliable data, we analyzed at least 10 fields in the microstructure for defining the content of each structural component. The experimental dendritic parameter data were used for evaluating the cooling rate in the alloy crystallization temperature range using a well-known empirical relationship [38]:

$$V_c = (A/d)^{1/n}, \tag{1}$$

where V_c—cooling rate upon solidification in K/s, d—dendritic parameter in μm and A, n—material-dependent constants.

Since we do not take determination of the precise material dependent constants as a mandatory, they were taken from [20] for high-strength Al-Zn-Mg-Cu alloys as $A = 100$ and $n = 1/3$.

Some samples were selectively subjected to a T5 heat treatment (aging at 200 °C for 1 h without prior quenching) and a T4 heat treatment (470 °C for 1 h). To control the properties evolved, a Vickers' hardness test at a load of 10 g (0.1 N) and 5 s dwell (for the MS sample), and at a load of 1 kg (10 N) and a dwell time of 10 s (for other samples) was used.

3. Results and Discussion

3.1. Computational Section

The polythermal section shown in Figure 2a indicates that under equilibrium conditions the nickel addition to the Al-8%Zn-3%Mg alloy significantly influences on the solidification path shifting to a hypereutectic manner at over 3.6%Ni. The Al-Zn-Mg-Ni system is convenient to study due to a lack of interaction between Al-Zn-Mg and Al-Ni systems. Thus, all of Zn and Mg are bonded into the T phase ($Al_2Mg_3Zn_3$) which is responsible for precipitation hardening of alloys with a Zn/Mg atomic mass ratio of less than 1 [39]. The liquidus line rises and the equilibrium solidus decreases slightly with an increase in nickel concentration in hypereutectic alloys. The Al8Zn7Ni3Mg alloy shows an experimental liquidus and solidus temperature of 665 and 520 °C, respectively. However, the Scheil–Goulliver simulation showed that the solidification range is much higher than the one in equilibrium condition and it ends with the proceeding of the [(Al) + Al₃Ni + T] reaction at a temperature of 478 °C (dotted line in the polythermal section). The temperatures, determined by thermal analysis using a single chromel–alumel thermocouple submerged into the melt, sufficiently agree with the calculated data that especially fair for non-equilibrium solidus. The reactions [L→Al₃Ni] at 659 °C, [L + Al₃Ni + (Al)] at 624 °C, and [L + Al₃Ni + (Al) + T] at 478 °C are detected. On the one hand, such a low solidus intrinsic to 7xxx alloys deteriorates their hot cracking tolerance in some RS methods, like SLM [40]. On the other hand, having a high volume of eutectic may lead to an improvement of brittleness due to cracking healing in the solid–liquid state [41].

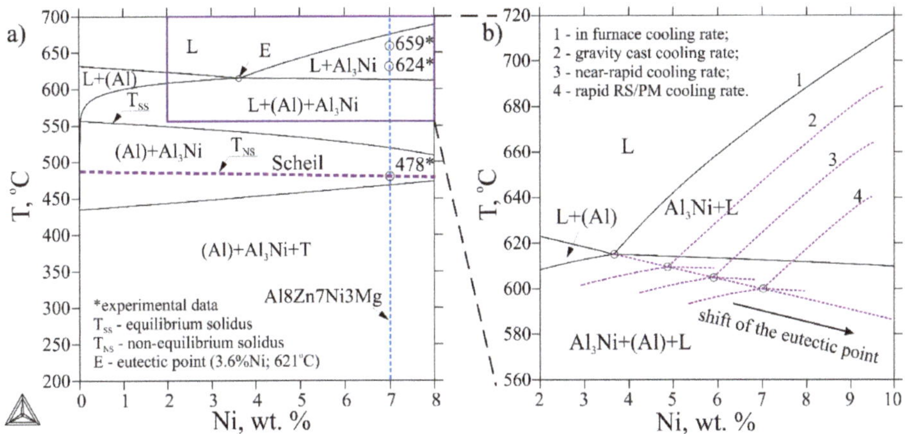

Figure 2. (a) Polythermal section of a Al-Zn-Mg-Ni system at 8%Zn and 3%Mg with marked non-equilibrium solidus calculated by Scheil–Goulliver simulation and (b) schematic representation of the solidification path shift as a result of an increase in the cooling rate.

In addition, Figure 2b schematically shows how the binary eutectic point shift depends on the cooling rate. According to [42], under high undercooling, diffusion in the liquid phase may be hindered causing local changes in composition and difference in local solidification rates. Under this condition, the sustainable extension of Zn and Mg solubility in the (Al) matrix can be expected [34]. In addition, the temperatures of liquidus and solidus may be lower than the equilibrium ones. It is anticipated that the experimental samples may solidify via different paths, causing a variety of structures from highly-hypereutectic to hypoeutectic, including eutectic manner if the two phases would solidify simultaneously in a diffusion coupled fashion. These microstructures can be qualified as quasi-structures, because the corresponding alloy is hypereutectic in general.

According to Figure 2a, the solvus temperature corresponding to the dissolution of the T phase (Al₂Mg₂Zn₃) increases along with nickel content and therefore the [(Al) + Al₃Ni] area is narrowing as well as the suitable range of quenching temperatures. We chose a temperature of 470 °C as a conventional temperature for a solid solution treatment of 7xxx alloys [1].

Figure 3a shows how the (Al) solid solution composition evolves at 470 °C depending on nickel content. It is assumed to be the same after rapid solidification. According to the (Al) matrix composition, we simulated its decomposition at the chosen aging temperature of 200 °C as a volume fraction of precipitation products both T′ and M′. The graph shows that nickel addition contributes to a gradual growth of Zn and Mg content in (Al) matrix, as well as precipitates volume, respectively. The solid solution of the experimental Al8Zn7Ni3Mg alloy composes 9.6%Zn and 3.6%Mg, promoting precipitations of 9.6 vol.% both T′ and M′ dispersoids responsible for matrix hardening. It is clearly seen that the (Al) solid solution composition and volume of dispersoids see a plateau (9.7%Zn and

3.7%Mg in (Al) and 10.1 vol.% of precipitates) at over 7.5%Ni. The further addition of nickel may cause an undesirable end of solidification via ternary eutectic [(Al) + Al$_3$Ni + T] leading to the appearance of an additional quantity of the T-phase in the structure undissolved in aluminum. That way, a higher amount of nickel seems to be unreasonable.

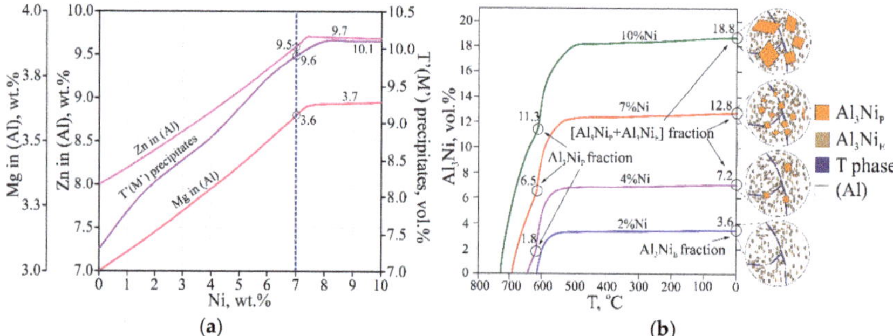

Figure 3. (**a**) Composition of the (Al) solid solution at 470 °C and corresponding volume fraction of T′ (M′) precipitates at 200 °C depending on nickel content in Al-8%Zn-3%Mg; (**b**) relationship between volume fraction of Al$_3$Ni phase and temperature at various nickel contents in Al-8%Zn-3%Mg alloy and schematic display of the structure evolution.

Figure 3b shows the volume fraction of the Al$_3$Ni phase in Al-8%Zn-3%Mg-xNi (x = 2%, 4%, 7%, 10%) alloy depending on the temperature under equilibrium conditions. In hypereutectic alloys, the formation of two types of intermetallics is possible. The primary phase (Al$_3$Ni$_P$) appears before the aluminum solid solution. Then it progresses with eutectic particles (Al$_3$Ni$_E$) formation along with aluminum. As we schematically represented, the Al$_3$Ni$_E$ type nucleates as disperse particles, while the Al$_3$Ni$_P$ type commonly has a faceted morphology. While at 2% and 4%Ni the volume of the reinforcement is significantly scarce (3.6% and 7.2%), at increasingly hypereutectic concentrations (10%Ni) the volume fraction is superior (18.8%), but the Al$_3$Ni$_P$ fraction is dominant (11.3%). On this occasion, the suppression of the pre-eutectic stage may be not accomplished in our experimental conditions. Barclay et al. [32], who first achieved a quasi-eutectic structure in Al-Ni hypereutectic alloys by the RS method, found that Al-10%Ni alloy requires a five times higher solidification rate than Al-7%Ni alloy for achieving suppression of the Al$_3$Ni primary crystallization.

Thus, the experimental Al8Zn7Ni3Mg alloy exhibits appropriate volume fraction of intermetallics (12.8%) composed of balanced eutectic (6.3%) and primary (6.5%) crystals. In addition to that, it shows the highest promise for precipitation hardening, a low inter-particles spacing is anticipated, taking into account positive results on binary alloys [32,33].

3.2. Solidification Path and Structure Analysis

The results of the cooling rate estimation are shown in Table 2, as well as the change in microstructural peculiarities pertaining to the dendritic parameter and Al$_3$Ni intermetallics size. As can be seen, an increase in the cooling rate drives a refinement of primary crystals up to the ultimate suppression of the pre-eutectic stage after cooling at more than 10^3 K/s.

The results are supported by a general view of microstructure evolution from a slow rate furnace-cooled sample to rapidly cooled melt-spun ribbons (Figure 4). In OM images, the Al$_3$Ni phase is visible as brown inclusions embedded into the light matrix of the (Al) solid solution, and otherwise, in the SEM image light intermetallics are incorporated into the dark matrix. Generally, the evolution allows to estimate the formation of a quasi-eutectic structure under an increase in the cooling rate.

Table 2. Microstructure peculiarities of the Al8Zn7Ni3Mg alloy depending on the cooling rate.

Experimental Sample	Dendritic Parameter (d) [1], μm	Cooling Rate (V_c) [2], K/s	Al$_3$Ni Size Range [1], μm	Al$_3$Ni Median Size [1], μm (Al$_3$Ni$_E$/Al$_3$Ni$_P$)
FC	198 ± 31	0.1	30–351	12/225
30 mm cast	39 ± 6	17	20–94	4/13
5 mm cast	19 ± 6	133	5–43	2/7
1 mm cast	9 ± 2	1.4×10^3	1–3	1.5/-
MS	1.6 ± 0.3	2.3×10^5	0.03–1.8	0.3/-

[1] determined using image analysis software; [2] calculated using the Equation (1).

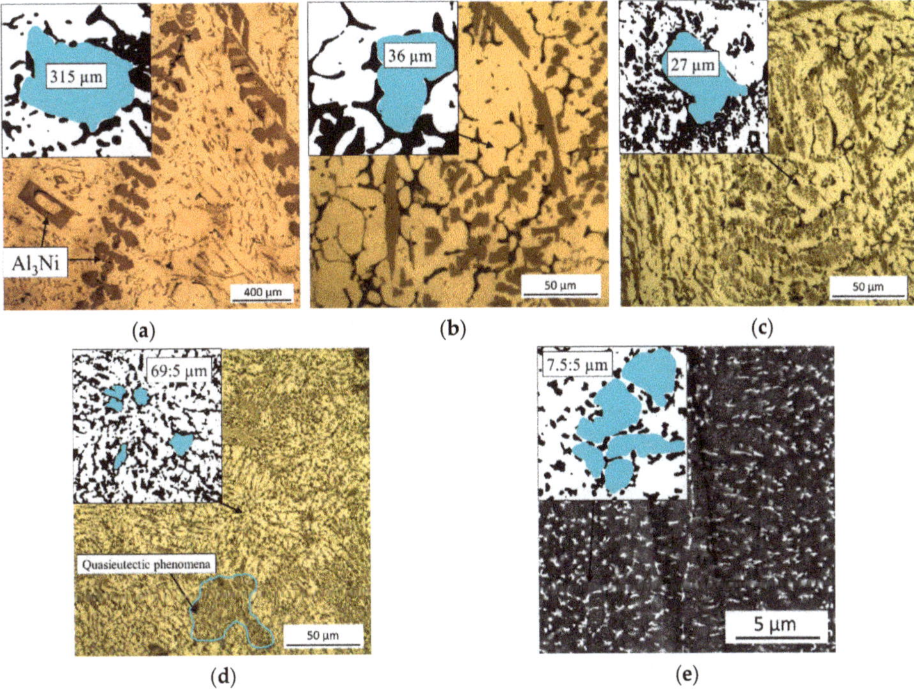

Figure 4. Microstructures of the Al8Zn7Ni3Mg alloy after solidification at different cooling rates, including binary imaging used for dendritic parameter determination: (**a**) furnace-cooled (FC) sample; (**b**) 30 mm cast sample; (**c**) 5 mm cast sample; (**d**) 1 mm cast sample; (**e**) melt spinning (MS) sample.

The microstructure of the FC sample (Figure 4a) can be qualified as the nearest to the equilibrium one. During in-furnace solidification, a large fraction of imposingly coarse primary crystals of up to 351 μm can be observed, which are probably responsible for a low ductility of material under such condition. The presented pattern reliably agrees with a schematic one shown in Figure 2b. The primary Al$_3$Ni$_P$ crystals are in an equilibrium with the eutectic mixture [(Al) + Al$_3$Ni$_E$] in which the eutectic-origin particles are relatively small (medium size 12 μm).

Considering the structure of gravity cast samples, we managed to achieve a significant difference among their solidification conditions and cooling rates as well. The cooling rate of 17 K/s for the 30 mm cast sample (Figure 4b) was calculated using equation 1 and a dendritic parameter of around 39 μm. The conception of the microstructure does not look modified in comparison to the furnace-cooled sample, but the refinement is obvious from the scale and increased number of dendritic cells reduced fivefold in size along with dramatically reduced intermetallics.

A further increase in the cooling rate resulted in intermetallic bands formation that indicates a closeness of the solidification path to the eutectic one. In the 5 mm cast sample (Figure 4c), there is still an extremely high number of Al_3Ni needle-like primary intermetallics, halved in size, obtained via near-rapid cooling ($d \sim 20$ µm, $V_c = 133$ K/s). Meanwhile, the 1 mm cast sample ($d \sim 10$ µm, $V_c = 1.4 \times 10^3$ K/s) contains wide areas where the coupled growth of Al_3Ni and the (Al) solid solution was provided (Figure 4d). This as qualified quasi-eutectic structure consists of micron-scale intermetallics, mixed with a hypoeutectic structure in vicinity, allowed to estimate the dendritic parameter. Such inhomogeneous structure is probably due to different crystallization front and related to a tough experimental condition. The quasi-eutectic structure is believed to develop via the following solidification path [43–45]. The Al_3Ni phase as part of the eutectic [(Al) + Al_3Ni] nucleates first due to its higher melting point. Next, the (Al) solid solution nucleates in the nickel depleted zone around the Al_3Ni particles preventing its growth. Therefore, the residual liquid phase is enriched with the nickel until its composition reaches the eutectic point. Thus, the reciprocal growth of both the Al_3Ni phase and the (Al) solid solution occurs.

In contrast, under melt spinning conditions, the highest cooling rate was achieved resulting in an ultrafine microstructure with visible dendritic cells of about 1.5 µm (Figure 4e). The estimated cooling rate is 2.3×10^5 K/s, which agrees with literature data [22]. The quasi-eutectic-origin intermetallics of submicron size are located along (Al) solid solution dendritic cells and the ultimate structure looks hypoeutectic. However, some areas are revealed to have very small in-bulks particles, which are to be studied using a high-magnification technique.

The microstructure of the 5 mm cast sample was studied in detail (Figure 5). According to EMPA analysis (Figure 5b), the white phase (in Figure 5a) corresponds to an insoluble Al_3Ni phase with a homogenous composition of 76% Al and 24%Ni. In the vicinity, T phase veins are clearly observed. It is striking that the most part of the Zn and Mg are dissolved in the (Al) matrix even under cooling at 133 K/s. Hence, it is reasonable to anticipate a dramatic extension of solid solubility in samples obtained at higher cooling rates.

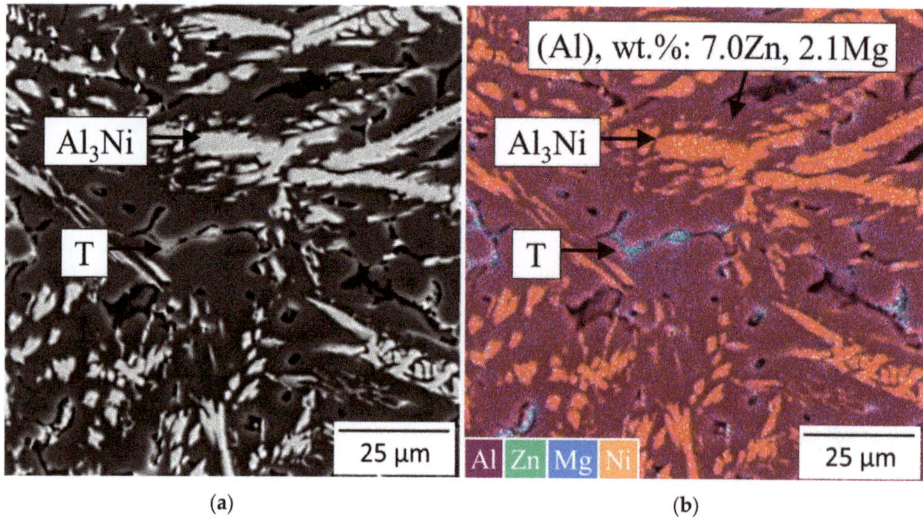

Figure 5. Microstructure of the 5 mm cast sample of the Al8Zn7Ni3Mg alloy: (**a**) SEM; (**b**) multilayer elemental map.

Ultimately, the analysis of the structure revealed a significant deviation of the solidification path from a local equilibrium condition. Microstructural peculiarities (dendritic cells and intermetallics) were far much refined under an increase in the cooling rate, and their characterization in the samples after a cooling rate of more than 10^3 K/s requires for detailed analysis using higher magnification. Preliminary, it is believed that a needle-like Al_3Ni phase, occurring in the structure of the FC, 30 mm, and 5 mm cast samples, is responsible for brittle behavior, while globular particles presented in other samples could play a reinforcing role in enhancing strength without a substantial loss of ductility. Moreover, the supersaturation of the (Al) solid solution is expected and will be discussed upon the hardness measurement results.

3.3. Characterization of the Quasi-Eutectic Structure

In the microstructure of the 1 mm cast sample (Figure 6), the tough casting condition resulted in the shrinkage cavities formation along the grains. There is no primary phase detected, and the structure seems to be homogeneous and quasi-eutectic in general. As can be seen, the Al_3Ni eutectic bands are in the bulk of the (Al) solid solution, and in their vicinities, Zn and Mg rich areas are presented with an Al–8.7%Zn–3.4%Mg composition. This result virtually agrees with a previously calculated (Al) matrix composition at 470 °C and corresponds to a supersaturated condition. The magnified section of the [(Al) + Al_3Ni] eutectic band shows that it has a rather fibrous morphology with a linear size of up to 3 μm. These microstructure features exhibit a correspondence to fiber-reinforced metal matrix composites, so it probably yields the best load transfer efficiency.

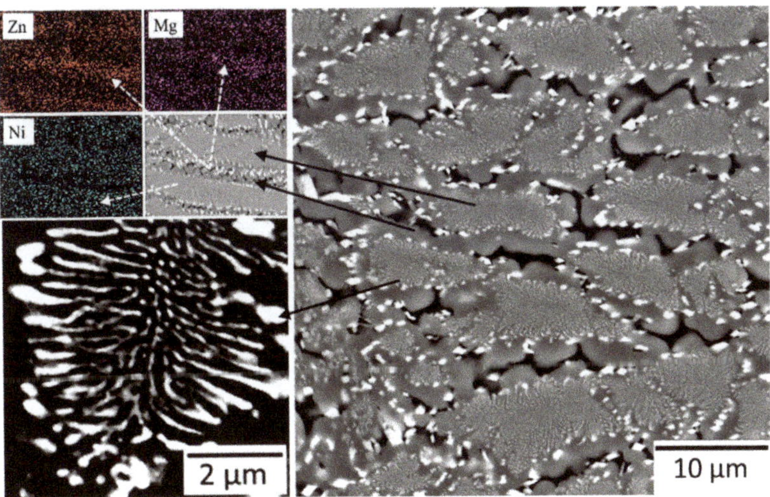

Figure 6. Microstructure of the Al8Zn7Ni3Mg alloy after solidification at 1.4×10^3 K/s.

Figure 7 displays the TEM characterization of the structure appearing in the melt-spun ribbons. As can be seen from Figure 7a, it is conceptually similar to the quasi-eutectic structure of the 1 mm cast sample, because it also shows wide bands of the fine intermetallics, surrounded by dendritic bulks of (Al) solid solution, which are clearly seen in Figure 7b. Since the dendritic cells presented are less than 300 nm in size, we can assume that the cooling rate achieved in some areas was much higher than $\sim 10^6$ K/s, probably provided by greater thermal conductivity of the aluminum matrix than intermetallics. The incoherent equiaxed Al_3Ni particles with a median linear size of 50 nm are located within the (Al) matrix (Figure 7c). The dark field image (Figure 7d) served as a more contrasting image for inter-particles spacing estimation. Its value is in the 10–50 nm range, which, in turn, shows good promise for a contribution into strength along with precipitate shearing due to possible naturally or artificially triggered decomposition of the (Al) solid solution.

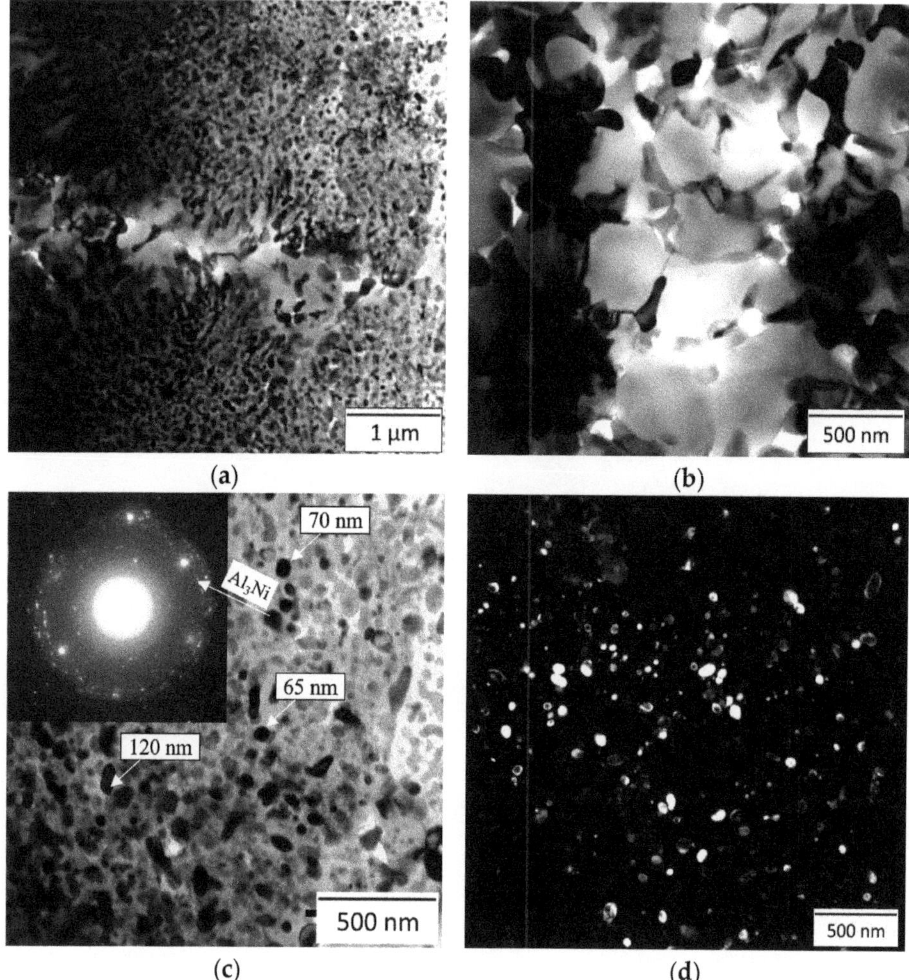

Figure 7. Microstructure of the melt-spun ribbons (MS sample, 2.3 × 105 K/s) of the Al8Zn7Ni3Mg alloy: (**a**) general view of the intermetallic bands; (**b**) a (Al) dendritic structure appeared in the bands' vicinities; (**c**) light field image of the globular intermetallcs' band; (**d**) dark field image.

3.4. Hardness and Influence of Heat Treatment

The results of the hardness measurement are shown in Figure 8. As it was earlier shown, the FC sample comprises of primary intermetallics of significantly giant linear size, and therefore, the hardness of 52 HV presented reflects the footprint after indentation into the (Al) matrix. Hence, the nickel contribution is relatively low, as well as the hardness value. A further increase in the cooling rate causes a strong visible effect on the hardness value. For comparison, it increased twofold in the 30 mm cast sample and threefold in the 5 mm and 1 mm samples. The most value of 195 HV was achieved in melt-spun ribbons, which have the finest virtually qualified nanocomposite microstructure. Furthermore, it is worth considering that the most uniform distribution of the hardness values was obtained in cast samples, while the inhomogeneous structure of the 1 mm cast sample provided high deviation, as well as for the MS sample measured using far lower load due to brittleness.

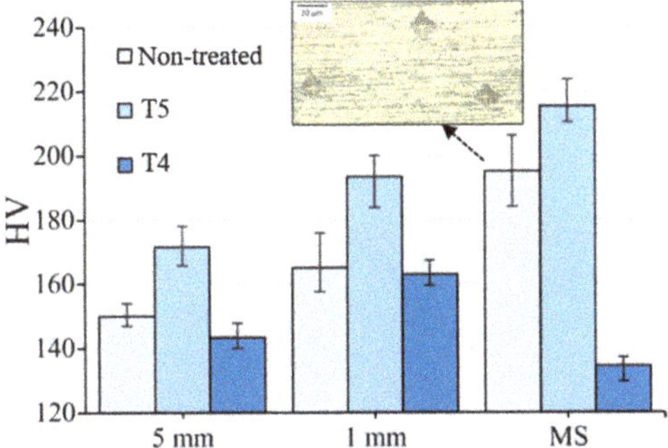

Figure 8. Influence of heat treatment on the hardness of experimental samples of the Al8Zn7Ni3Mg alloy.

Thermal treatment causes significant changes in microstructure and hardness, respectively. The bar chart demonstrates a relatively high aging response of around 20 HV at 200 °C without preliminary quenching in 5 mm and 1 mm cast samples as well as the MS sample. It is striking that the MS sample initially has a hardness of 200 HV, which is the same as an aged 1 mm cast sample. However, further aging leads to an increase in values of up to 220 HV, corresponding to ultrahigh-strength Al-based materials [10]. However, a further increase in temperature to 470 °C leads to a significant degradation of hardness. For all samples this is due to the (Al) matrix solutionization during alloy solidification. Moreover, for a composite-structured 1 mm cast sample and MS samples, this loss of properties is caused by a significant degradation of the intermetallics morphology.

The degradation of the Al$_3$Ni phase morphology after a 470 °C heat treatment is highly dependent on the initial microstructure. As the as-cast 1 mm sample microstructure contains fiber-like, slightly elongated inclusions, they were conjugated along a definite crystallographic plane, resulting in significant shape deformation as a mixture of needles and coalescenced particles, both blocky for load transfer (Figure 9a). Meanwhile, the MS sample initially contained equiaxed intermetallics, and heating to 470 °C caused advanced coalescence. The particles size is ranged in 0.2–2.5 μm, but their roundness is apparently appropriate to be 0.8–1 for 90% of the whole volume (Figure 9b). This factor seems to be advantageous in terms of hot consolidation of the melt-spun ribbons, hence, the microstructure of the bulk products may still be qualified as a reinforced composite.

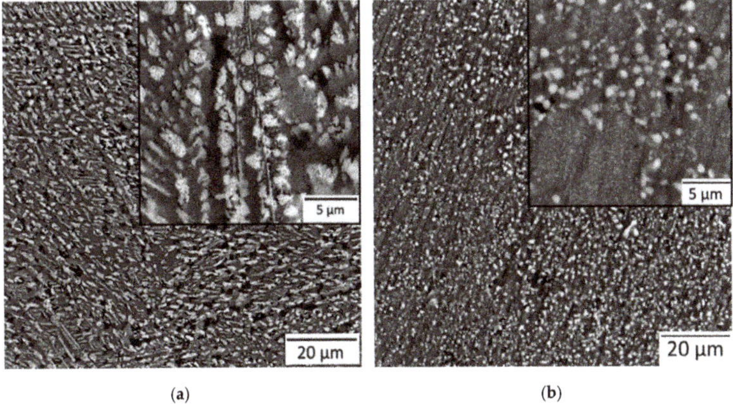

(a)   (b)

Figure 9. Microstructure of the Al8Zn7Ni3Mg alloy after 470 °C heat treatment: (**a**) 1 mm cast sample; (**b**) MS sample.

4. Conclusions

By using computational and experimental studies, the effect of different cooling rates (0.1 K/s, 17 K/s, 133 K/s, 1.4×10^3 K/s, and 2.3×10^5 K/s) on the phase composition, solidification manner, microstructure, and hardness of the Al8Zn7Ni3Mg aluminum alloy has been analyzed in details. The tremendous refinement of the microstructure along with extension in Zn and Mg solid solubility was accompanied with increase in hardness as a result of solidification path shift from a hypereutectic to a eutectic and hypoeutectic one. The results are believed to be beneficial for the development of new high-strength particulate reinforced composites which do not require ex-situ intervention for the input of reinforcements. Moreover, when considering eutectic forming element addition, a decrease in hot embrittlement is anticipated. Hence, the new composition may be highly recommended for laser additive manufacturing applications. The major conclusions are as follows:

(1) By using the CALPHAD approach, the concentration of nickel in the experimental Al8Zn7Ni3Mg alloy has been justified. While the eutectic point in the Al-8%Zn-3%Mg-Ni system corresponds to 3.6%Ni, the 7%Ni composition is highly hypereutectic. Taking into account the opportunity to shift the solidification path with further refinement caused by increase in cooling rates, the experimental alloy comprises 12.8 vol.% Al_3Ni intermetallics in which half corresponds to the primary phase;

(2) Composition of the (Al) solid solution after rapid solidification has been simulated similar to the one at the 470 °C solid solution temperature. The higher the nickel content, the more saturated the (Al) matrix. It is shown that a 7%Ni concentration is advantageous in terms of obtaining supersaturated solid solution containing 9.6%Zn and 3.6%Mg, promoting precipitation of 9.6 vol.% of T′ and M′ dispersoids. Meanwhile, a higher amount of nickel does not provide a significant change in these values;

(3) By OM, SEM, and TEM analysis, the increase in cooling rates on the microstructure was investigated, showing profound opportunities for microstructure tuning. A highly hypereutectic structure was observed after solidification at 0.1 K/s, 17 K/s, and 133 K/s accompanied with a refinement of the Al_3Ni phase from 50 to 7 μm in medium size. A cooling rate of 1.4×10^3 K/s appeared to be sufficient for providing quasi-eutectic solidification manner, and most of the structure area is covered with fiber-like composite microstructure of 1.5 μm intermetallics, while the melt spinning provided a cooling rate of 2.3×10^5 K/s resulting in a visible hypoeutectic structure with ultrafine equiaxed 50 nm intermetallics in the (Al) matrix bulk, beneficial for looping reinforcing;

(4) The hardness test revealed a substantial increase in strengthening as a result of structure refinement. While in slow and conventionally-cooled samples the hardness of 50–150 HV is relatively not contributed by Al_3Ni intermetallics appearance, the rapidly solidified samples showed a significant enhancement of 165 HV in the 1 mm cast sample and 195 HV in melt-spun ribbons. Moreover, these samples both showed a significant strengthening after low temperature annealing at 200 °C, achieving up to 220 HV;

(5) Nonetheless, a solid solution treatment at 470 °C resulted in significant degradation of hardness. While the 1 mm cast sample saw a decrease to the initial level, the melt-spun sample degraded to obtain hardness of around 140 HV. Such a loss in properties is caused by structure coarsening. Meanwhile, the fibrous-like intermetallics coalescenced to become needles and rods (up to 10 μm), and the globular particles evolved remaining a high roundness and relatively low size (0.2–2.5 μm), which is promising in terms of further consolidation processing.

Author Contributions: Conceptualization, P.S. and T.A.; investigation, N.K. and A.P.; methodology, A.P., A.B., and A.K.; resources, A.P., A.K., D.M., and A.B.; supervision, T.A.; writing—original draft, P.S.; writing—review & editing, P.S. and T.A. All authors have read and agreed to the published version of the manuscript.

Funding: This research was funded by the Russian Science Foundation (project No. 19-79-30025).

Metals **2020**, *10*, 762

Acknowledgments: The authors would like to thank Mikhail Gorshenkov for providing TEM investigations.

Conflicts of Interest: The authors declare no conflict of interest.

References

1. Polmear, I.; StJohn, D.; Nie, J.-F.; Qian, M. Physical metallurgy of aluminium alloys. In *Light Alloys*, 5th ed.; Elseiver: London, UK, 2017; pp. 31–107. [CrossRef]

2. Starke, E.A., Jr.; Staley, J.T. Application of modern aluminium alloys to aircraft. In *Fundamentals of Aluminium Metallurgy*; Lumley, R., Ed.; Woodhead Publishing Limited: Cambridge, UK, 2011; pp. 747–783. [CrossRef]

3. Ditta, A.; Weia, L.; Xub, Y.; Wua, S. Microstructural characteristics and properties of spray formed Zn-rich Al-Zn-Mg-Cu alloy under various aging conditions. *Mater. Charact.* **2020**, *161*, 110133. [CrossRef]

4. Zhao, J.; Liu, Z.; Bai, S.; Zeng, D.; Luo, L.; Wang, J. Effects of natural aging on the formation and strengthening effect of G.P. zones in a retrogression and re-aged Al-Zn-Mg-Cu alloy. *J. Alloys Compd.* **2020**, *829*, 154469. [CrossRef]

5. Chen, Z.; Yuan, Z.; Ren, J. The mechanism of comprehensive properties enhancement in Al–Zn–Mg–Cu alloy via novel thermomechanical treatment. *J. Alloys Compd.* **2020**, *828*, 154446. [CrossRef]

6. Tamarin, Y. *Atlas of Stress-Strain Curves*, 2nd ed.; ASM International: Novelty, OH, USA, 2002.

7. Peng, G.; Tietao, Z.; Xiaoqing, X.; Zhi, G.; Li, C. Refinement mechanism research of Al_3Ni phase in Ni-7050 alloy. *Rare Met. Mater. Eng.* **2013**, *42*, 6–13. [CrossRef]

8. Aoi, I.; Kuramoto, S.; Oh-ishi, K. Mechanical properties of Al-(8,10)%Zn-2%Mg-2%Cu base alloys processed with high-pressure torsion. *Light Met.* **2015**, 179–182. [CrossRef]

9. Ibrahim, M.F.; Samuel, A.M.; Alkahtani, S.A.; Samuel, F.H. A novel solution heat treatment of 7075-type alloy. *Light Met.* **2013**, 383–390. [CrossRef]

10. Wang, W.; Pana, Q.; Wanga, X.; Sun, Y.; Long, L.; Huang, Z. Mechanical properties and microstructure evolution of ultra-high strength Al-Zn-Mg-Cu alloy processed by room temperature ECAP with post aging. *Mater. Sci. Eng. A* **2018**, *731*, 195–208. [CrossRef]

11. Elliot, R. *Eutectic Solidification Processing: Crystalline and Glassy Alloys*, 1st ed.; Butterworth-Heinemann: Oxford, UK, 1983. [CrossRef]

12. Kim, C.S.; Cho, K.; Manjili, M.H.; Nezafati, M. Mechanical performance of particulate-reinforced Al metal-matrix composites (MMCs) and Al metalmatrix nano-composites (MMNCs). *J. Mater. Sci.* **2017**, *52*, 13319–13349. [CrossRef]

13. Zhang, Z.; Chen, D.L. Contribution of Orowan strengthening effect in particulate-reinforced metal matrix nanocomposites. *Mater. Sci. Eng. A* **2008**, *483–484*, 148–152. [CrossRef]

14. Garces, G.; Bruno, G.; Wanner, A. Load transfer in short fibre reinforced metal matrix composites. *Acta Mater.* **2007**, *55*, 5389–5400. [CrossRef]

15. Garg, P.; Jamwal, A.; Kumar, D.; Sadasivuni, K.K.; Hussain, C.M.; Gupta, P. Advance research progresses in aluminium matrix composites: Manufacturing & applications. *J. Mater. Res. Tech.* **2019**, *8*, 4924–4939. [CrossRef]

16. Chung, D.D.L. Metal-Matrix Composites. In *Carbon Composites*, 2nd ed.; Elseiver: London, UK, 2017; pp. 532–562. [CrossRef]

17. Dinaharan, I. Liquid metallurgy processing of intermetallic matrix composites. In *Intermetallic Matrix Composites*; Mitra, R., Ed.; Elseiver: London, UK, 2017; pp. 167–202. [CrossRef]

18. Gao, T.; Bian, Y.; Li, Z.; Xu, Q.; Yang, H.; Zhao, K.; Liu, X. Synthesis of a $(ZrAl_3 + AlN)/Al$ composite and the influence of particles content and element Cu on the microstructure and mechanical properties. *J. Alloys Compd.* **2019**, *791*, 730–738. [CrossRef]

19. Najarian, A.R.; Emadi, R.; Hamzeh, M. Fabrication of as-cast Al matrix composite reinforced by Al_2O_3/Al_3Ni hybrid particles via in-situ reaction and evaluation of its mechanical properties. *Mater. Sci. Eng. B* **2018**, *231*, 57–65. [CrossRef]

20. Bao, S.; Tang, K.; Kvithyld, A.; Engh, T.; Tangstad, M. Wetting of pure aluminium on graphite, SiC and Al_2O_3 in aluminium filtration. *Trans. Nonferrous Met. Soc. China* **2012**, *22*, 1930–1938. [CrossRef]

21. Cai, Z.; Zhang, C.; Wang, R.; Peng, C.; Wu, X.; Li, H. Microstructure, mechanical and thermo-physical properties of Al–50Si–xMg alloys. *Mater. Sci. Eng. A* **2018**, *730*, 57–65. [CrossRef]

22. Dobatkin, V.I.; Elagin, V.I.; Fedorov, V.M. Structure of rapidly solidified aluminium alloys. *Adv. Perform. Mater.* **1995**, *2*, 89–98. [CrossRef]

23. Uzun, O.; Karaaslan, T.; Gogebakan, M.; Keskin, M. Hardness and microstructural characteristics of rapidly solidified Al–8–16 wt.%Si alloys. *J. Alloys Compd.* **2004**, *376*, 149–157. [CrossRef]

24. Cai, Z.; Zhang, C.; Wang, R.; Peng, C.; Qiu, K.; Feng, Y. Preparation of Al–Si alloys by a rapid solidification and powder metallurgy route. *Mater. Des.* **2015**, *87*, 996–1002. [CrossRef]

25. Hatch, J.E. *Aluminium: Properties and Physical Metallurgy*; American Society for Metals: Cleveland, OH, USA, 1984.

26. The Aluminium Association, International Alloy Designations and Chemical Composition Limits for Wrought Aluminium and Wrought Aluminium Alloys. The Aluminium Association. Available online: https://www.aluminum.org/sites/default/files/Teal%20Sheets.pdf (accessed on 26 April 2020).

27. Fukui, Y.; Okada, H.; Kumazawa, N.; Watanabe, Y. Near-net-shape forming of Al-Al$_3$Ni functionally graded material over eutectic melting temperature. *Met. Mater. Trans. A* **2000**, *31*, 2627–2636. [CrossRef]

28. Nash, P.; Pan, Y.Y. The Al-Ni-Zr system (Aluminium-Nickel-Zirconium). *J. Phase Equilibria* **1991**, *12*, 105–113. [CrossRef]

29. Belov, N.A.; Zolotorevskiy, V.S. The Effect of nickel on the structure, mechanical and casting properties of aluminium alloy of 7075 type. *Mater. Sci. Forum* **2002**, *396–402*, 935–940. [CrossRef]

30. Belov, N.A.; Cheverikin, V.V.; Eskin, D.G.; Turchin, A.N. Effect of Al$_3$Ni and Mg$_2$Si eutectic phases on casting properties and hardening of an Al-7%Zn-3%Mg alloy. *Mater. Sci. Forum* **2006**, *519–521*, 413–418. [CrossRef]

31. Wang, S.H.; Uan, J.Y.; Lui, T.S.; Chen, L.H. Examination on the aging and tensile properties of Al-Zn-Mg/Al$_3$Ni eutectic composite. *Met. Mater. Trans. A* **2002**, *33*, 707–711. [CrossRef]

32. Barclay, R.S.; Kerr, H.W.; Niessen, P. Off-eutectic composite solidification and properties in Al-Ni and Al-Co alloys. *J. Mater. Sci.* **1971**, *6*, 1168–1173. [CrossRef]

33. Martínez-Villalobos, M.A.; Figueroa, I.A.; Suarez, M.A.; Rodríguez, G.A.L.; Peralta, O.N.; Reyes, G.G.; López, I.A.; Martínez, J.V.; Trujillo, C.D. Microstructural Evolution of Rapid Solidified Al-Ni Alloys. *J. Mex. Chem. Soc.* **2016**, *60*, 67–72. [CrossRef]

34. Lin, Y.; Mao, S.; Yan, Z.; Zhang, Y.; Wang, L. The enhanced microhardness in a rapidly solidified Al alloy. *Mater. Sci. Eng. A* **2017**, *692*, 182–191. [CrossRef]

35. Casati, R.; Coduri, M.; Riccio, M.; Rizzi, A.; Vedani, M. Development of a high strength Al–Zn–Si–Mg–Cu alloy for selective laser melting. *J. Alloys Compd.* **2019**, *801*, 243–253. [CrossRef]

36. Belov, N.A. Sparingly alloyed high-strength aluminium alloys: Principles of optimization of phase composition. *Mater. Sci. Heat Tr.* **2012**, *53*, 19–27. [CrossRef]

37. Thermo-Calc Software TTAL5 Al-Alloys. Available online: http://www.thermocalc.com (accessed on 17 April 2020).

38. Neikov, O.D.; Naboychenko, S.S.; Yefimov, N.A. *Handbook of Non-Ferrous Metal Powders: Technologies and Applications*, 2nd ed.; Elseiver: London, UK, 2018; p. 995. [CrossRef]

39. Nishi, M.; Matsuda, K.; Miura, N.; Watanabe, K.; Ikeno, S.; Yoshida, T.; Murakami, S. Effect of the Zn/Mg ratio on microstructure and mechanical properties in Al-Zn-Mg alloys. *Mater. Sci. Forum* **2014**, *794–796*, 479–482. [CrossRef]

40. Galy, C.; Le Guen, E.; Lacoste, E.; Arvieu, C. Main defects observed in aluminium alloy parts produced by SLM: From causes to consequences. *Addit. Manuf.* **2018**, *22*, 165–175. [CrossRef]

41. Glazoff, M.; Khvan, A.; Zolotorevsky, V.; Belov, N.; Dinsdale, A. *Casting Aluminium Alloys: Their physical and Mechanical Metallurgy*, 2nd ed.; Elsevier: Amsterdam, The Netherlands, 2018. [CrossRef]

42. Grandfield, J.; Eskin, D.G.; Bainbridge, I. *Direct-Chill Casting of Light Alloys: Science and Technology*, 1st ed.; John Wiley & Sons, Inc.: Hoboken, NJ, USA, 2013.

43. Todeschini, P.; Champier, G.; Samuel, F.H. Production of Al-(12–25) wt% Si alloys by rapid solidification: Melt spinning versus centrifugal atomization. *J. Mater. Sci.* **1992**, *27*, 3539–3551. [CrossRef]

44. Ma, P.; Prashanth, K.G.; Scudino, S.; Jia, Y.; Wang, H.; Zou, C.; Wei, Z.; Eckert, J. Influence of Annealing on Mechanical Properties of Al-20Si Processed by Selective Laser Melting. *Metals* **2014**, *4*, 28–36. [CrossRef]
45. Lekatou, A.; Sfikas, A.K.; Petsa, C.; Karantzalis, A.E. Al-Co alloys prepared by vacuum arc melting: Correlating microstructure evolution and aqueous corrosion behavior with Co content. *Metals* **2016**, *6*, 46. [CrossRef]

Article

Influence of Continuous Casting Speeds on Cast Microstructure and Mechanical Properties of an ADC14 Alloy

Yoon-Seok Lee [1,*], Yuya Makino [1], Jun Nitta [1] and Eunkyung Lee [2]

[1] Graduate School of Natural Science and Technology, Okayama University, Okayama 700-0082, Japan; po507fwv@s.okayama-u.ac.jp (Y.M.); p8l77t18@s.okayama-u.ac.jp (J.N.)

[2] Department of Ocean Advanced Materials Convergence Engineering, Korea Maritime and Ocean University, Busan 49112, Korea; elee@kmou.ac.kr

* Correspondence: y_lee@okayama-u.ac.jp; Tel.: +81-80-3332-4770

Received: 31 March 2020; Accepted: 8 May 2020; Published: 11 May 2020

Abstract: To improve the mechanical properties of the casting alloys, various attempts have been made to use alternative casting technologies. The Ohno continuous casting (OCC) process is a unidirectional solidification method, which leads to high-quality cast samples. In this study, the Al-Si-Cu-Mg alloy was cast at casting speeds of 1 mm/s, 2 mm/s, and 3 mm/s, by the OCC process. The aim of this study is to investigate the effects of the casting process parameters, such as casting speeds and cooling conditions, on the crystallization characteristics and mechanical properties of OCC-Al-Si-Cu-Mg alloy. Particularly, secondary dendrite arms spacing of α-Al dendrites in OCC samples significantly decreases with increasing casting speed. Moreover, the mean tensile strength of the samples, produced at the highest casting speed of 4.0 mm/s, is significantly higher than that for the samples produced at a casting speed of 1.0 mm/s.

Keywords: aluminum alloy; casting speed; solidification; Ohno continuous casting; gravity casting; dendritic spacing

---

1. Introduction

The reductions in automotive exhaust greenhouse gases, such as carbon dioxide (CO_2) and nitrogen oxides (NO_x), are strongly required for environmental reasons. Light-weighting of vehicles presents an opportunity for cutting greenhouse gas emissions. Aluminum (Al) alloys are one of the light weighting automotive materials, which meets vehicle safety and performance requirements [1]. Therefore, automotive parts made of heavy steel have been replaced with Al alloys. Particularly, the hypereutectic Al-Si alloy family is widely used in the automotive industry because of its high strength, good castability, and low density.

The Al-Si-Cu-Mg alloy, used in this study, is particularly the commercial hypereutectic $Al-Si_{16}-Cu_4-Mg_{0.6}$ alloy (JIS (Japanese Industrial Standards) ADC14). This Al-Si-Cu-Mg alloy has been also widely used for various automotive parts, such as linerless engine blocks, pistons, pumps, and compressors. Furthermore, there has been increasing use in the cast Al-Si-Cu-Mg alloy for automotive parts, over the world. The cast Al-Si-Cu-Mg alloy microstructure consists of mainly coarse α-Al, primary silicon (Si), and needle-shaped eutectic phases, which significantly affects the mechanical properties of Al-Si-Cu-Mg alloy [2]. In particular, it should be noted that the mechanical properties of Al-Si-Cu-Mg alloy are also directly affected by α-Al dendrite size, as indicated by dendrite arms spacing (DAS). However, the use of this Al-Si-Cu-Mg alloy, to replace heavy steel in automotive industry, has been restricted because of their lower strength and lower ductility.

Al-Si alloy family has been investigated using a unique continuous casting technique proposed by Ohno, which is known as the Ohno continuous casting (OCC) process. The OCC process is a unidirectional solidification method, and this casting process provides phase control and texture control [3], which can lead to easy control of microstructure size [4] and crystal orientation [5]. Some researchers have investigated the mechanical properties of OCC-Al alloys, showing excellent tensile and fatigue properties. This can be explained by that their unidirectional microstructures, low defect density, and uniformly oriented lattice structure [6–8]. Therefore, it is believed that the OCC process is useful in the automotive industry.

These days, a better understanding of how alloy solidifies at different casting conditions, involving cooling rates is required in order to improve mechanical properties of cast Al-Si-Cu-Mg alloys as automotive parts. However, very little investigation has been reported, which includes the interpretation of cooling conditions and mechanical properties for the OCC-Al-Si-Cu-Mg alloy on a metallurgical basis.

Therefore, the Al-Si-Cu-Mg samples were cast at casting speeds of 1 mm/s, 2 mm/s, 3 mm/s, and 4 mm/s via the OCC process in this study. In addition, the influences of microstructures, such as α-Al phases, primary Si, and DAS on the mechanical properties of cast Al-Si-Cu-Mg alloys were investigated. The aim of this work is to investigate the effects of casting speeds and cooling conditions on the α-Al dendritic grains-growing and tensile properties of OCC-Al-Si-Cu-Mg alloy.

2. Experimental Procedure

2.1. Materials

Table 1 shows the chemical compositions of Al-Si-Cu-Mg alloy used in this study. The Al-Si-Cu-Mg samples were prepared by the OCC process. In this study, the samples were also prepared by the gravity cast (GC) process for comparison with the OCC process. Figure 1a shows a schematic drawing of a horizontal-type OCC arrangement, consisting of a melting furnace, a heated graphite mold with a diameter of 5 mm, a graphite crucible, a cooling device, and a dummy rod for withdrawal of the cast sample. Approximately 0.4 kg of Al-Si-Cu-Mg ingot was placed in the graphite crucible for melting, with the graphite mold heated to approximately 910 K, which is just above the liquidus of the Al-Si-Cu-Mg alloy. The melted Al alloy in the crucible was fed continuously into the mold through a runner. A schematic drawing of conventional GC arrangement, consisting of an electric furnace, a crucible, and a metal mold, is shown in Figure 1b for comparison. The Al-Si-Cu-Mg ingot was also placed and melted in the crucible at approximately 910 K for 1 h, using the electric furnace. Then the molten alloy was solidified in the wide rectangular mold.

To obtain various microstructural characteristics, the casting operation of the OCC process was carried out at different speeds, from 1.0 mm/s to 4.0 mm/s. In this study, the round-rod samples were produced via the OCC process with a diameter of 5 mm and a length of approximately 1 m.

Table 1. Chemical composition (mass%) of the Al-Si-Cu-Mg alloy used in this study.

Element Alloy	Cu	Si	Mg	Zn	Fe	Mn	Al
Al-Si-Cu-Mg	4.20	16.49	0.61	0.37	0.72	0.31	bal.

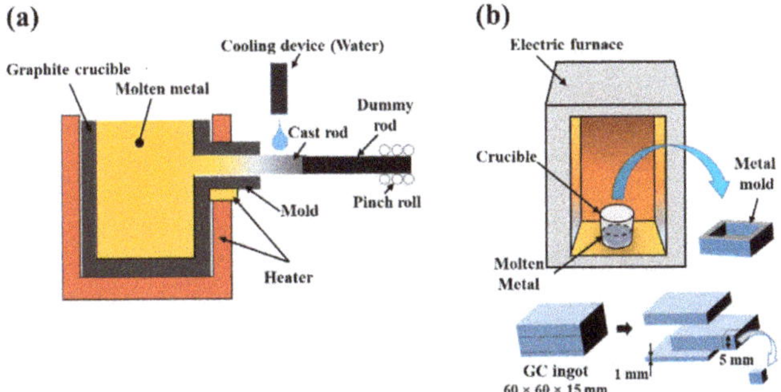

Figure 1. Schematic drawings of (**a**) vertical Ohno continuous casting (OCC) and (**b**) gravity cast (GC) process.

2.2. Metallurgical Analysis

Some of the OCC and GC samples were sectioned to observe microstructures using optical microscope (OM). The cut surfaces were first metallographically polished using up to a 4000 grit-sized silicon carbide paper and were subsequently buff-polished using an Al_2O_3 powder with an average diameter of 0.3 µm and a colloidal SiO_2 solution.

The DAS were also investigated from optical micrographs of cross-section of the cast samples. In particular, the secondary dendrite arms spacing (SDAS) was determined by several optical micrographs and applying an image analysis software. The total spacing from the first to the last arm in a certain area was firstly measured to calculate the SDAS. Consequently, the SDAS was obtained from the following equation:

$$SDAS = L / (n \times V),\tag{1}$$

where L is the measuring length in µm, n is the number of existing dendrite arms in a certain area, and V is the micrograph magnification. Then, an average value of the results was calculated after repeated measurements in different areas.

In addition, some of the cast rod was sectioned parallel to the withdrawal direction and scanning electron microscope (SEM)-based electron back-scattered diffraction (EBSD) analysis was performed to develop a more quantitative view of the microstructures. The cut surfaces were also buff-polished and the orientation data obtained from the EBSD scans were analyzed using an Orientation Imaging Microscopy (OIM^{TM}) analysis software (version 7.2, EDAX Inc., Mahwah, NJ, USA). Moreover, after the tensile tests, the fracture surfaces were also observed to analyze the material defects, using SEM.

2.3. Mechanical Properties

The hardness of longitudinal cross-sections of OCC samples was measured using a Vickers hardness tester at a load of 9.8 N for a holding time of 15 s. The hardness of GC samples was also measured under the same conditions. Nine different points were measured for each sample. The highest and lowest values were discarded, and the hardness of each sample was determined using the average of the remaining seven values.

Moreover, the tensile tests were carried out using an electro-servo-hydraulic system with a crosshead speed of 1 mm/min in air at room temperature. Dumbbell-shaped round specimens (4 mm in diameter and 30 mm in length) were used as shown in Figure 2. The three specimens were prepared and tested for each tensile test. Load and strain were measured using a load cell attached to the machine and a foil-type strain gage attached to the gage section of the specimens, respectively. The tensile strength and strain of the specimens were obtained from the tensile stress–strain curve.

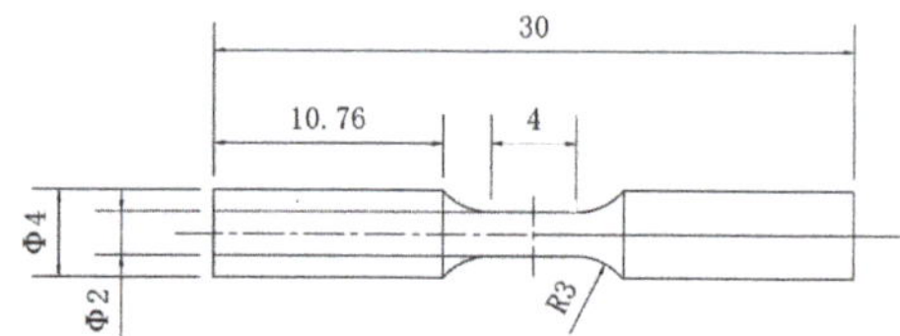

Figure 2. Schematic drawing of specimen used for tensile test.

3. Results

3.1. Microstructural Characteristics

Figure 3 shows optical micrograph of cross-section of sample produced by GC. The microstructure basically consists of large primary Si, embedded between primary α-Al grains. Moreover, Figure 4 shows optical micrographs of cross-sections of OCC samples produced at casting speeds of 1.0, 2.0, 3.0, and 4.0 m/s. The microstructures of OCC samples also consist of large primary Si, embedded between primary α-Al grains. It can be also clearly observed that the mean sizes of primary Si are the largest at a speed of 1.0 mm/s, regardless of the observed location.

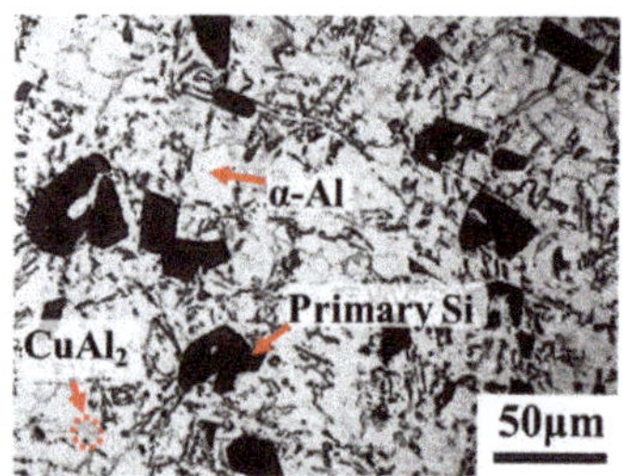

Figure 3. Optical micrograph showing microstructure of GC sample.

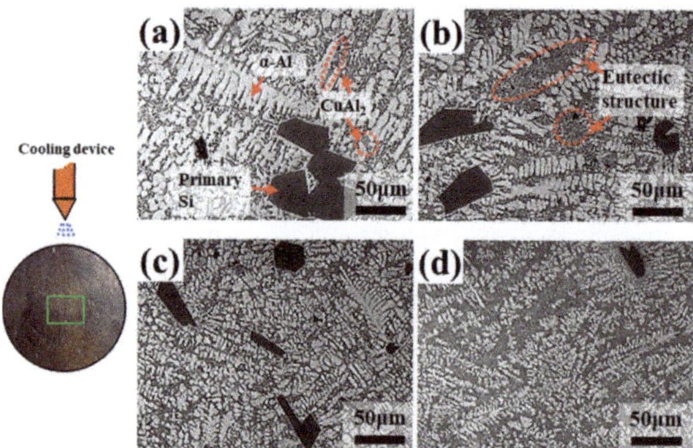

Figure 4. Optical micrographs of cross-sections of OCC samples; middle parts cast at speeds of (**a**) 1.0 mm/s, (**b**) 2.0 mm/s, (**c**) 3.0 mm/s, and (**d**) 4.0 mm/s.

Figure 4 shows middle parts of cross-sectioned OCC samples. In this study, no sharp difference is observed between upper and lower parts of the samples at any of the casting speeds, although upper parts were cooled directly by spray water and cooled at the fastest rate (Figure 1a). It can be explained by that the diameter of cast rod is too small to show large difference in the cooling rate, between upper and lower parts.

Moreover, it is observed that the grain structure transition occurs with increasing casting speeds, owing to the different cooling rates. The morphologies of primary Si phases show sharp edges and flat faces, indicating the facet characteristics. Furthermore, the distributions of primary Si phases are non-homogenous, regardless of the casting speeds. In particular, both the sizes and amount of primary Si are reduced between 1.0 mm/s and 2.0 mm/s. Larger primary Si particles (approximately 5–100 μm in equivalent diameter) are observed in the samples cast at a lowest speed (1.0 mm/s), while smaller primary Si particles (approximately 5–40 μm in equivalent diameter) are observed at the other speeds (2.0 mm/s, 3.0 mm/s, and 4.0 mm/s).

In addition, optical micrographs of longitudinal cross-sections are also shown in Figure 5. It is clear from the results at higher casting speeds (Figure 5b–d) that the α-Al dendrite cells grow in the withdrawal direction. No sharp difference was also observed between upper and lower parts of cast samples, regardless of the casting speeds. The grain structure transition is also observed in longitudinal cross-sections with changes in casting speeds. The quantitative analysis of dendritic structures can be conducted by measurements of DAS. As shown in Figure 5, it is clearly observed that the dendritic structures are refined with increasing casting speed. It is considered that higher cooling rates during solidification induce a higher degree of refinement in dendritic array.

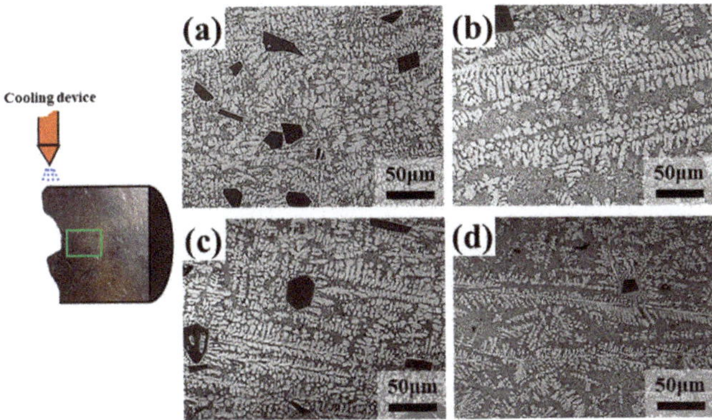

Figure 5. Optical micrographs of longitudinal cross-sections of OCC samples; middle parts cast at speeds of (**a**) 1.0 mm/s, (**b**) 2.0 mm/s, (**c**) 3.0 mm/s, and (**d**) 4.0 mm/s.

EBSD analysis was also carried out in order to analyze morphologies of the dendritic structures, including SDAS. All observations were obtained from longitudinal cross-sections of the samples as shown in Figure 6. It can be more clearly observed from both image quality (IQ) maps and corresponding grain boundary (GB) maps that SDAS of α-Al dendrites significantly decreases with increasing casting speed. SDAS is one of the most important factor, which affects the mechanical properties of Al-Si alloy family. The effects of SDAS and grain size on mechanical properties are similarly obtained for the related Al-Si alloys [9]. Figure 7 shows the relationship between the casting speeds and SDAS. SDAS of α-Al dendrites in OCC samples significantly decreases with increasing casting speed. Moreover, SDAS of OCC samples is smaller than that in GC samples, regardless of the casting speed. The mean SDAS is approximately 4.0 μm for the high casting speed of 4.0 mm/s and 8.7 μm for the low casting speed of 1.0 mm/s, resulting from the longitudinal cross-sections. It is considered that the changes in SDAS are due to a different cooling rate, which lead to different solidification behavior.

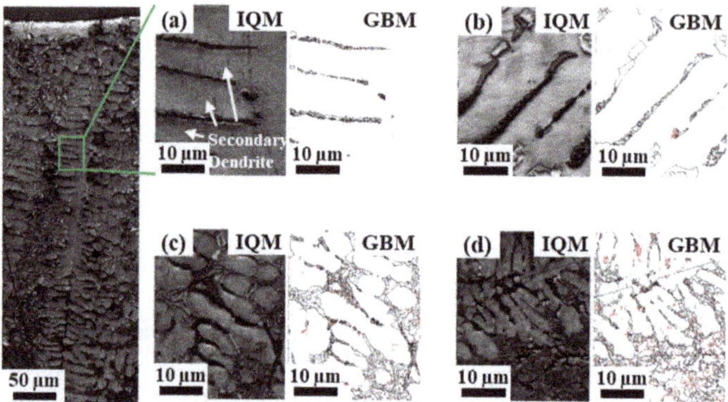

Figure 6. Electron back-scattered diffraction (EBSD) analysis of longitudinal cross-sections of OCC samples: image quality (IQ) maps and corresponding grain boundary (GB) maps of samples cast at speeds of (**a**) 1 mm/s, (**b**) 2 mm/s, (**c**) 3 mm/s, and (**d**) 4 mm/s.

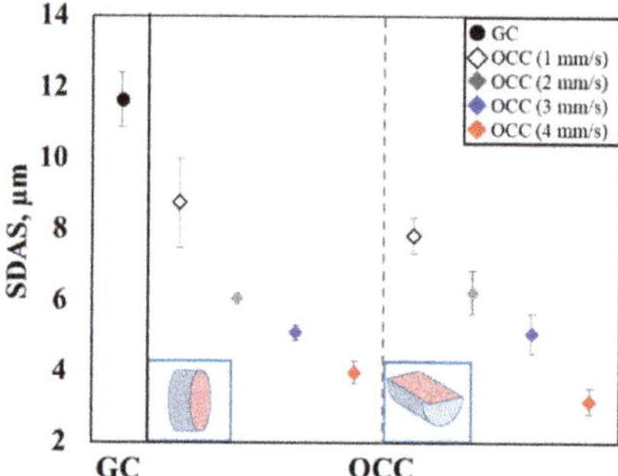

Figure 7. Secondary dendrite arms spacing (SDAS) of GC and OCC samples.

As shown in Figure 8, the cooling rates (CR) of the casting samples were also estimated from the following equation [10]:

$$CR = 2 \times 10^4 \times SDAS^{-2.67}$$

(2)

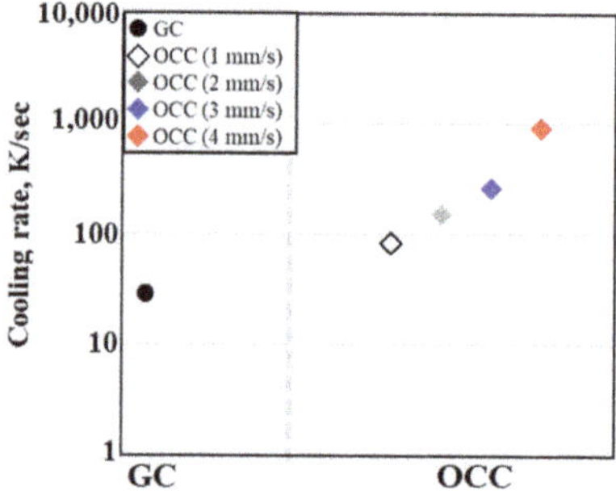

Figure 8. Relationship between casting rate and cooling rate for Al-Si-Cu-Mg alloys.

The cooling rates significantly increase with increasing casting speeds for OCC samples. Furthermore, the estimated cooling rates of OCC samples are higher than those of GC samples, regardless of the casting speed. It can be explained by that OCC samples have much smaller volumes with cooled directly with sprayed water.

3.2. Mechanical Properties

As shown in Figure 9, it is clearly observed that the hardness of OCC samples significantly increases with increasing casting speeds, showing similar behaviors with those of CR plotted in Figure 8. Grain boundaries generally act as barriers to dislocation motion. As mentioned in Section 3.1, the effects of SDAS and grain size on mechanical properties are similar for the Al-Si-Cu-Mg alloy. Consequently, it is considered that the refinement of SDAS, with increasing cooling rate, leads to increase in hardness. Moreover, the mean hardness of OCC samples cast at speeds over 3 mm/s exceeds that of GC sample.

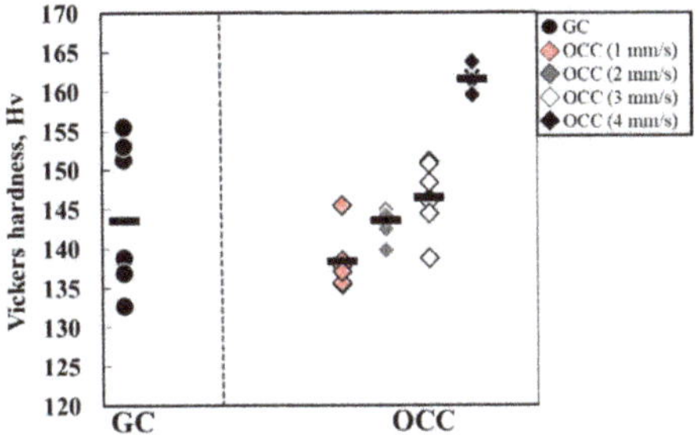

Figure 9. Vickers hardness of GC and OCC samples.

In addition, Figure 10 shows the variations of ultimate tensile strength as functions of casting speeds. It can be observed that the mean ultimate tensile strength increases at higher casting speeds. These results from the tensile tests are similar to the hardness results shown in Figure 9. The mean tensile strength shows a maximum for the samples produced at the highest casting speed of 4.0 mm/s, which is about 70% higher than that for the samples produced at a casting speed of 1.0 mm/s. Moreover, the mean tensile strength of OCC samples cast at 4.0 mm/s is also much higher (by more than 2 times) than that of the GC samples.

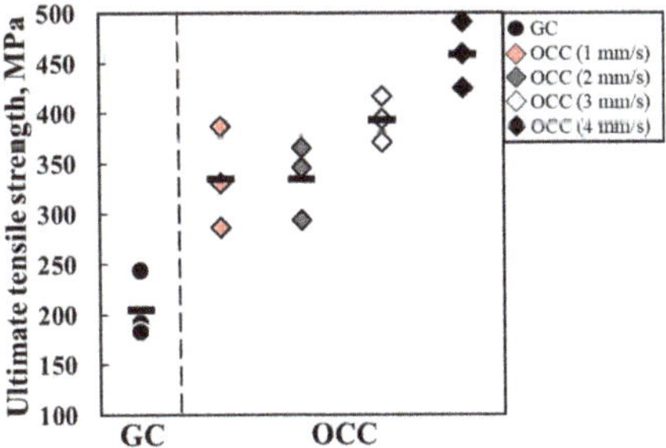

Figure 10. Ultimate tensile strength of GC and OCC samples.

The variations of fracture strain as functions of casting speeds are also shown in Figure 11. No sharp difference is observed with a changing casing speed. Nevertheless, the fracture strains of OCC samples are slightly higher compared to that of GC samples, regardless of the casting speeds.

It is believed that such improvements of mechanical properties of OCC samples can be explained by the presence of unidirectional microstructures and refined dendritic array. Particularly, the improvements of the tensile strength of OCC samples are owing to the refinement of SDAS, which may be the obstacles to the movement of dislocations. SDAS is particularly a significant factor determining the mechanical properties of cast Al-Si-Cu-Mg alloys due to grain-boundary hardening, although the mechanical properties are also affected by other microstructural characteristics, such as eutectic and Si phases.

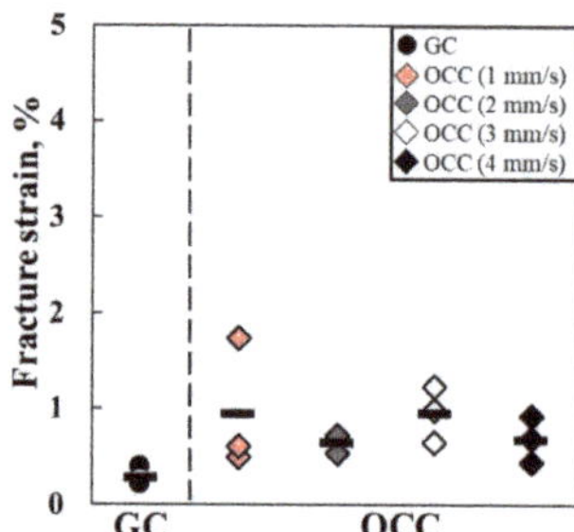

Figure 11. Fracture strain of GC and OCC samples.

SEM fractographs of the tensile specimens cast by GC and OCC are shown in Figure 12. Large pores (approximately 150 μm in diameter) can be observed only at the fracture surface of GC samples, as shown in Figure 12a. On the other hand, from the microstructural observations, the fine casting defects (pores or inclusions) are detected in all of OCC samples. The casting defects could give rise to the stress concentration, leading to the crack formation during the tensile test [11]. It should be noted that the large pore, among the casting defects, is the most responsible for crack formation. Hydrogen (H), as the only gas capable of dissolving to significant quantities in an aluminum melt, is the main factor influencing gas porosity. It is considered that the large pore of GC sample is generated by H gas during the solidification in this study. This is because the morphology of the large pore on the fracture surface does not seem to be a shrinkage pore with an irregular shape, but a gas pore of a spherical shape [11]. It is considered that the large gas pore in the GC sample is generated by the slower cooling rate compared to the OCC samples. Furthermore, not only SDAS, but also the casting defects are smaller and better distributed at a higher casting speed. Consequently, the lower strength and ductility of GC alloys are assumed to result from not only enlarged α-Al and SDAS but also internal defects in this study.

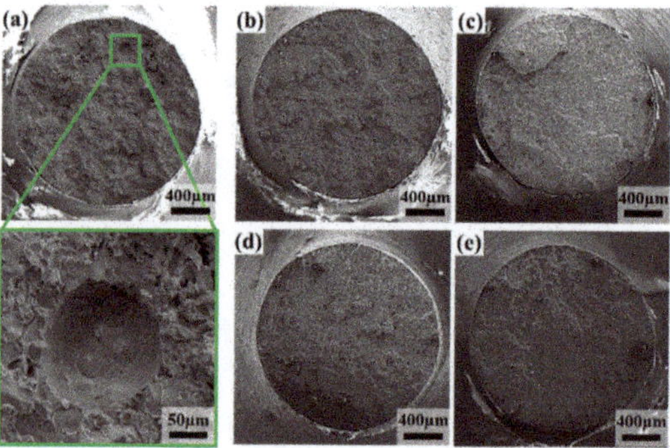

Figure 12. SEM fractographs of Al-Si-Cu-Mg samples cast by (**a**) GC and OCC at casting speeds of (**b**) 1.0 m/s, (**c**) 2.0 m/s, (**d**) 3.0 m/s, and (**e**) 4.0 m/s, after tensile tests.

4. Summary

In this study, the effects of casting speeds and cooling conditions on the cast microstructures and mechanical properties of Al-Si-Cu-Mg alloy were investigated. The primary findings of this study are the following:

1) The microstructures, particularly those of the sizes of α-Al grains, differ among the samples, which resulted from the different cooling rate. SDAS of α-Al dendrites in OCC samples significantly

decreases with increasing casting speed. Moreover, SDAS of α-Al dendrites in OCC samples is smaller than that in GC samples, regardless of the casting speed.

2) The improvements of mechanical properties of OCC samples are owing to the refinemennt of SDAS, which may be the obstacles to the movement of dislocations. The mean tensile strength shows maximum for the samples produced at the highest casting speed of 4.0 mm/s.

3) The large pore can be observed only at the fracture surface of GC samples. On the other hand, only fine casting defects are detected in any of OCC samples. It is believed that the presence of large pores may lead to crack formation during the tensile test.

4) The solidification conditions in continuous casting process, leading to the refinement of the dendritic structures and low porosity, are regarded as contributing to the higher quality of the products and mechanical properties of Al-Si-Cu-Mg casing alloy.

Author Contributions: Conceptualization, Y.-S.L.; methodology, Y.-S.L.; formal analysis, Y.-S.L. and Y.M.; investigation, Y.M. and J.N.; resources, Y.-S.L.; data curation, Y.-S.L. and Y.M.; writing—original draft preparation, Y.-S.L.; writing—review and editing, Y.-S.L. and E.L.; visualization, Y.-S.L.; supervision, Y.-S.L.; project administration, Y.-S.L. All authors have read and agreed to the published version of the manuscript.

Funding: This research received no external funding.

Acknowledgments: This work was supported in part by The Light Metal Educational Foundation, Inc., Japan.

Conflicts of Interest: The authors declare no conflict of interest.

References

1. Kim, H.-J.; Keoleian, G.A.; Skerlos, S.J. Economic assessment of greenhouse gas emissions reduction by vehicle lightweighting using aluminum and high-strength steel. *J. Ind. Ecol.* **2011**, *15*, 64–80. [CrossRef]

2. Okayasu, M.; Takeuchi, S.; Shiraishi, T. Crystallisation characteristics of primary silicon particles in cast hypereutectic Al–Si alloy. *Int. J. Cast Met. Res.* **2013**, *26*, 105–113. [CrossRef]

3. Soda, H.; Motoyasu, G.; Mclean, A.; Jen, C.K.; Lisboa, O. Method for continuous casting of metal wire and tube containing optical fibre. *Mater. Sci. Technol.* **1995**, *11*, 1169–1173. [CrossRef]

4. Sengupta, S.; Soda, H.; Mclean, A. Microstructure and properties of a bismuth-indium-tin eutectic alloy. *Mater. Sci.* **2002**, *37*, 1747–1758. [CrossRef]

5. Okayasu, M.; Ota, K.; Takeuchi, S.; Ohfuji, H.; Shiraishi, T. Influence of microstructural characteristics on mechanical properties of ADC12 aluminum alloy. *Mater. Sci Eng. A* **2014**, *592*, 189–200. [CrossRef]

6. Wang, Y.; Huang, H.-Y.; Xie, J.-X. Enhanced room-temperature tensile ductility of columnar-grained polycrystalline Cu–12 wt.% Al alloy through texture control by Ohno continuous casting process. *Mater. Lett.* **2011**, *65*, 1123–1126. [CrossRef]

7. Okayasu, M.; Yoshie, S. Mechanical properties of Al–Si$_{13}$–Ni$_{1.4}$–Mg$_{1.4}$–Cu$_1$ alloys produced by the Ohno continuous casting process. *Sci. Eng. A* **2010**, *527A*, 3120–3126. [CrossRef]

8. Okayasu, M.; Takasu, S.; Yoshie, S. Microstructure and material properties of an Al–Cu alloy provided by the Ohno continuous casting technique. *J. Mater. Process. Technol.* **2010**, *210*, 1529–1535. [CrossRef]

9. Ceschini, L.; Morri, A.; Gamberini, A.; Messieri, S. Correlation between ultimate tensile strength and solidification microstructure for the sand cast A357 aluminium alloy. *Mater. Des.* **2009**, *30*, 4525–4531. [CrossRef]

10. Okayasu, M.; Ohkura, Y.; Takeuchi, S.; Takasu, S.; Ohfuji, H.; Shiraishi, T. A study of the mechanical properties of an Al–Si–Cu alloy (ADC12) produced by various casting processes. *Mater. Sci. Eng. A* **2012**, *543*, 185–192. [CrossRef]

11. Teng, X.; Mae, H.; Bai, Y.; Wierzbicki, T. Pore size and fracture ductility of aluminum low pressure die casting. *Eng. Fract. Mech.* **2009**, *76*, 983–996. [CrossRef]

Article

Analysis of Different Solution Treatments in the Transformation of β-AlFeSi Particles into α-(FeMn)Si and Their Influence on Different Ageing Treatments in Al–Mg–Si Alloys

Florentino Alvarez-Antolin *, Juan Asensio-Lozano, Alberto Cofiño-Villar and Alejandro Gonzalez-Pociño

Materials Pro Group, Department of Materials Science and Metallurgical Engineering, University of Oviedo, Independencia 13, 33004 Oviedo, Spain; jasensio@uniovi.es (J.A.-L.); UO229780@uniovi.es (A.C.-V.); gonzalezpalejandro@uniovi.es (A.G.-P.)
* Correspondence: alvarezflorentino@uniovi.es; Tel.: +34-985181949

Received: 7 April 2020; Accepted: 7 May 2020; Published: 10 May 2020

Abstract: In the as-cast state, Al–Mg–Si alloys are not suitable for hot forming. They present low ductility due to the presence of intermetallic β-AlFeSi particles that form in the interdendritic regions during the solidification process. Homogenization treatments promote the transformation of these particles into α-(FeMn)Si particles, which are smaller in size and more rounded in shape, thus improving the ductility of the material. This paper analyses the influence of various solution treatments on the transformation of β-AlFeSi particles into α-(FeMn)Si particles in an Al 6063 alloy. Their effect on different ageing treatments in the 150–180 °C temperature range is also studied. An increase in the solution temperature favours greater transformation of the β-AlFeSi particles into α-(FeMn)Si, dissolving a greater amount of Si, thereby having a significant effect on subsequent ageing. We found that as the dwell time at a temperature of 600 °C increases, the rate of dissolution of the Fe atoms from α-(FeMn)Si particles exceeds the rate of incorporation of Mn atoms into said particles. This seems to produce a delay in reaching the peak hardness values in ageing treatments, which warrants further research to model this behaviour. The optimal solution treatment takes place at around 600 °C and the highest obtained peak hardness value is 104 HV after a 2 h solution treatment at said temperature and ageing at 160 °C for 12 h.

Keywords: Al–Mg–Si; α-Al$_8$(Fe$_2$Mn)Si particles; solution treatment; ageing; dissolution of Fe; Differential Scanning Calorimetry

1. Introduction

Aluminium alloys usually have iron as a common impurity. The maximum equilibrium solubility of Fe in solid aluminium is very low. Thus, most Fe forms Fe-rich intermetallic compounds together with other elements, which appear as needles or sharp edges in the microstructure. Some types of Fe-rich intermetallic compounds are very harmful to mechanical properties, especially ductility [1]. The particles present in Al–Mg–Si alloys during solidification are mostly β-AlFeSi (generally the β-Al$_5$FeSi phase), α-AlFeSi (generally the α-Al$_8$Fe$_2$Si phase) and Mg$_2$Si [1–3]. These alloys are not suitable for hot forming processes in the as-cast state or in processes of intensive deformation, as occurs in extrusion. Their ductility is too low, mainly due to the presence of intermetallic β-AlFeSi particles that locate at interdendritic regions, giving rise to the occurrence both inside the α-Al grains and at their grain boundaries after solidification is completed [4]. These particles usually have an acicular morphology in the polished plane. Hot ductility is impaired by the presence of these particles, decreasing significantly as their content increases. The purpose of the high-temperature homogenization process is the chemical homogenization of the dendrites, as well as the simultaneous fragmentation of the β-Al$_5$FeSi precipitates, thus avoiding their continuity along the grain boundaries and favouring their conversion into α-Al(FeMn)Si particles with rounded edges, which increase ductile behaviour. This conversion must be complete in hot extrusion processes, as they take place in a single stage with intensive deformation and in a very short time. Hot rolling processes that obtain flat rolled products and multi-pass hot strip mills enable the decrease in temperatures and homogenization times without completing the total transformation of β particles into α particles. β-AlFeSi particles can lead to local crack initiation and induce surface defects on the extruded material. The more rounded α-Al(FeMn)Si particles in the homogenized material improve the extrudability of the material as well as the surface quality of the extruded material [5]. The morphology of the Fe-rich phase is related to many factors, such as the composition of the alloy, the cooling rate, and the content of Fe. During solidification, for example, α-Al$_8$Fe$_2$Si tends to form at higher cooling rates than β-Al$_5$FeSi [6]. The presence of Mn can change the morphology of the acicular Fe-rich phase and form a granular intermetallic α-Al(MnFe)Si phase [7,8]. The formation of the β-AlFeSi particles is prior to, or at least concurrent with, the solidification of the alloy, which is why these precipitates are found in the interdendritic regions. The same grain can eventually encompass several dendrites. Therefore, these particles can appear inside the grains. As a result, the above processes lead to an increase in hot ductility [9]. A homogenization heat treatment promotes the transformation of these needle-like particles into smaller particles with more rounded shapes denoted as α-(FeMn)Si, α-Al$_8$(FeMn)$_2$Si or α-Al$_{12}$(FeMn)$_3$Si [5], which allows an enhancement in the ductility of the material [9–14]. The main force inducing the transformation of the β particles into α particles is found in the difference of Fe concentration between the Al matrix and the β particles themselves. It has been found that an Fe content in the alloy between 0.15% and 0.25% favours the appearance of smaller volume fractions of β intermetallics during solidification and promotes fine acicular morphologies, favouring a faster β → α transformation [15]. Moreover, the Mn in turn modifies this inducing force by causing variations in the concentration of Fe in the matrix and β interfaces. Kuijpers et al. determined that the optimal rate of transformation of β particles into α particles occurs when the manganese concentration falls within the 0.02% to 0.2% range [16]. With an increase in solution time, the acicular morphology of the β-AlFeSi particles can be transformed into rounded shapes and the α-Al8(FeMn)$_2$Si phase can be formed. A minimal amount of Mn is needed to promote this transformation. The desirable Mn/Fe ratio could be around 0.5. The α-(FeMn)Si particles present an Fe/Si atomic ratio that is higher than that of its homologous β phase. Therefore, the solid solution of Si in the matrix phase accordingly increases, favouring subsequent hardening via an ageing treatment [12,17]. Lui et al. concluded that the (Fe + Mn)/Si atomic ratio remains constant in homogenization treatments within the 550–580 °C temperature range [18]. In addition to the aforementioned objectives, the homogenization treatment also aims to reduce the microsegregation produced during non-equilibrium solidification and endow the material with a fine, homogeneous structure of precipitates that ensures its subsequent manufacture. The final mechanical properties are determined by the solution heat treatment, quench rate and ageing heat treatment employed. In the solution heat treatment, the aim is to dissolve the phases containing Mg and Si. The temperatures employed in these alloys usually fall within the 460–530 °C range [19]. Chen et al. placed the optimal treatment at 520 °C for 3.5 h [20]. The ageing of these alloys begins with the formation of clusters in the Guinier–Preston (GP) zones followed by sequential precipitation of the β″ phase and other metastable phases until precipitation of the β equilibrium phase is achieved [21]. The high level of saturation and high concentration of voids that the α-Al phase presents after quenching promotes the rapid formation of Si clusters and/or GP zones. These clusters have a high concentration of solute and maintain absolute coherence with the matrix, although elastic stresses are induced around them due to the difference in size between the atoms of the solute and those of the solvent. At this point, there is an increase in hardness, as these clusters/zones constitute a major obstacle to the displacement of dislocations [22,23]. Transition precipitates that are coherent with the matrix may nucleate from these zones. The increase in strength is defined by the size of the precipitates, their distribution, and their coherence with the matrix. In these alloys, the precipitate responsible for hardening is the β″ (associated with the Mg_2Si phase) [22–25]. It was found that the width of the precipitate-free zones (PFZs) was significant and optimal precipitation was associated with low thickness [26,27]. The temperatures used in the solubilization treatment and the dwell times at these temperatures seem to have a significant influence on the strength obtained after the ageing treatment. An increase in the temperature or dwell times at the aforementioned temperatures seems to favour an increase in strength after ageing [19,28]. Nandy et al. managed to obtain a peak hardness value of 90 HV by ageing at 175 °C for around 8 h, after a solution treatment at 525 °C for 2 h [29]. Yuksel obtained a maximum hardness value of nearly 100 HV by means of a solution treatment at 535 °C for 2 h and ageing at 204 °C for 2 h [25]. Note that the greatest decrease in hardness due to over-ageing was obtained at this same temperature compared to lower ageing temperatures. Several authors concluded that the rate of cooling prior to ageing and after the solubilization treatment seems to influence the peak hardness value after the ageing treatment. This cooling rate should be as high as possible in order to achieve the maximum supersaturation of Si and Mg in solid solution [24,30–32]. The most sensitive temperature range at industrial cooling rates seems to be between 450 and 250 °C [24,31,33]. Prolonged solubilization treatments are necessary in order to ensure the total dissolution of the Mg_2Si formed in the cooling after chemical homogenization. As a form of validation, it is likewise necessary to employ a cooling rate which is fast enough to avoid the premature precipitation of Mg_2Si. This would allow the full potential of structural hardening to be available in the subsequent ageing process.

The aim of this paper is to analyse the influence of various solution treatments, employing quenching in water, on the transformation of the β-AlFeSi particles in an Al 6063 alloy into α-(FeMn)Si particles and their possible influence on different ageing treatments carried out in the 150–180 °C range.

2. Materials and Methods

One hundred twenty specimens were taken from the intermediate zone of the radius of a number of 200-mm-diameter slabs in the as-cast state. Table 1 shows the chemical composition of the as-cast material. These slabs were manufactured by continuous casting in discontinuous processes, with cooling and lubrication by demineralized water in the die. The round aluminium products had an approximate diameter of 228 mm and a length of 12 m. The cooling rates measured at the centre of these products were 1 K/s, and 20 K/s at the surface. The specimens had a cubic geometry measuring 12–15 mm per side. One hundred sixteen of these specimens were subjected to four solution treatments, employing three different ageing temperatures for each. Quenching in water at 15 °C was used after the solution treatments. Figure 1 outlines the experimental work carried out. The 120 specimens were distributed as follows:

(1) 108 specimens were used to obtain the hardness profiles: 4 different solution treatments were carried out, employing 3 ageing treatments per solution treatment, with a total of 9 specimens in each ageing treatment.

(2) 8 specimens had no ageing. There were 2 specimens after each solution treatment, one was used for metallographic inspection and calculation of the material's hardness, while the other was used for differential scanning calorimetry (DSC).

(3) The remaining 4 specimens were analysed in the as-cast state.

Table 1. Chemical composition (% by weight).

%Si	%Fe	%Cu	%Mg	%Mn	%Ti
0.42	0.22	0.018	0.47	0.031	0.019

Metallographic inspection was carried out using an optical microscope and a scanning electron microscope. Preparation of the metallographic specimens was carried out by cutting with a SiC disc, cold mounting with an epoxy resin, grinding with SiC paper of different abrasive grain sizes ranging from grit 240 to 600 then finally polishing in three steps with textile cloths using different types of abrasives in each step. In the first step, 6 and 1 μm diamond paste was used, while a 0.5 μm alumina solution in distilled water was used in the second step. In the third stage, a 0.05 μm colloidal silica solution, which was also in distilled water, was utilized.

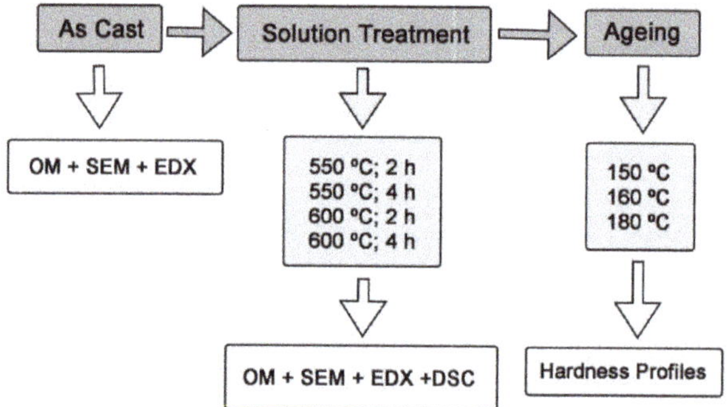

Figure 1. Outline of the experimental work carried out. OM: optical microscopy; DSC: differential scanning calorimetry; EDX: energy dispersive X-ray spectroscopy; SEM: scanning electron microscopy.

The etching reagents used in the optical microscopy analysis were:

(1) 2 mL HF, 3 mL HCl, 5 mL HNO_3 and 190 mL distilled water.
(2) A solution made of 4 g $KMnO_4$, 1 g NaOH and 100 mL distilled water. This reagent reveals the chemical heterogeneities derived from dendritic microsegregation.

The optical microscope used was a NIKON Epiphot 200 (Nikon, Tokyo, Japan) and the images were obtained using Beuhler Omnimet Enterprise image analyser software (5.0, Beuhler, Lake Bluff, USA). The scanning electron microscope employed was a JEOL JSM-5600 (JEOL, Nieuw-Vennep, The Netherlands). The Fe/Si atomic ratio in the α-Al(FeMn)Si particles was determined on metallographic samples in the polished state, without etching with a chemical reagent. Energy dispersive X-ray (EDX) microanalysis was used for this purpose, randomly analysing 30 particles per specimen (20 kV with reflections of up to 10 keV).

Vickers hardness values were obtained under the application of a 300 N load. The results correspond to the mean value obtained from 12 indentations per specimen.

The differential scanning calorimeter used was a Mettler Toledo DSC 822e (Metter Toledo, Schwerzenbach, Switzerland). The heating rate was 10 °C/min, and the test was conducted between 25 and 500 °C. The results obtained by DSC will not provide us with the isothermal temperature at which the precipitation of the metastable phases that cause structural hardening takes place. During the DSC test, these temperatures are obtained by means of a heating ramp and are therefore always higher than the temperatures used isothermally for artificial ageing. However, when the aim is to compare various ageing treatments, this test allows us to compare whether said precipitation of structural hardening will occur sooner or later.

3. Results

Figure 2 shows the microstructure obtained in the as-cast state. It can be seen that the microstructure presents dendritic segregation and precipitation of β-Al_5FeSi particles in the segregated zones, preferably at the grain boundaries. Figure 3 shows images of this microstructure obtained using electron channelling contrast imaging (ECCI) [34]. The elongated, acicular morphology of the β-Al_5FeSi particles can be observed, forming a more or less continuous network through the grain boundaries network of the α-Al phase. It can also be seen that the grain size is around 200–250 μm. Figure 4 shows the microstructure obtained following the solution treatments with cooling in water. A homogenized microstructure can be seen in all cases, with no dendritic segregation. A decrease in the volume fraction of AlFeSi particles and loss of their continuity were likewise observed. Figure 5 shows the transformation of the microstructure during the homogenization treatment. Figure 5a shows the microstructure obtained in the as-cast state. The β-Al$_5$FeSi particles and the Mg$_2$Si phase can be observed, both located at a grain boundary. Figure 5b shows the microstructure after solubilization at 600 °C, employing cooling in water. This treatment also allows the solubilization of the Mg$_2$Si phase and the transformation of the β-Al$_5$FeSi particles into α-Al$_8$(FeMn)$_2$Si particles. Cooling in water prevented solid-state precipitation of Mg$_2$Si.

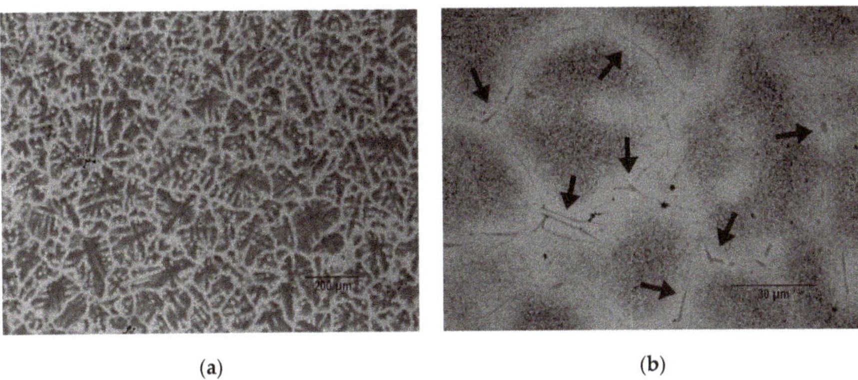

(a) (b)

Figure 2. Microstructure in the as-cast state. (**a**) Dendritic segregation is observed; (**b**) β-Al$_5$FeSi particles at the grain boundaries. The etching reagent consisted of a solution of 100 mL distilled water with 4 g KMnO$_4$ and 1 g NaOH.

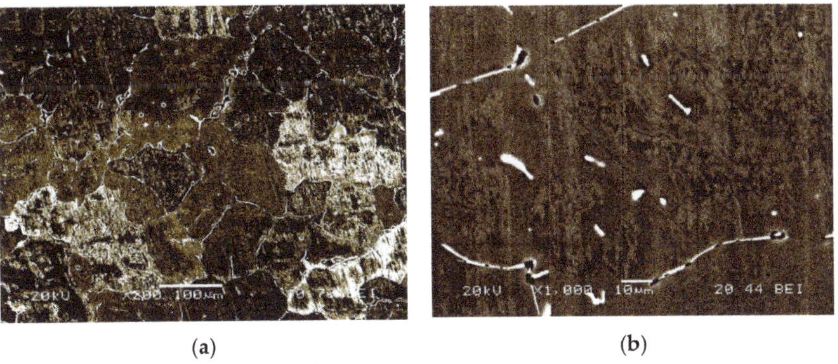

(a) (b)

Figure 3. Microstructure in the as-cast state obtained under a scanning electron microscope (SEM) using electron channelling contrast imaging (ECCI). (**a**) It can be seen that the grain size can be estimated to be around 200–250 μm. (**b**) The β-Al$_5$FeSi particles are in white and Mg$_2$Si particles are in black.

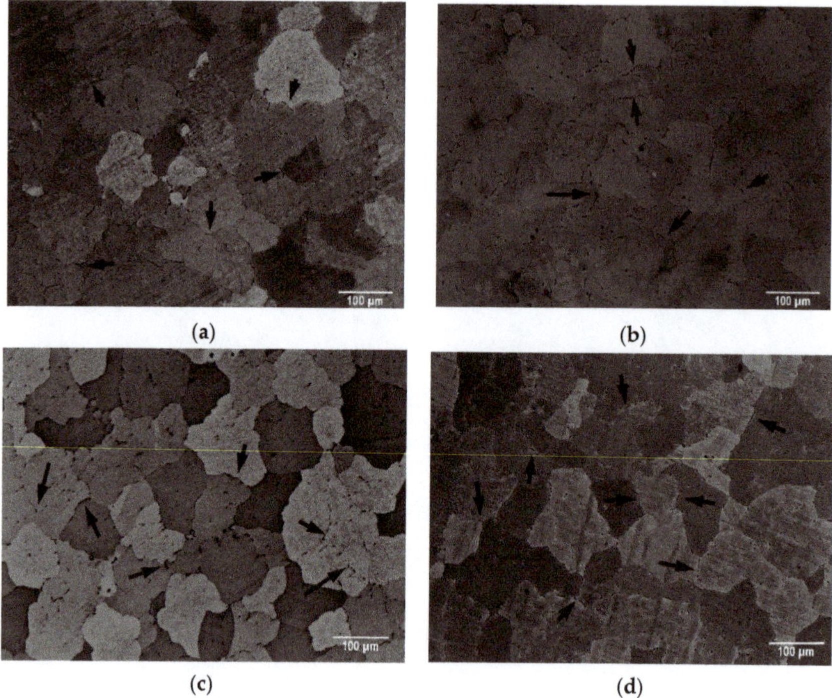

Figure 4. Microstructure after solution treatments, employing quenching in water. (**a**) Treatment at 550 °C for 2 h. (**b**) Treatment at 550 °C for 4 h. (**c**) Treatment at 600 °C for 2 h. (**d**) Treatment at 600 °C for 4 h.

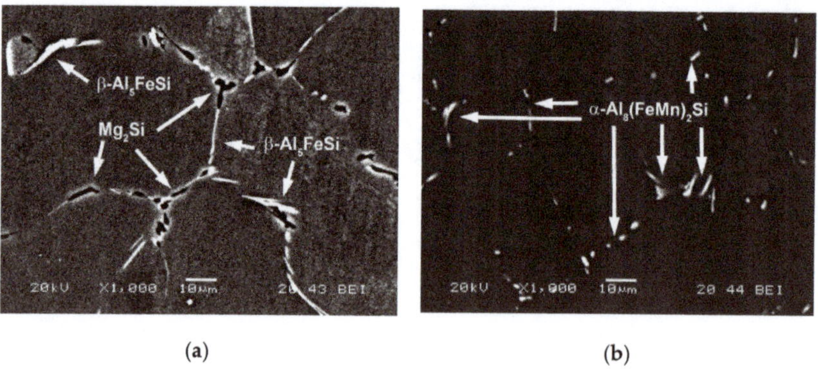

Figure 5. Microstructures obtained by scanning electron microscopy (SEM). (**a**) The as-cast state. The β-Al$_5$FeSi particles and the Mg$_2$Si phase can be observed, located at a grain boundary. (**b**) After a solubilization treatment at 600 °C with water cooling, which enabled the maximum solubilization of the Mg$_2$Si phase and the transformation of the β-Al$_5$FeSi particles into α-Al$_8$(FeMn)$_2$Si particles.

Figure 6 shows the Fe/Si atomic ratio and the atomic percentage of Mn in the α-Al(FeMn)Si particles, while Figure 7 shows the (Fe + Mn)/Si ratio, along with the atomic percentages of Si and Fe. It can be observed that the Fe/Si ratio was higher in treatments at 600 °C than in treatments at 550 °C, a finding that could be justified by the greater dissolution of Si atoms at 600 °C. This would allow an increase in the potential for structural hardening. However, this ratio decreased when the duration of the treatment was 4 h. Note that the atomic % of Fe in the α particles was much lower in the treatments at 600 °C, being lower with dwell times of 4 h versus 2 h. Moreover, the β-AlFeSi particles in the as-cast state do not contain Mn atoms. However, as the dwell times of the treatment temperature increased, the atomic content of Mn in the α-Al(FeMn)Si particles also increased.

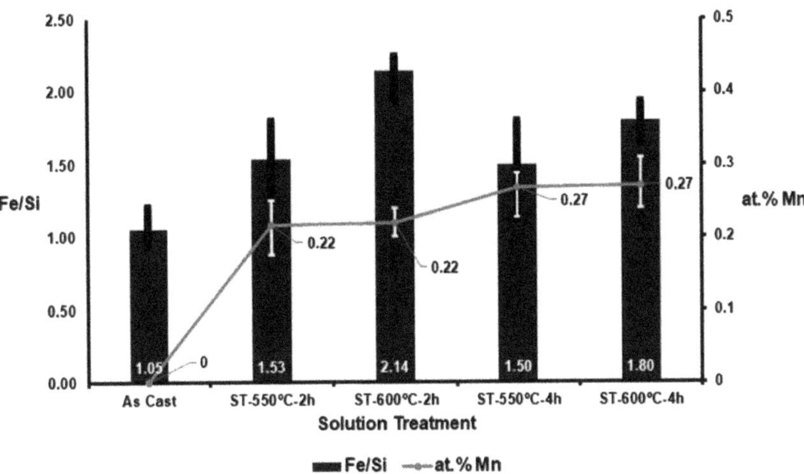

Figure 6. Mean Fe/Si atomic ratio of AlFeSi particles after the different solution treatments (ST). The results are correlated with the atomic % of Mn in these particles. The "As Cast" state refers to the β-Al$_5$FeSi particles, while the state after the solution treatment refers to the α-Al$_8$(FeMn)$_2$Si particles. The error bars show the distance between the mean value and the maximum and minimum values of the 30 particles analysed in each specimen.

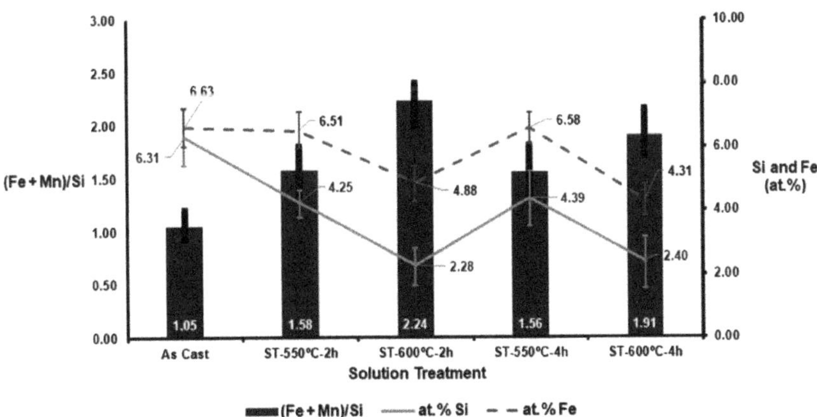

Figure 7. Mean (Fe + Mn)/Si atomic ratio of AlFeSi particles after the different solution treatments (ST). The atomic percentages of Si and Fe are also shown. The "As Cast" state refers to β-Al$_5$FeSi particles, while the state after the solution treatment refers to the α-Al$_8$(FeMn)$_2$Si particles. The error bars show the distance between the mean value and the maximum and minimum values of the 30 particles analysed in each specimen.

The hardness value in the as-cast state was 40 HV and the hardness value obtained after these treatments was 45 HV, with no differences in hardness being found between the different solution treatments. The hardness test was carried out immediately after these treatments.

Figure 8 shows the results obtained after the DSC analysis. Two main exothermic peaks were detected, designated as A and C. Peak A corresponds to precipitation of the β″and β′ phases. In between peaks A and C is the endothermic peak B that corresponds to the dissolution of these phases. Lastly, peak C corresponds to the precipitation of the β phase [35–42]. The endothermic peak prior to the exothermic peak A shows the dissolution of the GP zones formed during natural ageing. The results of the DSC test provide the temperatures at which the structural hardening precipitations are verifiable when this temperature is reached by means of a heating ramp. In our case, the heating rate is 10 °C/ min. This means that these temperatures cannot be compared with the isothermal ageing temperatures, the latter being necessarily lower. However, when the aim is to compare various ageing treatments, this test allows us to compare whether the structural hardening will occur sooner or later. The solution treatment at 600 °C with a dwell time of 4 h produced the greatest delay in the formation of the metastable transition phases (Peak A) and their subsequent dissolution (Peak B). This delay could be related to the dissolution of the Fe atoms that were observed in the previous solution treatment at this temperature with prolonged dwell times, which could be grounds for further investigation in this regard.

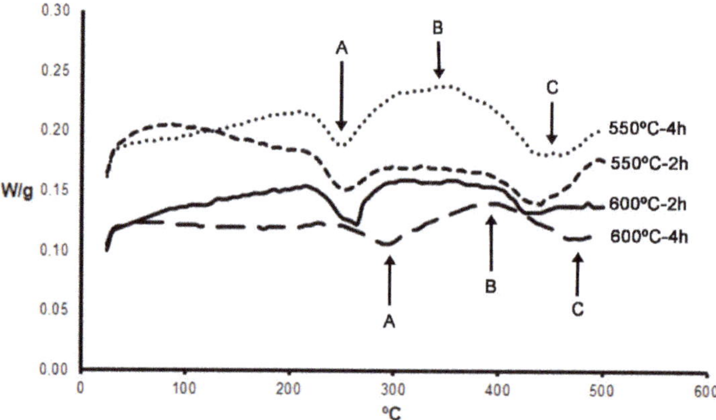

Figure 8. Continuous heating DSC at 10 °C/min. The samples were previously aged naturally. The exothermic peaks (A and C) indicate precipitation of the β″-β′ and β phases, respectively. The endothermic peak (B) reflects the dissolution of the β″-β′ phases, precipitated in A.

Figure 9 shows the hardness profiles obtained after the ageing treatments. The peak hardness values were obtained with solution treatments at 600 °C, which are the ones that dissolved the highest % Si from the β-AlFeSi particles. This would allow more Si to be available in solid solution for the subsequent precipitation of the metastable β″ phase during ageing. The peak hardness value was 104 HV after the solution treatment at 600 °C for 2 h and ageing at 160 °C for 12 h. Very similar hardness profiles following the solution treatment at 550 °C were obtained with dwell times of 2 and 4 h. However, the peak hardness values obtained after a dwell time of 4 h at 600 °C were delayed in the three ageing temperatures compared to the 2 h dwell time. This fact may be conditioned by the dissolution of Fe atoms observed after the 4 h dwell time at 600 °C.

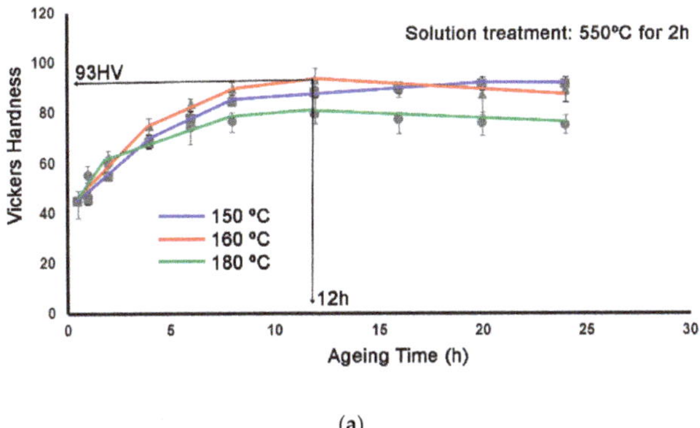

(**a**)

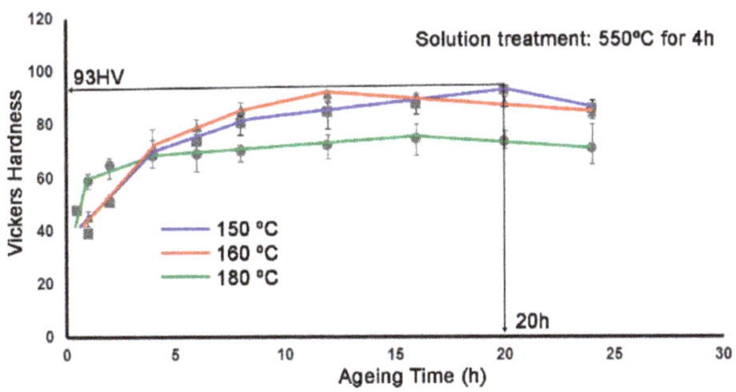

(**b**)

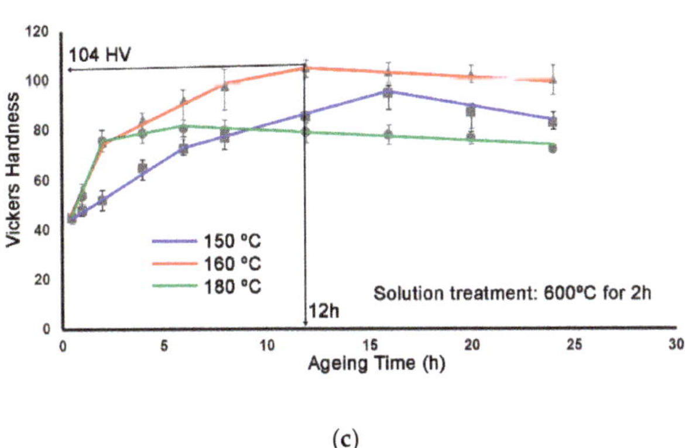

(**c**)

Figure 9. *Cont.*

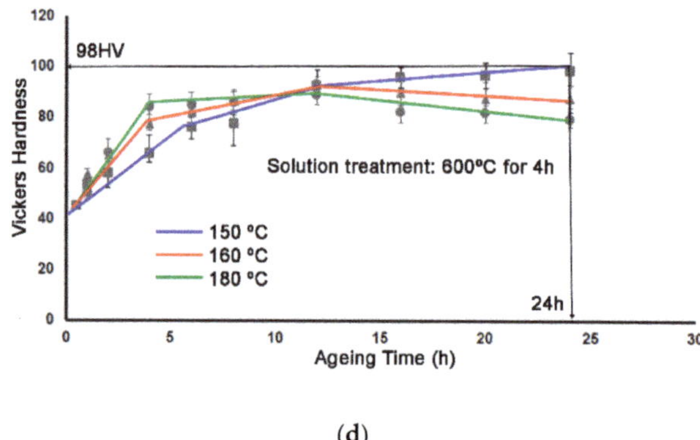

(**d**)

Figure 9. Hardness profiles after the ageing treatments. (**a**) Treatment at 550 °C for 2 h. (**b**) Treatment at 550 °C for 4 h. (**b**) Treatment at 600 °C for 2 h. (**d**) Treatment at 600 °C for 4 h.

4. Conclusions

This paper analyses the influence of solution treatments on the transformation of β-AlFeSi particles into α-(FeMn)Si particles and their possible influence on different ageing treatments carried out in the 150–180 °C range. The main conclusions are:

(1) The Fe/Si atomic ratio increased with increasing solution treatment temperature from 550 to 600 °C. This reflects a greater degree of transformation of β-Al$_5$FeSi particles into α-Al$_8$(FeMn)$_2$Si particles, and a greater potential for structural hardening.

(2) In the transformation of β-Al$_5$FeSi particles into α-Al$_8$(FeMn)$_2$Si, a greater dissolution of both Si and Fe atoms was observed in the matrix when the solution treatment was carried out at 600 °C. The dissolution of Fe was somewhat more pronounced when the dwell times were increased from 2 to 4 h.

(3) At a solution temperature of 550 °C, the (Fe + Mn)/Si atomic ratio remained practically constant. However, at 600 °C, this ratio decreased when the dwell time was increased from 2 to 4 h. This suggests that the rate of dissolution of Fe atoms exceeded the rate of incorporation of Mn atoms. This could lead to a delay in reaching peak hardness values during ageing at temperatures between 150 and 180 °C.

(4) The peak hardness value obtained was 104 HV, following a solution treatment at 600 °C for 2 h and ageing at 160 °C for 12 h.

Author Contributions: J.A.-L. conceived and designed the investigation; A.C.-V. and A.G.-P. performed all laboratory work; F.A.-A. led the investigation, analysed the data, and wrote the paper. All authors have read and agreed to the published version of the manuscript.

Funding: This research received no external funding.

Conflicts of Interest: The authors declare no conflicts of interest.

References

1. Chanyathunyaroj, K.; Patakham, U.; Kou, S.; Limmaneevichitr, C. Microstructural evolution of iron-rich intermetallic compounds in scandium modified Al-7Si-0.3Mg alloys. *J. Alloy. Compd.* **2017**, *692*, 865–875. [CrossRef]

2. Hsu, C.; O'Reilly, K.A.Q.; Cantor, B.; Hamerton, R. Non-equilibrium reactions in 6xxx series Al alloys. *Mater. Sci. Eng. A-Struct. Mater. Prop. Microstruct. Process.* **2001**, *304*, 119–124. [CrossRef]

3. Sha, G.; O'Reilly, K.; Cantor, B.; Worth, J.; Hamerton, R. Growth related metastable phase selection in a 6xxx series wrought Al alloy. *Mater. Sci. Eng. A-Struct. Mater. Prop. Microstruct. Process.* **2001**, *304*, 612–616. [CrossRef]

4. Li, H.; Wang, J.Y.; Jiang, H.T.; Lo, Z.F.; Zhu, Z.F. Characterizations of precipitation behavior of Al-Mg-Si alloys under different heat treatments. *China Foundry* **2018**, *15*, 89–96. [CrossRef]

5. Kuijpers, N.C.W.; Vermolen, F.J.; Vuik, K.; van der Zwaag, S. A model of the beta-AlFeSi to alpha-Al(FeMn)Si transformation in Al-Mg-Si alloys. *Mater. Trans.* **2003**, *44*, 1448–1456. [CrossRef]

6. Gorny, A.; Manickaraj, J.; Cai, Z.H.; Shankar, S. Evolution of Fe based intermetallic phases in Al-Si hypoeutectic casting alloys: Influence of the Si and Fe concentrations, and solidification rate. *J. Alloy. Compd.* **2013**, *577*, 103–124. [CrossRef]

7. Fan, C.; Long, S.Y.; Yang, H.D.; Wang, X.J.; Zhang, J.C. Influence of Ce and Mn addition on alpha-Fe morphology in recycled Al-Si alloy ingots. *Int. J. Miner. Metall. Mater.* **2013**, *20*, 890–895. [CrossRef]

8. Wu, X.Y.; Zhang, H.R.; Zhang, F.X.; Ma, Z.; Jia, L.N.; Yang, B.; Tao, T.X.; Zhang, H. Effect of cooling rate and Co content on the formation of Fe-rich intermetallics in hypoeutectic A17Si0.3Mg alloy with 0.5%Fe. *Mater. Charact.* **2018**, *139*, 116–124. [CrossRef]

9. Zajac, S.; Gullman, L.O.; Johansson, A.; Bengtsson, B. Hot ductility of some Al-Mg-Si alloys. *Mater. Sci. Forum* **1996**, *217*, 1193–1198. [CrossRef]

10. Bayat, N.; Carlberg, T.; Cieslar, M. In-situ study of phase transformations during homogenization of 6060 and 6063 Al alloys. *J. Phys. Chem. Solids* **2019**, *130*, 165–171. [CrossRef]

11. Sirichaivetkul, R.; Wongpinij, T.; Euaruksakul, C.; Limmaneevichitr, C.; Kajornchaiyakul, J. In-situ study of microstructural evolution during thermal treatment of 6063 aluminum alloy. *Mater. Lett.* **2019**, *250*, 42–45. [CrossRef]

12. Couto, K.B.S.; Claves, S.R.; Van Geertruyden, W.H.; Misiolek, W.Z.; Goncalves, M. Effects of homogenisation treatment on microstructure and hot ductility of aluminium alloy 6063. *Mater. Sci. Technol.* **2005**, *21*, 263–268. [CrossRef]

13. Kumar, S.; Grant, P.S.; O'Reilly, K.A.Q. Evolution of Fe Bearing Intermetallics During DC Casting and Homogenization of an Al-Mg-Si Al Alloy. *Metall. Mater. Trans. A-Phys. Metall. Mater. Sci.* **2016**, *47A*, 3000–3014. [CrossRef]

14. Zajac, S.; Hutchinson, B.; Johansson, A.; Gullman, L.O. Microstructure control and extrudability of al-mg-si alloys microalloyed with manganese. *Mater. Sci. Technol.* **1994**, *10*, 323–333. [CrossRef]

15. Asensio-Lozano, J.; Suarez-Pena, B. Quantitative analysis and morphological characterization of 6063 alloy. Microestructural and mechanical comparison between periphery and center of semi-continuous casting round billets. *Rev. De Metal.* **2012**, *48*, 199–212. [CrossRef]

16. Kuijpers, N.C.W.; Vermolen, F.J.; Vuik, C.; Koenis, P.T.G.; Nilsen, K.E.; van der Zwaag, S. The dependence of the beta-AlFeSi to alpha-Al(FeMn)Si transformation kinetics in Al-Mg-Si alloys on the alloying elements. *Mater. Sci. Eng. A-Struct. Mater. Prop. Microstruct. Process.* **2005**, *394*, 9–19. [CrossRef]

17. Osada, Y. EPMA mapping of small particles of alpha-AlFeSi and beta-AlFeSi in AA6063 alloy billets. *J. Mater. Sci.* **2003**, *38*, 1457–1464. [CrossRef]

18. Liu, C.L.; Azizi-Alizamini, H.; Parson, N.C.; Poole, W.J.; Du, Q. Microstructure evolution during homogenization of Al-Mg-Si-Mn-Fe alloys: Modelling experimental results. *Trans. Nonferrous Met. Soc. China* **2017**, *27*, 747–753. [CrossRef]

19. Gavgali, M.; Totik, Y.; Sadeler, R. The effects of artificial aging on wear properties of AA 6063 alloy. *Mater. Lett.* **2003**, *57*, 3713–3721. [CrossRef]

20. Chen, B.; Wu, Y.J.; Zhu, T.; Li, X.Y. The effects of Solid-solution on properties and microstructure of 6063 aluminum alloy. *Adv. Mater. Res.* **2014**, *881–883*, 1346–1350. [CrossRef]

21. Lang, P.; Povoden-Karadeniz, E.; Falahati, A.; Kozeschnik, E. Simulation of the effect of composition on the precipitation in 6xxx Al alloys during continuous-heating DSC. *J. Alloy. Compd.* **2014**, *612*, 443–449. [CrossRef]

22. Sjolander, E.; Seifeddine, S. The heat treatment of Al-Si-Cu-Mg casting alloys. *J. Mater. Process. Technol.* **2010**, *210*, 1249–1259. [CrossRef]

23. Siddiqui, R.A.; Abdullah, H.A.; Al-Belushi, K.R. Influence of aging parameters on the mechanical properties of 6063 aluminium alloy. *J. Mater. Process. Technol.* **2000**, *102*, 234–240. [CrossRef]

24. Milkereit, B.; Wanderka, N.; Schick, C.; Kessler, O. Continuous cooling precipitation diagrams of Al-Mg-Si alloys. *Mater. Sci. Eng. A-Struct. Mater. Prop. Microstruct. Process.* **2012**, *550*, 87–96. [CrossRef]

25. Yuksel, B. Effect of artificial aging on hardness and intergranular corrosion of 6063 Al alloy. *Pamukkale Univ. J. Eng. Sci. -Pamukkale Univ. Muhendis. Bilimleri Derg.* **2017**, *23*, 395–398. [CrossRef]

26. Liu, S.H.; Wang, X.D.; Pan, Q.L.; Li, M.J.; Ye, J.; Li, K.; Peng, Z.W.; Sun, Y.Q. Investigation of microstructure evolution and quench sensitivity of Al-Mg-Si-Mn-Cr alloy during isothermal treatment. *J. Alloy. Compd.* **2020**, *826*, 11. [CrossRef]

27. Saito, T.; Marioara, C.D.; Royset, J.; Marthinsen, K.; Holmestad, R. The effects of quench rate and pre-deformation on precipitation hardening in Al-Mg-Si alloys with different Cu amounts. *Mater. Sci. Eng. A-Struct. Mater. Prop. Microstruct. Process.* **2014**, *609*, 72–79. [CrossRef]

28. Meyveci, A.; Karacan, I.; Caligulu, U.; Durmus, H. Pin-on-disc characterization of 2xxx and 6xxx aluminium alloys aged by precipitation age hardening. *J. Alloy. Compd.* **2010**, *491*, 278–283. [CrossRef]

29. Nandy, S.; Abu Bakkar, M.; Das, D. Influence of Ageing on Mechanical Properties of 6063 Al Alloy. *Mater. Today-Proc.* **2015**, *2*, 1234–1242. [CrossRef]

30. Zohrabyan, D.; Milkereit, B.; Kessler, O.; Schick, C. Precipitation enthalpy during cooling of aluminum alloys obtained from calorimetric reheating experiments. *Thermochim. Acta* **2012**, *529*, 51–58. [CrossRef]

31. Li, H.Y.; Zeng, C.T.; Han, M.S.; Liu, J.J.; Lu, X.C. Time-temperature-property curves for quench sensitivity of 6063 aluminum alloy. *Trans. Nonferrous Met. Soc. China* **2013**, *23*, 38–45. [CrossRef]

32. Cavazos, J.L.; Colas, R. Quench sensitivity of a heat treatable aluminum alloy. *Mater. Sci. Eng. A-Struct. Mater. Prop. Microstruct. Process.* **2003**, *363*, 171–178. [CrossRef]

33. Giersberg, L.; Milkereit, B.; Schick, C.; Kessler, O. In-situ isothermal calorimetric measurement of precipitation behaviour in Al-Mg-Si alloys. *Mater. Sci. Forum* **2014**, *794–796*, 939. [CrossRef]

34. Luoh, T.; Bow, J.S.; Peng, A.; Tsai, S.Y.; Tseng, M.R. Observation of recording marks in phase-change media using scanning electron microscopy channelling contrast image. *Jpn. J. Appl. Phys. Part 1-Regul. Pap. Short Notes Rev. Pap.* **1999**, *38*, 1698–1700. [CrossRef]

35. Frock, H.; Kappis, L.V.; Reich, M.; Kessler, O. A Phenomenological Mechanical Material Model for Precipitation Hardening Aluminium Alloys. *Metals* **2019**, *9*, 1165. [CrossRef]

36. Hamdi, I.; Boumerzoug, Z.; Chabane, F. Study of precipitation kinetics of an Al-Mg-Si alloy using differential scanning calorimetry. *Acta Metall. Slovaca* **2017**, *23*, 155–160. [CrossRef]

37. Vedani, M.; Angella, G.; Bassani, P.; Ripamonti, D.; Tuissi, A. DSC analysis of strengthening precipitates in ultrafine Al-Mg-Si alloys. *J. Therm. Anal. Calorim.* **2007**, *87*, 277–284. [CrossRef]

38. van de Langkruis, J.; Vossenberg, M.S.; Kool, W.H.; van der Zwaag, S. A study on the beta ′ and beta ″ formation kinetics in AA6063 using differential scanning calorimetry. *J. Mater. Eng. Perform.* **2003**, *12*, 408–413. [CrossRef]

39. Ohmori, Y.; Doan, L.C.; Matsuura, Y.; Kobayashi, S.; Nakai, K. Morphology and crystallography of beta-Mg2Si precipitation in Al-Mg-Si alloys. *Mater. Trans.* **2001**, *42*, 2576–2583. [CrossRef]

40. Edwards, G.A.; Stiller, K.; Dunlop, G.L.; Couper, M.J. The precipitation sequence in Al-Mg-Si alloys. *Acta Mater.* **1998**, *46*, 3893–3904. [CrossRef]

41. Tsao, C.S.; Chen, C.Y.; Jeng, U.S.; Kuo, T.Y. Precipitation kinetics and transformation of metastable phases in Al-Mg-Si alloys. *Acta Mater.* **2006**, *54*, 4621–4631. [CrossRef]

42. Murayama, M.; Hono, K. Pre-precipitate clusters and precipitation processes in Al-Mg-Si alloys. *Acta Mater.* **1999**, *47*, 1537–1548. [CrossRef]

Article

Simulation on the Effect of Porosity in the Elastic Modulus of SiC Particle Reinforced Al Matrix Composites

Jorge E. Rivera-Salinas [1,*], Karla M. Gregorio-Jáuregui [2], José A. Romero-Serrano [2], Alejandro Cruz-Ramírez [2], Ernesto Hernández-Hernández [3], Argelia Miranda-Pérez [4] and Víctor H. Gutierréz-Pérez [5]

[1] Catedrático CONACyT—Department of Plastics Transformation Processing, Centro de Investigación en Química Aplicada—CIQA, 25294 Saltillo, Coahuila, Mexico
[2] Metallurgy and Materials Department, Instituto Politécnico Nacional, Escuela Superior de Ingeniería Química e Industrias Extractivas—ESIQIE, UPALM, 07738 México D.F., Mexico; kaarlaa@gmail.com (K.M.G.-J.); romeroipn@hotmail.com (J.A.R.-S.); alex73ipn@gmail.com (A.C.-R.)
[3] Department of Advanced Materials. Centro de Investigación en Química Aplicada—CIQA, 25294 Saltillo, Coahuila, Mexico; ernesto.hernandez@ciqa.edu.mx
[4] Corporación Mexicana de Investigación en Materiales S.A. de C.V., Ciencia y Tecnología—COMIMSA, 25290 Saltillo, Coahuila, Mexico; argelia.miranda@comimsa.com
[5] Profesional Specific Training Department. Instituto Politécnico Nacional—Unidad Profesional Interdisciplinaria de Ingeniería campus Zacatecas (UPIIZ), 98160 Zacatecas, Mexico; metalurgico2@hotmail.com
* Correspondence: enrique.rivera@ciqa.edu.mx; Tel.: +52-844-389-830

Received: 17 February 2020; Accepted: 14 March 2020; Published: 19 March 2020

Abstract: Although the porosity in Al-SiC metal matrix composites (MMC) can be diminished; its existence is unavoidable. The purpose of this work is to study the effect of porosity on Young's modulus of SiC reinforced aluminum matrix composites. Finite element analysis is performed based on the unit cell and the representative volume element approaches. The reliability of the models is validated by comparing the numerical predictions against several experimental data ranging in low- and high-volume fractions and good agreement is found. It is found that despite the stress transfer from the soft matrix to the reinforcement remains effective in the presence of pores, there is a drop in the stress gathering capability of the particles and thus, the resulting effective elastic modulus of composite decreases. The elastic property of the composite is more sensitive to pores away the reinforcement. It is confirmed, qualitatively, that the experimentally reported in the literature decrease in the elastic modulus is caused by the presence of pores.

Keywords: Al/SiC composite; porosity in composites; finite element analysis

1. Introduction

Particles embedded in a matrix are commonly encountered in metal matrixes since they arise during melt processing by non-controlled phase changes, mechanical interaction of the melt with its surroundings, or they are added intentionally as filler material. Stiff and soft particle inclusions in a matrix have effects that could be considered adverse or beneficial in the physical and mechanical properties of the bulk matrix. For example, hard particles in steel are considered harmful and they often originate the variability of steel properties [1]. On the other hand, the incorporation of ceramic stiffer particles—e.g., SiC, SiO_2, Al_2O_3, and WC—to matrixes such as aluminum alloys reinforce the bulk matrix and metal matrix composites (MMC) arise. Due to their high specific strength and modulus, reinforced materials can be tailored, finding applications in multiple industries such as naval, aeronautics, automotive, and machining-tools. Moreover, MMC exhibit isotropic material properties [2].

The elastic properties of MMC depend principally on geometrical parameters: volume fraction, size distribution, shapes and stiffness of the particles, porosity, and bonding state at the interface. Studies have been conducted to elucidate the effects of particles embedded in metal matrixes [3,4], concluding that the shape of the particles influences the overall elastic response of the composite. While the particle size has no significant influence on the elastic modulus at low content of reinforcement [5,6].

For the microstructures having different phases like hard ceramic particles and pores, it is desirable to predict strength properties for the development of high-performance materials. Porosity in MMC can greatly influence the material properties and is mainly caused by manufacturing processes, thermal cycles, and both distribution and percentage particles [7–9]. For instance, enhancement of strength is affected in the interface causing crack initiation sites, which was evidenced in Hassani's study at varying milling times with different particle sizes and amounts. The visualization of porosity and voids employed specialized techniques such as nondestructive laser-ultrasonic spectroscopy where Podymova et al. [10] proposed a calibration curve for porosity evaluation as this defect compromise composite strength. Other techniques, such as neutron and synchrotron diffraction, were voids and pores are easily observed and micro stresses can be detected. Previous investigations conclude that thermal fatigue damage is produced by the porosity presence and can influence the coefficient of thermal efficiency behavior in MMC, tensile and fatigue properties are also affected [8,11]. Consequently, porosity content decreases mechanical properties of MMC—such as tensile strength, Poisson ratio, and Young's modulus.

Modeling of inclusion-matrix interaction or pore-matrix interaction at the microscale is usefully in accurately predicting effective properties (e.g., Young's modulus). In the literature, several constitutive models (analytical, semi-empirical, differential, and variational methods) for the prediction of mechanical properties with specified volume fractions can be found elsewhere. Among these models, the ones proposed by Voigt and Reuss (V–R) [12,13] and by Hashin–Shtrikman (H–S) [14] have proven to be effective means to obtain upper and lower bounds for the elastic moduli of the MMC from the known properties of its constituents. The H–S model represents the tightest bounds. This model is based on variational principles of linear elasticity and assumes isotropic composite. None of the previous methods (V–R, H–S) consider microstructural information.

Micromechanical analytical models also have been established to predict the mechanical properties of porous solids. Ramakrishnan and Arunachalam (R–A) [15] derived a useful analytical model using the principles of statistical continuum mechanics, where the effective elastic moduli of bulk material are predicted as a function of the volume fraction of the pores, the variation of the effective Poisson's ratio, and the elastic modulus of the corresponding dense material. This model considers an assemblage of hollow spheres. Additional information about other theoretical models for porous solids and its validity can be found in [16]. It is difficult to derive analytical solutions for particle–pore interaction systems because the stress distribution around particles and pores start interfering with each other.

The finite element modeling of micromechanics is an alternative means to analytical models for the simulation of mechanical properties. Finite element analysis has become increasingly more popular and an accurate approach to study different phases and complex morphologies of reinforcement particles in the microstructures [1]. Approaches are the unit cell model (UC) and the representative volume element (REV). In the UC approach, one or two particles are embedded in the metal matrix to represent a periodical array since this approach assumes that the material microstructure is periodic. Because of the periodic assumption, the UC approach is computationally efficient. The REV approach assumes random microstructures. Studies on numerical modeling in MMC have simplified the morphology of the particle reinforcement to that of a circle, rectangle, or ellipse to study the elastic response of the composite [17,18]. Good correlation between the numerical predictions and the experimental data was obtained. Available theoretical studies have shown the effect of reinforcement volume fraction on the stress distribution and modulus of MMC. However, such an understanding still lacks when particles interact with pores in MMC, where the presence of pore close to particle or even within the particle produces inhomogeneous stress distributions and the load-bearing capacity of MMC is affected.

On the other hand, in polymer matrix composites containing fibers, numerical studies have shown that the porosity has a much larger influence in transversal properties than the longitudinal ones and a critical volume fraction exists below when the strength is unaffected by pores [19]. For solders, the effects of the morphology of intermetallic and porosity volume fraction on the modulus of the solder were studied by Chawla et al. [20]. They used a 2D microstructure based finite element modeling to study the effect of the morphology of Ag_3Sn intermetallic and porosity on the Young's modulus of Sn–Ag solders. The Ag_3Sn morphology was controlled by cooling rate; spherical and needle-like particles were obtained. They conclude that the morphology of the particles should not have a significant influence on Young's modulus of the Ag-solder. Rather, the porosity is responsible for the variability of the elastic moduli. However, in this study the porosity effect was not included in the numerical model, instead, it analyzed with the (R–A) analytical model.

MMC porosity arises generally by causes of gas entrapment and the evolution of dissolved gases [21]. Although the porosity in MMC can be diminished, its existence is unavoidable [22]. As porosity causes structural weakness in particulate composites, this factor remains as one of the focuses to attract researchers' attention in the development of high-performance materials. Ray [23] classified porosity of composites in two types: (a) at the boundary of the matrix phase and the reinforcement, and (b) away from the filler particles in the matrix alloy. In the works of Sun et al. [24] and Podymova et al. [10], it was found that pores coexist with the clustered and large-sized particles, and they significantly decrease the mechanical properties of the composites. The porosity in MMC arises from different origins, therefore it is of great interest to screening the inhomogeneous stress field around pores interacting with hard particles, and how the discontinuity produced by the pores has a damaging effect on the load-bearing capacity of the MMC. Finite element analysis is a means of properly interpreting the effect of porosity on the elastic properties of porous materials [16,19]. This paper aims to investigate the effect of porosity on the elastic modulus of SiC particle reinforced Al matrix composites. Finite element models based on the UC and REV approaches were developed. Porosity is considered within the Al matrix and within SiC particle to determine how the load-bearing capacity of the composite is affected by the different regions of stress concentration origins. The shapes of the reinforcement particle are angular and circular in the UC approach. Whereas, in the REV approach, the microstructures were developed to incorporate the mechanical interaction between the reinforcing particles. Firstly, the numerical methodology is validated against experimental data available in the literature ranging in a low and high-volume fraction of reinforcement. Also, the constitutive models of H–S, V–R, and R–A are used for reference.

2. Numerical Modeling

Square and circular morphologies of the particle reinforcement were selected to elucidate the effect of the inclusion shape in the elastic moduli. Two validation studies are conducted first by comparing the FEA results within experimental data reported for the system Co-WC [25] and experimental results for porous refractory spinel [16]. In the case of matrix strengthening, the composite matrix is made of cobalt (Co), whose elastic modulus $E_2 = 207$ [GPa], Poisson's ratio $\eta_2 = 0.31$ and shear modulus $G_2 = 79$ [GPa]. The inclusion is tungsten carbide (WC) whose elastic modulus $E_1 = 700$ [GPa], Poisson's ratio $\eta_1 = 0.23$, and shear modulus $G_1 = 278$ [GPa]. For the Co-WC composite, experimental data of the elastic moduli as a function of the volume fraction of reinforcement ranging at high concentrations of filler are available in the literature [25]. Therefore, this data allows examining the reliability of the numerical model when high volume fractions are considered. As the volume fraction of the reinforcement increases, advanced micromechanical models are needed [26].

In the case of the porous material, the elastic modulus and Poisson's ratio of $MgAl_2O_4$ are $E_{spinel} = 268.2$ [GPa] and $v_0 = 0.262$, respectively. For modeling the pore, zero-moduli and zero-Poisson's ratio are assigned appropriately to avoid singularities in the numerical model. For computing the effective elastic moduli of the materials, two-dimensional (2D) UCs models of 8 × 8 mm with a different volume fraction of the reinforcement and porosity were created. Plane strain behavior is assumed. In the case of the UCs models the center of mass of the reinforcement particles coincides with that of the square representing the matrix. Pores are introduced explicitly in the metal matrix and within the particle reinforcement. Since the model is 2D, the volume fraction is equal to that of area fraction. The sizes of the particles are varied to obtain different volume fractions. For the REV models, the volume fraction and particle sizes are fixed. The unit cell is under tensile loading.

2.1. Linear Elasticity Equations

The governing equations for the static mechanic's problem are given by the equilibrium equation

$$\nabla \cdot \sigma + f = 0 \tag{1}$$

where, σ and f are the stress tensor and body force, respectively. For the stress-strain relation, it is assumed a plane strain model. Therefore, strain normal to the plane is zero, and shear strains that involve angles normal to the plane are assumed to vanish. Constituent materials are isotropic, so the constitutive equation ($\sigma = D\varepsilon$) is given by [27]

$$D = \frac{E}{(1+v)(1+2v)} \begin{bmatrix} 1-v & v & 0 \\ v & 1-v & 0 \\ 0 & 0 & (1-2v)/2 \end{bmatrix} \tag{2}$$

where E is the Young's modulus and v the Poisson's ratio. Macroscopically, the heterogeneous material can be assumed as a homogenous medium. The average strains, stress and strain in the composite are related to the boundary displacements of the UC and the REV by Gauss theorem. Therefore, using appropriate boundary conditions to produce uniform stress and strain in the homogeneous medium, the relation between the actual heterogeneous composite and the homogeneous medium is given by averaging the stress and strain tensors over the area of the unit cell [28].

$$\bar{\sigma}_{ij} = \frac{1}{A} \int_A \sigma_{ij}(x,y) dA = \frac{1}{A} \int_\Gamma \left(E u_i n_j + E u_j n_i \right) d\Gamma \tag{3}$$

$$\bar{\varepsilon}_{ij} = \frac{1}{A} \int_A \varepsilon_{ij}(x,y) dA = \frac{1}{A} \int_\Gamma \left(u_i n_j + u_j n_i \right) d\Gamma \tag{4}$$

The effective elastic moduli of the metal matrix composite (E_c), is given by substitution of the two previous equations into the Hook's law $\bar{\sigma}_{ij} = E_c \bar{\varepsilon}_{ij}$.

2.2. Finite Element Implementation

The 2D numerical models are developed using the mechanic's module in the software COMSOL Multiphysics. Boundary conditions assume plane strain-tension to evaluate the elastic modulus of the composites. Figure 1 shows the boundary conditions and the periodic microstructures that are considered for the UCs models. The mesh refinement is applied to ensure mesh independent solutions for each microstructure. The inclusion volume fraction is varied by changing the inclusion size in a parametric study, where the circular reinforcement is only able of producing volume fraction up to 0.71% while the square reinforcement a maximum volume fraction of 0.91%. The external load **F** is set to 0.1 Pa. The microstructures used in the REV approach are shown in Section 3.2.5. This latter study was performed to compare the mechanical behavior predicted by the UCs models, with that predicted by the structure of real composite material. The boundary conditions are the same as those used in the UCs models.

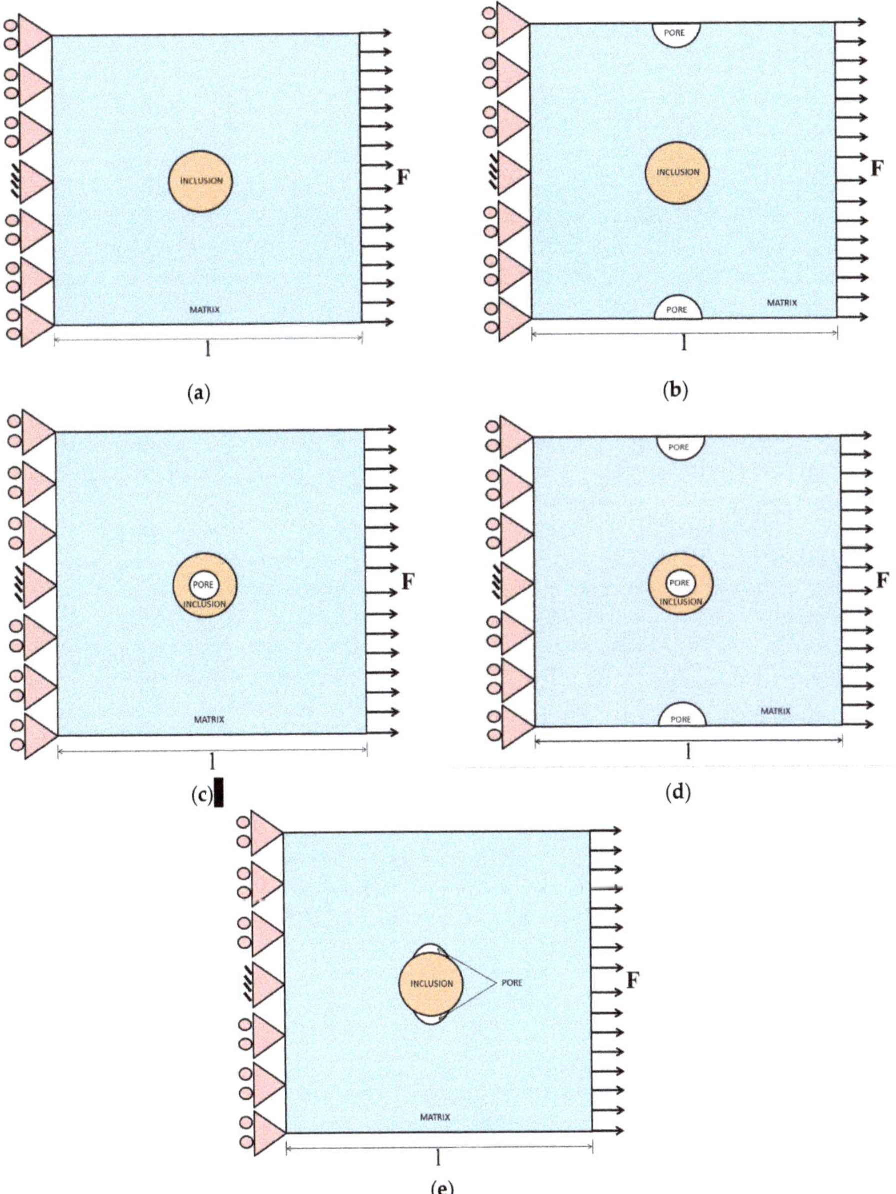

Figure 1. Schematic representation of the unit cell model and its boundary conditions. Particle reinforcements are considered square and circular. (**a**) Pore free matrix and pore-free particle. (**b**) Porous matrix and pore-free particle. (**c**) Pore free matrix and porous particle and (**d**) porous matrix and porous particle. (**e**) Porosity at particle–matrix interface. The triangles on the left side represent constraints displacement in the node.

3. Results and Discussions

3.1. Comparison of FEA Predictions with Experimental Data

The numerical methodology was validated by comparing the numerical effective elastic modulus of the composites against experimental data published for the cemented carbides, which consist of WC particles distributed in cobalt [25]. This study considers small and large granules, with reinforcement volume fraction ranging from 50–90% and the elastic modulus were determined by resonant ultrasound spectroscopy (RUS) and impulse excitation. The elastic modulus predicted from the bounds of Hashin–Shtrikman (H–S) and Voigt–Reuss (V–R) is also used as a reference. The H–S micromechanical model is perhaps the most widely used because of its accuracy. Bounds are the tightest possible ones from a range of composite moduli; the two-phase material moduli can be obtained as

$$E^{\pm} = E_2 + \frac{\phi_1}{1/(E_1 - E_2) + \left(3\phi_2/(3E_2 + 4G_2)\right)} \tag{5}$$

For the computation of the upper bound $E_2 > E_1$, meanwhile, the lower bound is computed by interchanging the subscripts $(1, 2)$ in the previous equation. Here, E is the Young's modulus, G is the shear moduli, and ϕ is the volume fraction.

Figure 2 shows the comparison between H–S bounds on the Young's modulus for the Co/WC system, experimental results [25] and numerical data obtained by simplifying the actual particle morphology to a square and a circle. V–R bounds are also plotted for reference. It is seen that the numerical predictions fall within the H–S bounds and that the square capture the angular nature of the WC particles since these data are closer to the experimental values (WC particles are angular). However, at the volume fraction of filler of 91%, the numerical prediction lies above the H–S upper bound. The numerical error was evaluated as $\left|\left(E_{HS}^{+} - E_{numerical}\right)/E_{HS}^{+}\right|$, which is lower than 3% at the volume fraction of 91% with $E_{numerical} = 655$ [GPa] and $E_{HS}^{+} = 638$ [GPa]. Whenever the numerical predictions do not overtake the (V–R) bounds, the predictions are considered valid [29]. The upper bound of the V–R bounds is just 655 [GPa] at the volume fraction of 91%, so the numerical prediction lies at the upper bound of the V–R model and does no overtake this bound. This confirms that the predictions for the elastic moduli from the UC models developed are reliable. Results also show that the square particle has a little greater stiffening effect than the circular particles. Since the square particle values are close to the experimental ones, no additional effort was undertaken to consider a polygonal shape of the inclusion.

The information plotted in Figure 2 shows that the numerical model correctly predicts the stiffening effect. The validity of the developed numerical methodology was evaluated for the prediction of the effective Young's moduli when one phase of the composite has a zero-moduli and zero-Poisson's ratio. The analytical model for micromechanics of porous mediums proposed by Ramakrishnan and Arunachalam (R–A) was used as a reference. Experimental data reported in [16] was used to determine the accuracy of the numerical methodology in capturing the presence of pores in the composite. The R–A model was established considering that the dispersed phase of the composite has zero moduli, which means that in the porous area, the stress is zero but not the strain, the equations that describe this behavior are the following

$$E^{*} = E(1 - \theta)^2 / (1 + b_{\theta}\theta) \tag{6}$$

$$b_{\theta} = 2 - 3v_0 \tag{7}$$

where E is the Young's modulus of the fully dense material, θ the porosity, and b_{θ} is a constant that depends on the Poisson's ratio v_0 of fully dense material.

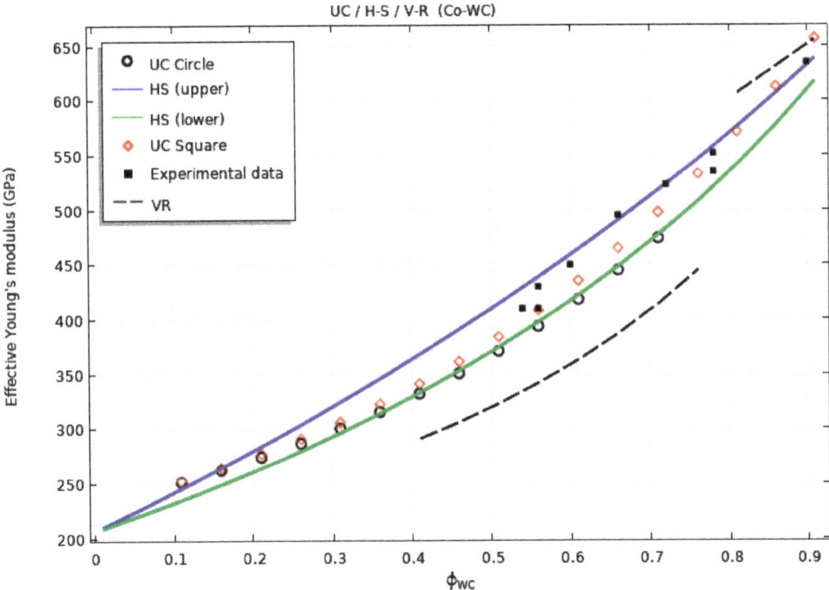

Figure 2. H–S bounds (——), experimental data (■) [25], numerical data (square ◆, circle ●) and V–R bounds (– – –) for the Young's modulus for Co-WC system.

Numerical results were compared to the experimental data for the porous refractory spinel, where the effective elastic modulus was determined in [16] as a function of the porosity ranging from a fully dense material up to 40% porosity. Figure 3 shows this comparison when the pore shapes are circular and square to reflect the effect of pore shape on the Young's moduli. The circular porous geometry fit the experimental data better than that of the square one. Since the simulation results with a circular pore show a close match to experimental data, the pore geometry considered in the UC models is only circular. The comparison between numerical results and experimental data shows that the present numerical models developed are also reliable when zero moduli and zero-Poisson's ratio are assigned for one phase in the heterogeneous material. The computed results show that the UC model calculations are in good agreement with experimental data available in the literature [16,25] which proves the validity of the presented approach for microstructures of two phases. However, it is necessary to be careful when considering a three-phase material since it is necessary to place an inclusion (particle or pore) out from the center of mass of the UC model. In this work, pores are placed in the boundary, at the center of the UC and both the boundary and the center of the UC. In order to take into care that the composite behavior is not dominated by the boundary conditions (Saint-Venant's principle), the influence of pore location on the computed elastic modulus was first established. As mentioned above, when the pore is in the center of mass of the UC, the numerical results are close to the experimental data in a wide range of values of pore content (Figure 3). On the other hand, when the pores are placed on the boundaries of the unit cell, the model predictions are reliable only when the pore content is ranging in low volume fraction as shown in Figure 4. Besides, it is observed that the model predictions are reliable when considering porosity up to 14% with a maximal deviation $((E_{exp} - E_{FEM})/E_{exp})$ of 6.6%. Thus, to interpret the numerical results properly when the pore is out from the center of mass of the UC, pore content should be lower than 14%. It is assumed that similar accuracy pertains when the numerical methodology is used to find the local stress and strain for the porous (Al/SiC) composites.

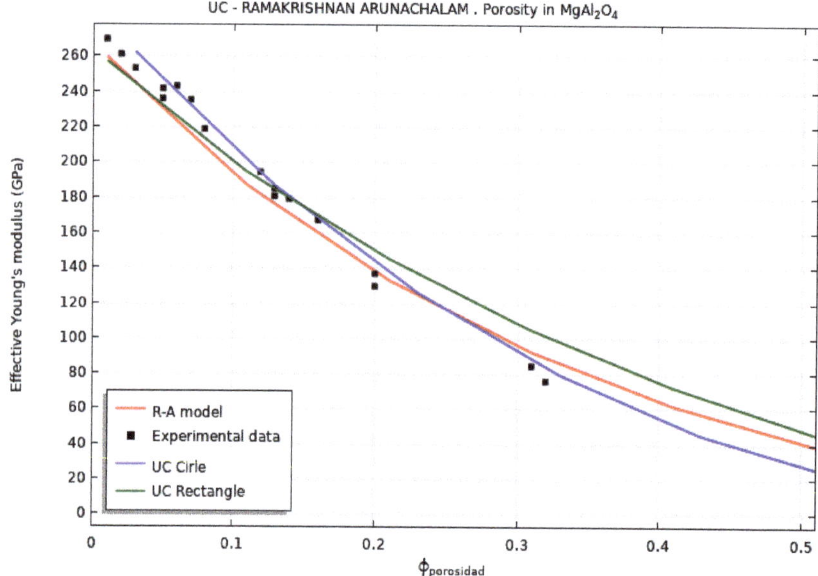

Figure 3. Variation of the effective elastic moduli for porous spinel. R–A model (——), experimental data (■) [16] and numerical data (square ——, circle ——).

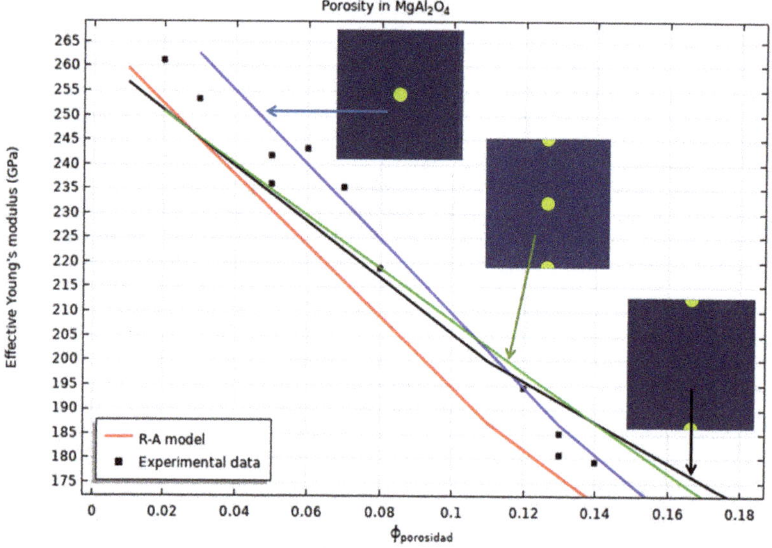

Figure 4. Influence of pore location on the numerical predictions of the elastic modulus. When the pore is placed at the center of mass, the model is reliable for a wide range of porous volume fraction. On the other hand, when the pores are placed in the boundaries of the UC, the model is reliable for porosity content ranging from 0% to 14% with an error lower to 7%.

3.2. SiC Particle Reinforced Aluminum Composites

Aluminum matrix composites possess both structural and functional applications, as a result, they are being increasingly used in several industries such as automotive; however, with the current manufacturing process, it is difficult to avoid the porosity in MMC. Moreover, the porosity arises from different origins [10,23,24]. Therefore, the porosity has different destructive effects on the strength of the MMC. This factor has attracted the interest of the current research to investigate the effect of the porosity using finite element analysis on the load-bearing capacity of Al/SiC composites. The numerical models based on the microstructures shown in Figure 1 were constructed. The shape of the SiC particle is considered square and circular to determine the strengthening effect due to the particle's shape. The mechanical properties of the particle and matrix are shown in Table 1 [30]. The results are shown in the next section.

Table 1. Material properties of the SiC particle and Al matrix [30].

Material	Young's Modulus (GPa)	Poisson's Ratio
Al	74	0.33
SiC	410	0.19

3.2.1. Pore-Free Al-Matrix and Pore-Free SiC-Particle

Figure 5 illustrates the surface plots of von Mises stress in the microstructure when the particle and matrix are fully dense. Both unit cells (square and circular SiC-particle) are under the same external load and the volume fraction of reinforcement is 0.26%. For the SiC square particle (Figure 5a) the maximum stress value in the particle is 0.4 Pa, this stress is mainly concentrated at the corners of the particle. In the case of the circular SiC particle (Figure 5b), its maximum stress value is 0.11 Pa, which is concentrated at the contour of the particle. A comparison of both the square and circle particles shows that the stress gradient at the interface of the particle-matrix is larger when the reinforcement shape is angular (Figure 5a) than when is circular (Figure 5b). In both cases, the stress in the particle is larger than the stress in the matrix. Regarding the deformation in the composite, the strain in the matrix is larger than that of the particle because the particles are stiffer than the matrix. Based in the fact that the external load must equal the sum of the volume-averaged loads borne by the constituents (matrix and reinforcement). Then, if the reinforcement carries a relatively high proportion of the externally applied load, it is regarded as acting efficiently ($\sigma_c = \sigma_m(1 - \phi) + \sigma_p\phi$) [31]. Then, the load is better transferred from the soft Al-matrix to the hard SiC particles when angular particles are used as reinforcement. These results agree with that reported in [17], where the load transfer by a shear lag type of mechanism is more effective across planar interfaces than a spherical interface.

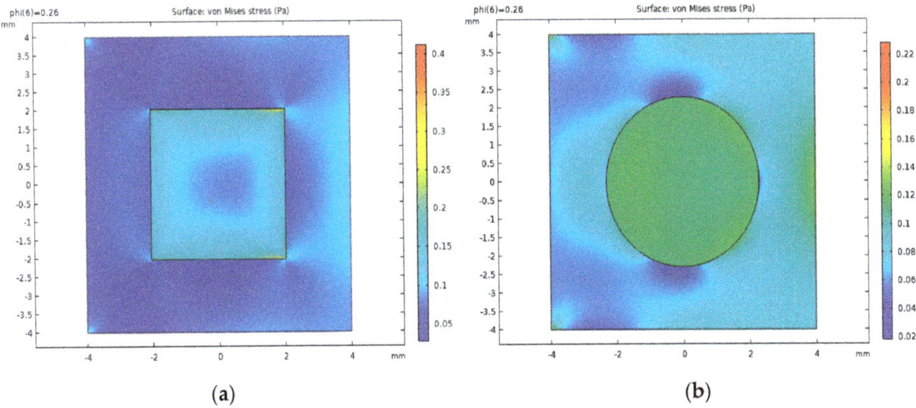

(a)   (b)

Figure 5. Surface plots of von Mises stress for the composite reinforced by SiC particles of (**a**) square shape and (**b**) circular shape. The volume fraction of reinforcement is 0.26%.

Figure 6 shows that, at low volume fractions of particle reinforcement, the effective elastic modulus of the circular particles composite is like that obtained with angular particles (red and black dashed lines). Then, the amount of transfer external-load from the matrix to the particle reinforcement at low volume fractions of SiC particles is similar for angular than circular particles. Therefore, at low volume reinforcement content, the particle shape has little effect on the elastic modulus of the composite. Figure 6 also shows the comparison between H–S bounds on Young's modulus for the Al/SiC system with data from various Al/SiC MMCs experimental results [32–37] and the numerical data obtained in this study. In the plot, the gathered experimental data—as well as the numerical predictions on the Young's modulus of the composite—lies near the lower bound of the H–S model. Therefore, the prediction on the strength of the Al-SiC composites can be done with a high level of confidence using the lower bound of the H–S micromechanical model. Numerical predictions of the elastic modulus based on the square and circular SiC particles are very close to the reported measured data for the Al alloy-SiC systems. As the volume fraction of SiC increases, angular particles strengthen the composite better than the circular ones do. According to the H–S lower bound, the circular shape in the numerical model underpredicts the elastic modulus of the Al-SiC composite. Though these predictions (with circular particles) fall off from the H–S lower bound, the data reported by Qu et al. [34] demonstrates that the numerical predictions are reliable and the condition proposed in [29], is confirmed; while the V–R bounds are not overtaken, the numerical model is reliable. However, the accuracy of the numerical results must be checked by comparing the results to experimental data.

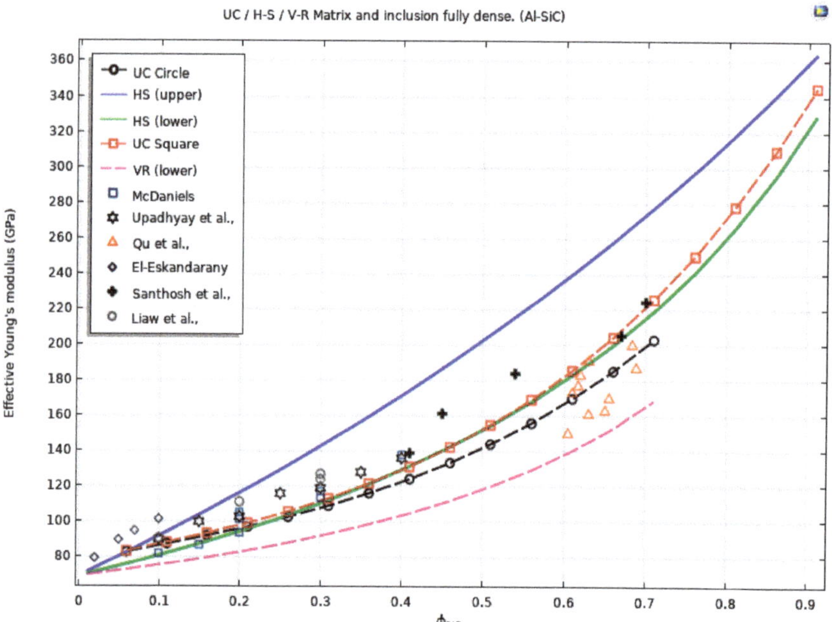

Figure 6. Effect of reinforcement content on the elastic modulus of Al-SiC composite. H–S bounds (━), experimental data [32–37], numerical data (square ■, circle ●) and V–R bounds (━━).

The SiC square or angular particle reinforced Al matrix composites endure more load than reinforced with SiC round particles as shown in Figure 7. This plot shows the stress ratio (stress in the particle and the stress in the matrix, σ_p/σ_m) as a function of the reinforcement volume fraction. It is considered that the composite remains elastic independently of the applied load [31]. It is found that the square and circular particles endure the same load to the volume fraction up to 11%. However, above this reinforcement content, the angular particle endures more load. Then, the effective elastic modulus of the MMC is independent of the particle morphology when the reinforcement content is lower than 11%.

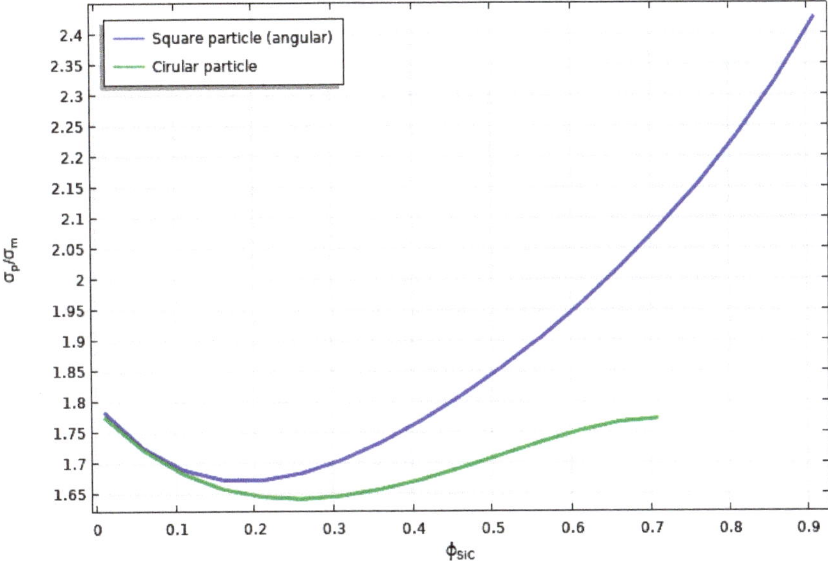

Figure 7. The portion of the external load borne by each of the constituents of the composite. Square particle (blue) and circular particle (green).

3.2.2. Porous Al-Matrix and Pore-Free SiC Particle

The porosity significantly affects the mechanical performance of the composites. The porosity increases as the weighting percentage of particles increases. In [21] it was found that the porosity content of the Al composites varied from 1.01% to 2.69% with the increased SiC wt % from 0% to 25%. A parametric study was performed, where a similar increase-relationship between the amount of porosity and the content of SiC particles is maintained; where the porosity increases as the volume fraction of reinforcement do. Porosity content (volume fraction) varies from 0% to 3.1% with SiC particle added from 1% to 31%. Figure 8 shows the surface plots of von Mises stress in the microstructure when the Al-matrix is porous, and the SiC-particle is fully dense. These results are based on the microstructure shown in Figure 1b. The maximum stress obtained in the square particle is about 0.5 MPa (Figure 8a). Whereas, for the circular particle, the maximum stress in the particle is 0.14 MPa (Figure 8b). In both Al-matrixes (square and circular particle) high-stress concentration around the pore is observed, with a value of about 0.2 MPa, however, the stress in the matrix is bigger than the stress in the particle in the case of the circular reinforcement. As the stress in the SiC square particle is higher than that of the Al-matrix, the load transfer remains effective despite the presence of pores in the Al-matrix. In Figure 8b it is shown that the stress gradient at the interface of the SiC circular particle and Al-matrix is low; namely at some points of the interface, the matrix and the reinforcement endure the same load.

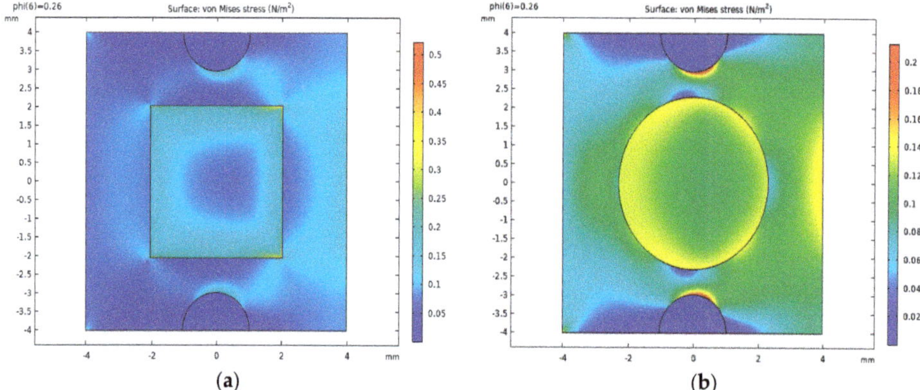

Figure 8. Von-Mises stress distribution of 26% SiC particulate filled porous-aluminum matrix under tensile loading. (**a**) Square shape and (**b**) circular shape.

3.2.3. Pore Free Al-matrix and Porous SiC Particle

Figure 9 shows the surface plots von Mises stress in the microstructure when the Al-matrix is fully dense, and the SiC-particle is porous. These results are based on the microstructure shown in Figure 1c. For the square porous particle (Figure 9a), the stress in the particle (maximum stress 0.44 Pa) is greater than the stress in the matrix, which shows that strengthening of the composite is by transfer load from the soft Al-matrix to the SiC hard particle. The stress concentration around the pore within the particle is 0.35 Pa, while the stress located in the angles of the particle is 0.44 Pa. In the case of the SiC circular particle (Figure 9b), the maximum stress in the particle is around the pore with a value of 0.42 Pa, while in the Al-matrix the stress is lower, so transfer load is also effective in this case. Stress concentration around the pore is higher for the circular particles than square particles, which means that circular porous particles are more prone to fracture than angular porous particles when pores coexist inside the particles.

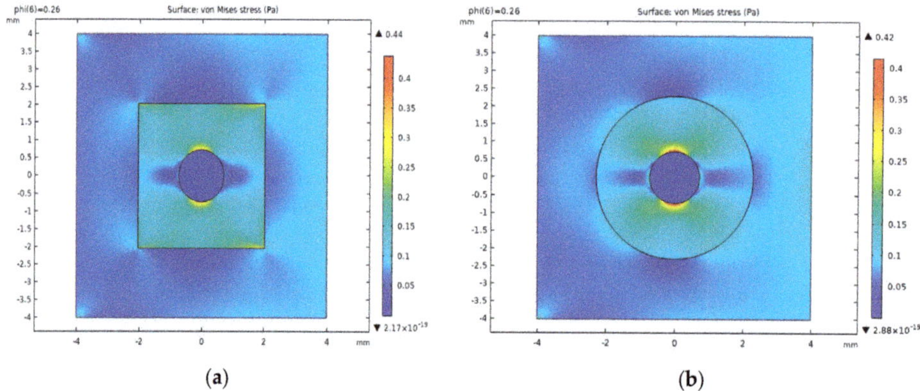

Figure 9. Von Mises stress distribution of SiC porous-particulate filled aluminum matrix under tensile loading. (**a**) Square shape and (**b**) circular shape. The volume fraction of SiC particles and pore content is 26% and 2.6%, respectively.

3.2.4. Porous Al-Matrix and Porous SiC Particle

Figure 10 shows the surface plots of von Mises stress in the microstructure when the Al-matrix and the SiC-particle are both porous. These results are based on the microstructure shown in Figure 1d. In this case, the total porosity is divided into an equal content in the matrix and the particle; so, the pore content in the matrix is 1.3%, while the pore content in the particle is 1.3%. The total pore content in the composite is 2.6%. Figure 10a,b show that the stress concentration around the pore in the Al-matrix is about 0.2 Pa in both cases (the square particle and circular one). However, the stress around the pore in the SiC particle reinforcement is bigger for the circular particle (0.39 Pa) than the stress in the pore located in the square particle (0.31 Pa). Therefore, when the matrix is porous, the stress gathering around the pore within the particles is lower than when the matrix is fully dense. That means that the pore in the matrix reduces the transfer load mechanism and more load is endured by the matrix lowering the strengthening of the composite. For the square SiC particle, the maximum stress (in the corners) is 0.5 Pa, while the stress in the Al-matrix is 0.2 Pa, this shows that the transfer load mechanism from the matrix to the reinforcement remains despite the pore presence at the matrix and particle.

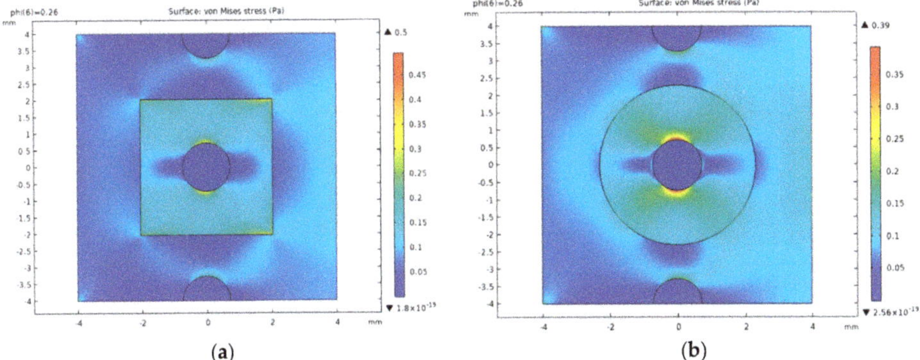

(**a**) (**b**)

Figure 10. Von Mises stress distribution of SiC porous-particulate filled porous aluminum matrix under tensile loading. (**a**) Square shape and (**b**) circular shape. The volume fraction of SiC particles and pore content is 26% and 2.6%, respectively.

3.2.5. Porosity at Matrix–Particle Interface

Figure 11 shows the surface plots of von Mises stress in the microstructure when the pore is located at the particle-matrix interface. For the square particle (Figure 11a), the stress in the particle (0.3 Pa) is greater than the stress in the matrix (0.09 Pa), which shows that strengthening of the composite is by transfer load from the soft Al-matrix to the SiC hard particle. The stress is concentrated at the particle corners, but the highest stress concentration (0.5 Pa) is located at some contact points between the particle, matrix, and pore. In the case of the SiC circular particle (Figure 11b), the stress concentration is at all the contact points among the particle, matrix, and pore, and when compared to the square particle, the stress intensity is two-fold. Moreover, the stress level in the particle (0.13 Pa) is a little bit greater than that in the matrix (0.11 Pa). Therefore, in the case of the circular particles, the pores located at the particle–matrix interface significantly affect the transfer load mechanism, whereas in the case of the angular particles, the transfer load remains effective still because a relatively high proportion of the externally applied load is endured by the particle. As mentioned in Ray [23] this type of porosity aid the debonding of particles from the matrix under low stress, because the interface damage starts to develop since the contact points among the particle matrix and the pores act as a stress concentrator, as shown here.

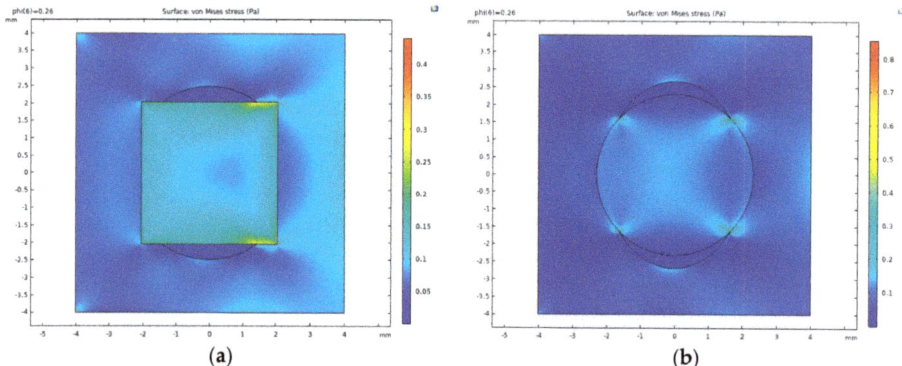

Figure 11. Surface plots of the von Mises stress for the composite reinforced by SiC particles of (**a**) square shape and (**b**) circular shape. The volume fraction of SiC particles and pore content is 26% and 2.6%, respectively. Porosity at particle–matrix interface.

It is found that square particles carry much higher stress than matrix in the case of porous matrix and fully dense particle (Figure 8a). Whereas, in the case of the circular particle, it carries much higher stress than matrix when the particle is porous, and the matrix is fully dense (Figure 9b). From the previous results, it can be seen that despite the presence of pore in the composite, stress transfer from the soft matrix to the hard particle is effective in all cases, except in the case of fully dense circular particle and porous matrix (Figure 8b), where the matrix endures more load than the particle. Then, the transfer load is lowered, and this is reflected as a reduction in the elastic modulus of the composite (Figure 12, black line with circular markers). Figure 12 shows the results gathered from the parametric study to demonstrate the effect of the porosity in the elastic modulus of the SiC particle reinforced Al matrix. In this parametric study, porosity content varies from 0% to 3.1% with SiC particle added from 1% to 31%. For the studied cases that consider coexistence of pores inside the particle and the matrix, the greater impairing on the Young's modulus of the composite was found to be in the case of the fully dense particle and porous matrix. The lower impairing on the elastic modulus of the composite is for the porous particle and fully dense matrix. Intermediate values on the impairing of the Young's modulus of the composite are for the case of porous particle and porous matrix, as well as the pore located at particle–matrix interface when the particle is circular. For the square or angular particles, the pore within the particle affects the load transfer mechanism in a similar manner to the pore located at the particle matrix interface. These results mean that the elastic property of the composite is more sensitive to the porosity in the matrix because the particles are stiffer than the matrix. The declining trend for the elastic moduli is consistent in both the angular and circular particle shapes. In all cases, the square or angular particles strengthen the composite when compared to the circular particles. From the results, it is concluded that the disrupting of continuity in the composite by the presence of pores causes elastic relaxation of the matrix, which leads to the reinforcement particles (angular or circular) undergoing increased loads. Consequently, this leads to a relaxation in the particle, and there is a drop in its stress gathering capability, and thus the resulting effective elastic modulus of the Al/SiC composite decreases. Therefore, the data indicate that the porosity in the composite matrix is the main cause in the impairing of the elastic modulus in MMC. In practice, the scatter of experimental data on the elastic modulus is caused mainly by the porosity in the composite because the difficult of experimentally reproducing the same porosity in the composite every test. Table 2 shows how the % increment in porosity in the matrix reduces the elastic modulus of the composite. Moreover, despite the volume fraction of the reinforcement is increased, if the porosity increase, the impairing on the Young's modulus is increased also. Therefore, the porosity plays a significant role in the strength of the composites, and for a given porosity in the matrix, it impairs the Young's modulus in the same way independently of the shape of the SiC-reinforcement.

Table 2. Porosity increase in the Al-matrix and its effect on the Young's modulus of the composite.

SiC Volume Fraction	10%	-	20%	-	30%	-
Shape of the Particle	1% in porosity	% reduction	2% in porosity	% reduction	3% in porosity	% reduction
Square	88.2 GPa * 80.8 GPa †	8.3	99.3 GPa * 85.1 GPa †	14.3	113.15 GPa * 94.7 GPa †	16.3
Circular	87.3 GPa * 80.0 GPa †	8.3	97.1 GPa * 82.9 GPa †	14.6	109.0 GPa * 90.2 GPa †	17.2

* Pore-free matrix † Porous matrix.

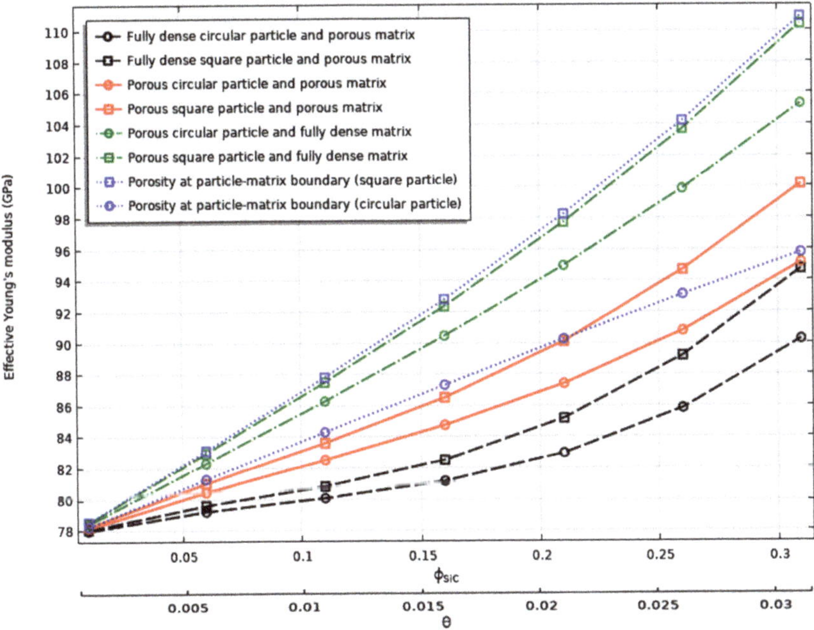

Figure 12. Effect of porosity in the elastic modulus of the Al-SiC composite. ϕ_{SiC} is the reinforcing content and θ is the porosity content in the composite. Square markers are for the square particle while circular markers are for the circular particle.

3.2.6. REV Models with SiC Particles Representing 2D Real Microstructures

The effective material properties are linked to the microstructure of composite. Therefore, the global behavior of the material depends on the microstructure, and modeling of the microstructure is as important as the prediction of global behavior. To visualize the difference between the material properties predictions conducted by assuming a single SiC particle of simple geometry in a unit cell, and those predictions of a cell model with several SiC particles exhibiting statistical homogeneity and isotropy as the real microstructures. Thus, two cell models (REV approach) that display distribution and orientation of particles of different sizes and shapes are established. The interactions between the particles produce a complicated local stress field with high stress concentrations which depends on the particle morphology. Figure 13 shows the developed microstructures. By comparing these microstructures with the experimentally obtained microstructures for SiC reinforced Al matrix composites reported elsewhere, they are quite similar in the context that the position of the polygonal particles are randomly defined, the particles present different sizes, shapes, and orientation. Hence, these microstructures capture the same complicated mechanical interactions as the experimental ones will do. The volume fraction of reinforcement is 12% (Figure 13a) whereas the pore content is 1.2% in volume fraction (Figure 13b). The finite element implementation of these unit cell models is the same followed in Section 2.2, with the main difference that in this section, a parametric study is not needed, due to the volume fraction of the constituents is fixed. It is important to note that despite the shape of some pores look like a circle, they are not perfectly circular.

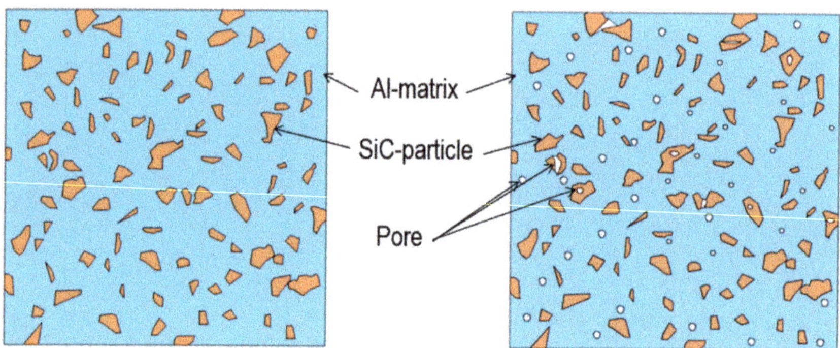

Figure 13. Developed microstructures for Al-SiC composite. (**a**) Fully dense materials and (**b**) porous materials.

Figure 14a,b show the surface plots of von Mises stress in the Al-matrix and the SiC particles, respectively. The stress in the particles is larger than in the matrix, and the stress in the particles is mainly concentrated in the sharp angular corners. These results are in agreement with the stress fields shown previously in Sections 3.2.1–3.2.5. Moreover, the stress field magnitudes are similar for the matrix as well as for the particles. The Young's modulus of the composite using this microstructure is 91.06 GPa, while Young's modulus predicted by the unit cell (used in Section 3.2.1) is 89.35 GPa. Evaluating the difference, this value is lower than 2%.

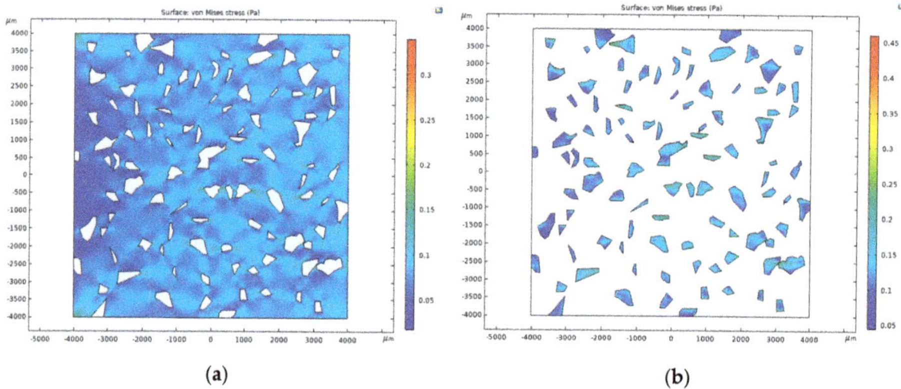

(**a**) (**b**)

Figure 14. Von Mises stress distribution in (**a**) matrix and (**b**) particles. Constituents are fully dense.

Figure 15a,b show the surface plots of von Mises stress in the porous Al-matrix and the SiC particles (some particles are porous), respectively. The presence of the pores within the composite produces higher stress in the matrix as well as in the particles, but the particles are under higher stress when compared to the fully dense composite, indicating that the particles undergoing increased loads. The stress is concentrated in the poles of the pores, in the sharp angular corners of the particles and at the contact points among the particle, matrix, and pore. The Young's modulus of this porous composite is 86.6 GPa. The pore content of 1.2% (volume fraction), reduced the elastic modulus by about 5% (from 91.06 to 86.6 GPa) despite the load transfer is increased. The mechanical behavior observed within the structure of a real composite material under load, is similar to that observed in the results of Sections 3.2.1–3.2.5. Thus, as the results computed with the UCs models are coincident with experimental data, as well as with the results computed by the REV models, the conclusions drawn by this research are confirmed. Therefore, the presence of the pores leads to a relaxation of the composite and reduces its stress gathering capability.

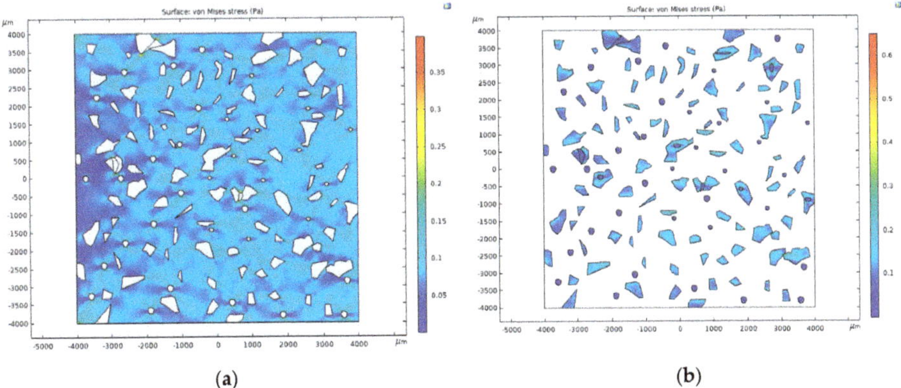

Figure 15. Von Mises stress distribution in (**a**) matrix and (**b**) particles. Matrix is porous while particles are both porous and pore free.

4. Conclusions

In the present work, 2D models (UC and REV approaches) and finite element analysis are used to investigate the porosity effect on the SiC particle reinforced Al matrix composites. The reliability of the numerical methodology was validated by comparing the numerical results against several experimental data, micromechanical constitutive models, and structures that represent real composite materials. The main results to be highlighted are as follows:

(1) In the case of fully dense SiC particles and fully dense Al-matrix, the square and circular particles endure the same load to the volume fraction up to 11%. However, above this reinforcement content, the angular particle endures more load. Therefore, the effective elastic modulus of the MMC is independent of the morphology when the reinforcement content is lower to 11%.

(2) Despite the presence of pore in the composite, stress transfer from the soft matrix to the hard particle is effective in all cases, except in the case of fully dense circular particle and porous matrix.

(3) For pores within the particles, the stress concentration around the pore is higher for the circular particles than the square ones.

(4) The elastic modulus is more sensitive to porosity in the matrix, which is a common defect in MMC.

(5) For the square or angular particles, the pore within the particle affects the load transfer mechanism in the same way to the pore located at the particle matrix interface.

(6) For the porosity at the particle–matrix interface, the contact points among the particle, matrix, and the pore acts as a stress concentrator.

(7)   The porosity plays a significant role in the strength of the composites, and for a given porosity in the matrix, it impairs the Young's modulus in the same way independently of the shape of the SiC-reinforcement.

Author Contributions: Conceptualization, J.E.R.-S. and K.M.G.-J.; methodology, J.E.R.-S. and K.M.G.-J.; software, J.E.R.-S.; validation, J.E.R.-S. and K.M.G.-J.; formal analysis, J.E.R.-S.; investigation, J.E.R.-S.; resources, A.C.-R., J.A.R.-S., V.H.G.-P., A.M.-P., and E.H.-H.; data curation, J.E.R.-S.; writing—original draft preparation, J.E.R.-S. and K.M.G.-J.; writing—review and editing, J.E.R.-S., K.M.G.-J., A.C.-R. and A.M.-P.; visualization, J.E.R.-S., K.M.G.-J., A.C.-R., J.A.R.-S., A.M.-P., E.H.-H., and V.H.G.-P.; supervision, J.E.R.-S., A.C.-R., and J.A.R.-S.; project administration, J.E.R.-S.; funding acquisition, J.E.R.-S, K.M.G.-J., A.C.-R., J.A.R.-S, E.H.-H., A.M.-P. and V.H.G.-P. All authors have read and agreed to the published version of the manuscript.

Funding: This research was funded by CONACYT, grant number "A1-S-9692" and "The APC was funded by CONACYT A1-S-9692".

Acknowledgments: The authors wish to thank the Institutions CONACYT, SNI, Centro de Investigación en Química Aplicada, and Instituto Politécnico Nacional, for their permanent support. J.E. Rivera gratefully acknowledges the financial support received from Cátedras CONACYT project 809.

Conflicts of Interest: The authors declare no conflict of interest.

References

1. Arreola-Herrera, R.; Cruz-Ramírez, A.; Rivera-Salinas, J.E.; Romero-Serrano, J.A.; Sánchez-Alvarado, R.G. The effect of non-metallic inclusions on the mechanical properties of 32 CDV 13 steel and their mechanical stress analysis by numerical simulation. *Theor. Appl. Fract. Mech.* **2018**, *94*, 134–146. [CrossRef]

2. Kulkarni, S.G.; Achchhe, L.; Menghani, J.V. Effect of reinforcement type and porosity on strength of metal matrix composite. *Int. J. Comput. Mater. Sci. Eng.* **2016**, *5*, 1650006. [CrossRef]

3. Shen, L.; Finot, M.; Needleman, A.; Suresh, S. Effective plastic response of two-phase composites. *Acta Metall. Mater.* **1995**, *43*, 1701–1722. [CrossRef]

4. Chawla, N.; Chawla, K.K. Microstructure-based modeling of the deformation behavior of particle reinforced metal matrix composites. *J. Mater. Sci.* **2006**, *41*, 913–925. [CrossRef]

5. Nasher, N.; Hori, M. *Micromechanics: Overall Properties of Heterogeneous Materials*, 2nd ed.; Elsevier: North-Holland, The Netherlands, 1999.

6. Gentieu, T.; Catapano, A.; Jume, J.; Broughton, J. Computational modeling of particulate-reinforced materials up to high volume fractions: Linear elastic homogenization. *J. Mater. Des. Appl.* **2019**, *233*, 1101–1116.

7. Hassani, A.; Bagherpour, E.; Qods, F. Influence of pores on workability of porous Al/SiC composites fabricated through powder metallurgy + mechanical alloying. *J. Alloys Compd.* **2014**, *591*, 132–142. [CrossRef]

8. Schöbel, M.; Requena, G.; Fiedler, G.; Tolnai, D.; Vaucher, S.; Degischer, H.P. Void formation in metal matrix composites by solidification and shrinkage of an AlSi7 matrix between densely packed particles. *Compos. Part A Appl. Sci. Manuf.* **2014**, *66*, 103–108. [CrossRef]

9. Arif, M.N.; Bukhari, M.Z.; Brabazon, D.; Hashmi, M.S.J. Coefficient of thermal expansion (CTE) study in metal matrix composite of CuSiC vs. AlSiC. *IOP Conf. Ser. Mater. Sci. Eng.* **2019**, *701*, 012057. [CrossRef]

10. Podymova, N.B.; Kalashnikov, I.E.; Bolotova, L.K.; Kobeleva, L.I. Laser-ultrasonic nondestructive evaluation of porosity in particulate reinforced metal-matrix composites. *Ultrasonics* **2019**, *99*, 105959. [CrossRef]

11. Schöbel, M.; Altendorfer, W.; Degischer, H.P.; Vaucher, S.; Buslaps, T.; Di Michiel, M.; Hofmann, M. Internal stresses and voids in SiC particle reinforced aluminum composites for heat sink applications. *Compos. Sci. Technol.* **2011**, *71*, 724–733. [CrossRef]

12. Voigt, W. Uber die Beziehung zwischen den beiden Elastizitatskonstanten Isotroper Korper. *Wied. Ann.Phys.* **1889**, *38*, 573–587. [CrossRef]

13. Reuss, A. Berechnung der Fliessgrense von Mischkristallen auf Grund der Plastizitätsbedingung für Einkristalle. *J. Appl. Math. Mech.* **1929**, *9*, 49–58.

14. Hashin, Z.; Shtrikman, S. A variational approach to the theory of the elastic behavior of multiphase materials. *J. Mech. Phys. Solids* **1963**, *11*, 127–140. [CrossRef]

15. Ramakrishnan, N.; Arunachalam, V. Effective elastic moduli of porous solids. *J. Mater. Sci.* **1990**, *25*, 3930–3937. [CrossRef]

16. Ramakrishnan, N.; Arunachalam, V. Effective elastic moduli of porous ceramic materials. *J. Am. Ceram. Soc.* **1993**, *76*, 2745–2752. [CrossRef]

17. Chawla, N.; Shen, Y. Mechanical behavior of particle reinforced metal matrix composites. *Adv. Eng. Mater.* **2001**, *6*, 357–370. [CrossRef]
18. Chawla, N.; Ganesh, V.; Wunsch, B. Three-dimensional (3D) microstructure visualization and finite element modeling of the mechanical behavior of SiC particle reinforced aluminum composites. *Scr. Mater.* **2004**, *51*, 161–165. [CrossRef]
19. Dong, C. Effects of process-induced voids on the properties of fibre reinforced composites. *J. Mater. Sci. Technol.* **2016**, *32*, 597–604. [CrossRef]
20. Deng, X.; Koopman, M.; Chawla, K.K.; Scarritt, S. Measurement and prediction of Young's modulus of a Pb-free solder. *J. Mater. Sci. Mater. Electron.* **2004**, *15*, 385–388.
21. Kukshal, V.; Gangwar, S.; Patnaik, A. Experimental and finite element analysis of mechanical and fracture behavior of SiC particulate filled A356 alloy composites: Part I. *J. Mater. Des. Appl.* **2015**, *229*, 91–105. [CrossRef]
22. Hashim, J.; Looney, M.; Hashmi, M.S.J. The enhancement of wettability of SiC particles in cast aluminum matrix composites. *J. Mater. Process. Technol.* **1999**, *1*, 329–335.
23. Ray, S. Review Synthesis of cast metal matrix particulate composite. *J. Mater. Sci.* **1993**, *28*, 539–5413. [CrossRef]
24. Sun, C.; Song, M.; Wang, Z.; He, Y. Effect of particle size on the microstructures and mechanical properties of SiC-reinforced pure aluminum composites. *J. Mater. Eng. Perform.* **2011**, *20*, 1606–1612. [CrossRef]
25. Koopman, M.; Chawla, K.K.; Coffin, C.; Patterson, B.R.; Deng, X.; Patel, B.V.; Fang, Z.; Lockwood, G. Determination of elastic constants inWC/Co metal matrix composites by resonant ultrasound spectroscopy and impulse excitation. *Adv. Eng. Mater.* **2002**, *4*, 37–42. [CrossRef]
26. Eroshkin, O.; Tsukrov, I. On micromechanical modeling of particulate composites with inclusions of various shapes. *Int. J. Solids Struct.* **2005**, *42*, 409–427. [CrossRef]
27. Fish, J.; Belytschko, T. *A Fist Course in Finite Elements*; John Wiley and Sons, Ltd.: Hoboken, NJ, USA, 2007.
28. Sun, C.T.; Vaidya, R.S. Prediction of composite properties from a representative volume element. *Compos. Sci. Technol.* **1996**, *56*, 171–179. [CrossRef]
29. Tillmann, W.; Klusemann, B.; Nebel, J.; Svendsen, B. Analysis of the mechanical properties of an Arc-sprayed WC-FeCSiMn coating: Nanoindentation and simulation. *J. Therm. Spray Technol.* **2011**, *20*, 328–335. [CrossRef]
30. Ganesh, V.V.; Chawla, N. Effect of particle orientation anisotropy on the tensile behavior of metal matrix composites: Experiments and microstructure-based simulation. *Mater. Sci. Eng. A* **2005**, *391*, 342–353. [CrossRef]
31. Clayne, T.W.; Withers, P.J. *An Introduction to Metal Matrix Composites*; Cambridge University Press: Cambdrige, UK, 1995.
32. McDaniels, D.L. Analysis of stress-strain, fracture, and ductility behavior of aluminum matrix composites containing discontinuous silicon carbide reinforcement. *Metall. Trans. A* **1985**, *16*, 1105–1115. [CrossRef]
33. Upadhy, A.; Ramvir, S. Prediction of effective elastic modulus of biphasic composite materials. *Mod. Mech. Eng.* **2012**, *2*, 6–13. [CrossRef]
34. Qu, S.G.; Lou, H.S.; Li, X.Q. Influence of particle size distribution on properties of SiC particles reinforced aluminum matrix composites with high SiC particle content. *J. Compos. Mater.* **2016**, *50*, 1049–1058. [CrossRef]
35. Sherif El-Eskandarany, M. Mechanical solid state mixing for synthesizing of SiCp/Al nanocomposites. *J. Alloys Compd.* **1998**, *279*, 263–271. [CrossRef]
36. Santhosh Kumar, S.; Seshu Bai, V.; Rajkumar, K.V.; Sharma, G.K.; Jayakumar, T.; Rajasekharan, T. Elastic modulus of Al–Si/SiC metal matrix composites as a function of volume fraction. *J. Phys. D Appl. Phys.* **2009**, *42*, 175504. [CrossRef]
37. Liaw, P.K.; Shannon, R.E.; Clark, W.G.; Harrigan, W.C.; Jeong, H.; Hsu, D.K. Nondestructive characterization of material properties of metal-matrix composites. *Mater. Chem. Phys.* **1995**, *39*, 220–228. [CrossRef]

Article

Pure Aluminum Structure and Mechanical Properties Modified by Al₂O₃ Nanoparticles and Ultrasonic Treatment

Ilya A. Zhukov [1], Alexander A. Kozulin [1], Anton P. Khrustalyov [1,2,*], Nikolay I. Kahidze [1], Marina G. Khmeleva [1], Evgeny N. Moskvichev [1,2], Dmitry V. Lychagin [1] and Alexander B. Vorozhtsov [1]

[1] National Research Tomsk State University, Lenin av. 36, 634050 Tomsk, Russia; gofra930@gmail.com (I.A.Z.); kzln2015@yandex.ru (A.A.K.); nick200069@yandex.ru (N.I.K.); khmelmg@gmail.com (M.G.K.); em_tsu@mail.ru (E.N.M.); lychagindv@mail.ru (D.V.L.); abv1953@mail.ru (A.B.V.)

[2] Institute of Strength Physics and Materials Science of the Siberian Branch of the Russian Academy of Sciences, pr. Akademicheskiy 2/4, 634055 Tomsk, Russia

* Correspondence: tofik0014@mail.ru; Tel.: +7-952-155-55-68

Received: 13 October 2019; Accepted: 6 November 2019; Published: 7 November 2019

Abstract: This paper examines dispersion hardened alloys based on commercial-purity aluminum obtained by permanent mold casting with the addition of aluminum oxide nanoparticles. Ultrasonic treatment provides a synthesis of non-porous materials and a homogeneous distribution of strengthening particles in the bulk material, thereby increasing the mechanical properties of pure aluminum. It is shown that the increase in the alloy hardness, yield stress, ultimate tensile strength, and lower plasticity depend on the average grain size and a greater amount of nanoparticles in the alloy.

Keywords: aluminum; alumina nanoparticles; microstructure; mechanical properties; elastic limit; strength

---

1. Introduction

The development of aerospace and automotive industries is accompanied by increased requirements for the materials applied, in particular, high specific strength, workability, and low cost [1,2]. Conventional aluminum alloys do not fully meet these requirements. For example, zirconium- and scandium-containing alloys currently used in industry possess the appropriate level of mechanical properties but are rather expensive [3–5]. Consequently, new, lightweight metal-based materials with the required properties are being intensively developed.

The use of metal matrix composites seems to be relevant [1,6,7] because the mechanical properties of such composites can be improved by the addition of high-melting particles, including ceramic nanoparticles [8–13]. There are different methods of manufacturing such materials, but casting technologies are the most universal and efficient [14,15]. At the same time, the production of composite materials using casting technology is connected with a spectrum of problems concerning particle agglomeration and flotation due to low wettability with melt [16–19]. As a result, the composite density lowers, and the structure becomes non-homogeneous, leading to a decrease in the mechanical properties of the material. This problem can be solved by several methods, namely the coating deposition onto particles to enhance wettability [20], the addition of a master alloy, and the exposure of molten metal to external effects. One such effect is the ultrasonic melt treatment. As is known, the ultrasonic treatment provides melt degassing and also allows the introduction and distribution of nanoparticles owing to their wettability and deagglomeration. These uniformly distributed nanoparticles promote structural refinement and harden the material. Structural refinement occurs due to the melt overcooling around the introduced particles on which grains start to grow. Dispersion hardening is achieved by the particle's ability to resist dislocation motion during deformation. This requires additional energy to overcome these barriers.

In most publications focused on the investigation of the nanoparticle effect on the structure and properties of aluminum, a small amount (not over 0.5 wt.%) of additives was used alongside base alloys with complex chemical compositions. For example, Vorozhtsov et al. [21] introduced non-metallic nanoparticles via an A356 aluminum alloy. This allowed them to significantly increase the yield stress, ultimate tensile strength, and plasticity of the alloy. The difference in the thermal-expansion coefficients of the matrix and nanoparticles made the most important contribution to the improvement of the mechanical properties of the alloy. At the same time, the presence of other elements in alloys prevents one from fully appreciating the contribution of nanoparticles in matrix deformation and failure.

The aim of this work is to explore the influence of different amounts of aluminum oxide (Al_2O_3) nanoparticles on the structure and mechanical properties of pure aluminum.

2. Materials

2.1. Preparation of Nanoparticles

Al_2O_3 nanoparticles were used in this experiment as a hardener. They were obtained by the electrical explosion of wire [22]. The structure, particle distribution histogram, and phase composition of the initial Al_2O_3 nanoparticles are shown in Figure 1.

Al_2O_3 nanoparticles and aluminum powder (20 μm) were mixed with the addition of 200 mL petroleum ether and 1.5 wt.% octadecanoic acid as a superficially active substance to provide deagglomeration and homogeneous distribution of the nanoparticles. After 20 min of mixing, the powder composition was air-dried and then sifted. The powder composition was wrapped in foil to achieve a cylindrical package. The prepared powder mixture was then introduced to the melt.

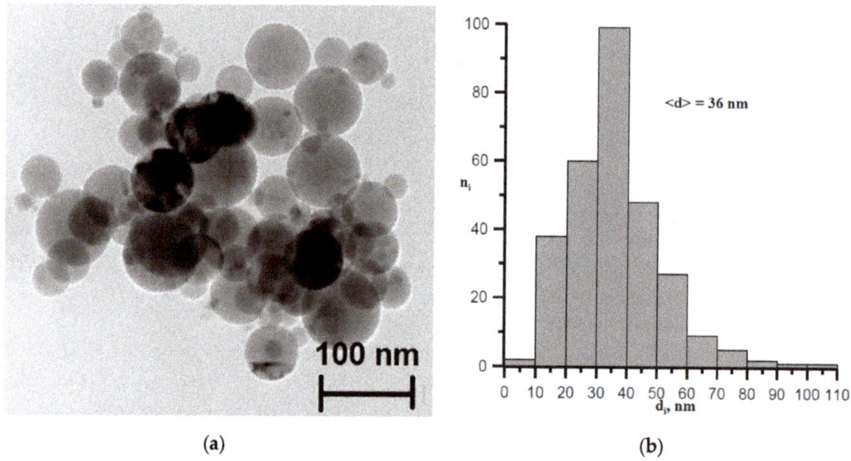

(a)   (b)

Figure 1. *Cont.*

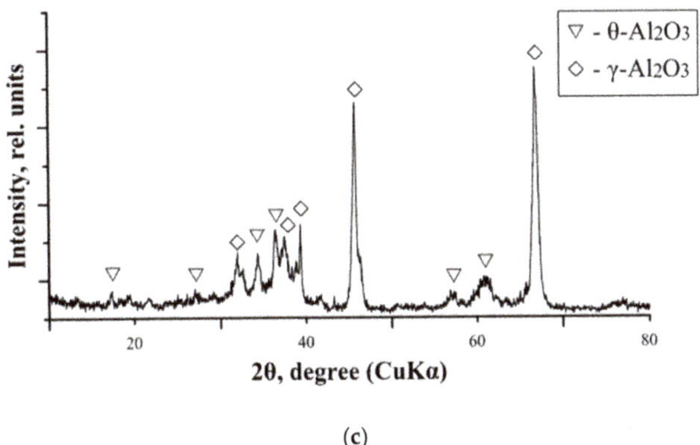

(c)

Figure 1. TEM image of the structure (**a**), particle distribution histogram (**b**), and powder X-ray diffraction [22] (**c**) of the Al_2O_3 nanoparticles.

2.2. Alloy Casting

The 1100 aluminum alloy (99.0 wt.% Al, 0.8 wt.% Si, 0.04 wt.% Mg) castings were prepared in a laboratory-scale melter with the addition of master alloy containing ceramic Al_2O_3 nanoparticles. The ultrasonic treatment (UST) for 2 min, and the addition of the master alloy was carried out simultaneously at 730 °C. The aluminum alloy melt was prepared at 1003 K in an electric furnace, in a clay graphite crucible. The permanent mold casting in a $100 \times 150 \times 10$ mm^3 steel die was then used to obtain the cast products. Prior to experimental research, the cast products were annealed at 300 °C for one hour. Alloys with nanoparticles of aluminum oxide were also obtained without using ultrasonic treatment (no UST).

3. Methods

The electron backscatter diffraction (EBSD) technique for a Tescan Vega II LMU scanning electron microscope (SEM) (TESCAN ORSAY HOLDING, Brno, Czech Republic) was used to study the grain structure and orientation of the alloy surface. The surface was prepared by mechanical polishing and ion milling on a SEMPrep2 multifunction device (Technoorg Linda Co. Ltd., Budapest, Hungary). The EBSD analysis of the obtained data was performed via HKL Channel 5 software (Oxford Instruments, High Wycombe, UK).

In accordance with the ASTM E-8M-08 standard, the plate-like samples had 25 mm gauge length, 2 mm thickness, 6 mm width, and 14 mm spherical radius and were obtained via electrical discharge machining. The samples were used for uniaxial tension testing on an Instron 3369 Dual Column Tabletop Testing System (Instron European Headquarters, HighWycombe, UK) conducted at 24 °C with a 0.001 s^{-1} strain rate.

The Brinell hardness test method was used to determine the Brinell hardness, as defined in the ASTM E103 standard, using a wide sample surface in different places. For hardness measurements, a Duramin 500 (Struers GmbH, Ballerup, Denmark) hardness tester was used. Measurements were carried out at a 2500 N indentation load over 30 s. More than 20 measurements were performed for each sample. The sample surface was prepared in accordance with the standard procedures for abrasive machining.

4. Experimental Results

The grain structure and the grain size distribution in the studied alloys are illustrated in Figure 2. The average grain size of the initial Al 1100 alloy is 200 μm. The grain size of 112 and 69 μm belongs to the dispersion-hardened alloys with 0.5 and 1 wt.% Al_2O_3, respectively.

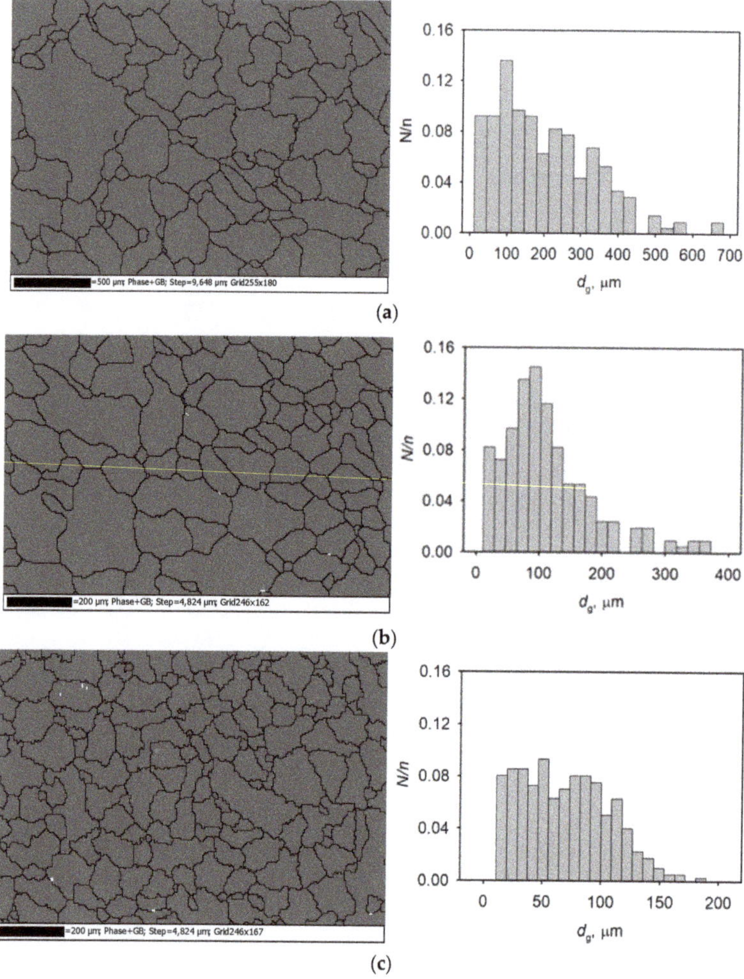

Figure 2. EBSD images of the structure and block diagrams of the grain size distribution in the alloy samples: (**a**) initial state, (**b**) 0.5 wt.% Al_2O_3, (**c**) 1 wt.% Al_2O_3. d_g is the grain size, and N/n is the ratio between the number of measurements and the number of grains of a certain size.

The results of the Brinell hardness tests are shown in Figure 3.

As can be seen from Figure 3, the Brinell hardness increases with an increasing content of hardening nanoparticles and decreasing average grain size. Its maximum value of 21.7 HBW is observed after the addition of 1 wt.% Al_2O_3. The increased content of 1.5 wt.% Al_2O_3 does not change the alloy hardness. The addition of 0.1 wt.% Al_2O_3 without ultrasonic treatment produces an insignificant hardness increase from 18.5 to 19.6 HBW. A further increase in Al_2O_3 content lowers the alloy hardness because the nanoparticle distribution in aluminum cannot be reached without ultrasonic treatment.

The stress–strain curves resulting from the uniaxial tensile tests are presented in Figure 4.

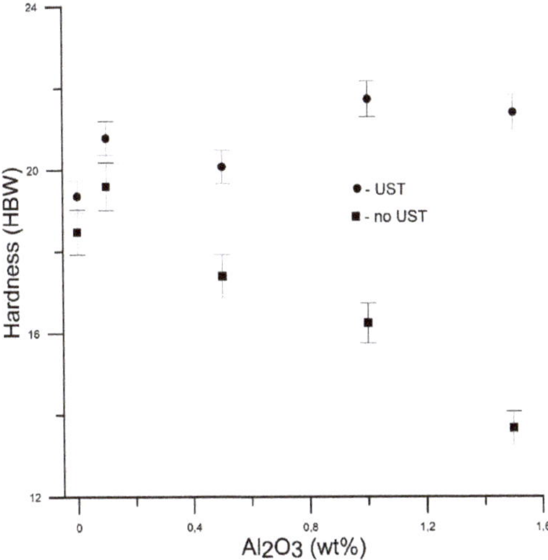

Figure 3. The dependence between the Brinell hardness and Al$_2$O$_3$ content in the 1100 aluminum alloy.

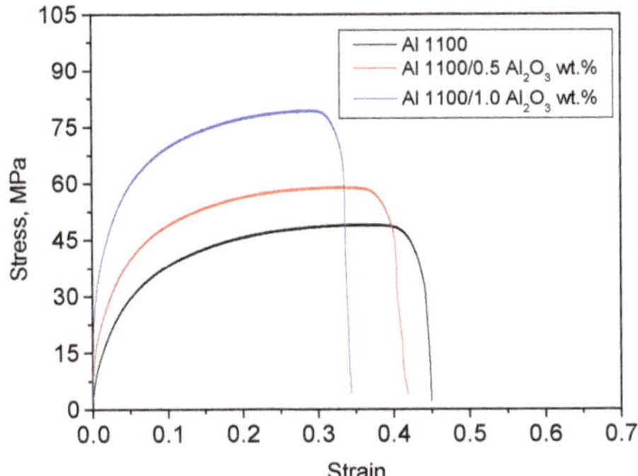

Figure 4. Stress–strain curves obtained for the tensile test samples from investigated alloys: Al 1100: Initial state; Al 1100/0.5 wt.% Al$_2$O$_3$; Al 1100/1 wt.% Al$_2$O$_3$.

Table 1 presents the mechanical properties of aluminum alloys with different amounts of Al$_2$O$_3$ nanoparticles. According to this table, when the content of Al$_2$O$_3$ nanoparticles increased to 1 wt.%, the yield stress (YS) and the ultimate tensile strength (UTS) grew from 12 to 27 MPa and from 48.75 to 79.1 MPa, respectively. The growth of YS and UTS led the plasticity to reduce from 45% to 34.4%.

Table 1. Mechanical properties of the investigated alloys.

Alloy Composition	$P \times 10^3$, g/mm^3	YS, MPa	UTS, MPa	ε_{max}, %	Brinell Hardness
Initial state	2.68	12.08	48.75	45	19.36
0.5 wt.% Al$_2$O$_3$	2.66	16.36	58.85	41.8	20.08
1 wt.% Al$_2$O$_3$	2.69	27.40	79.10	34.4	21.73

Figure 5 shows the yield stress (YS) and the ultimate tensile stress (UTS), which are dependent on the concentration of Al$_2$O$_3$ nanoparticles in the 1100 aluminum alloy.

As shown in Figure 5, the addition of Al$_2$O$_3$ nanoparticles with ultrasonic treatment in the amount of 0.1 wt.% significantly increased both the yield stress and the ultimate tensile stress. Al$_2$O$_3$ nanoparticles introduced in an amount over 0.1 wt.% decreases the yield stress. The ultimate tensile stress is the same. The addition of Al_2O_3 nanoparticles without ultrasonic treatment does not allow enhancement of the mechanical properties of the alloy. One can see the reduction in the yield stress from 8 to 4 MPa, the ultimate tensile stress from 35 to 16 MPa, and the plasticity from 35% to 5%.

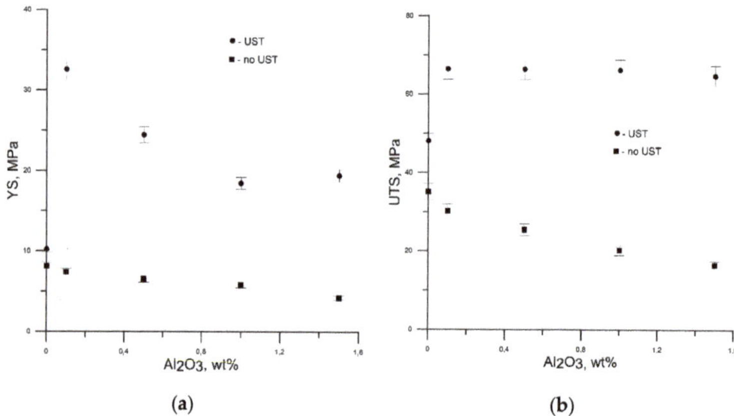

Figure 5. The dependencies of (**a**) the yield stress (YS) and (**b**) the ultimate tensile stress (UTS) depending on the Al_2O_3 nanoparticle concentration in the 1100 aluminum alloy.

5. Discussion

The investigation results of the alloy microstructure show that the presence of Al_2O_3 nanoparticles in the melt affects the grain size. This is because the melt–nanoparticle system is in a low-stability state. Therefore, any thermal effects on the grain nucleation from inoculant particles modify the aggregate state of the melt on their surface. As a result, the matrix starts to crystallize, and decreases the size of the grain nucleus, thus forming a lower grain size. A homogeneous distribution of inoculant particles in the melt provides the formation of a uniform structure in the bulk material.

The significant improvement of the mechanical properties of pure aluminum can be explained by the influence of four hardening mechanisms: The Hall–Petch relationship, Orowan strengthening, the particle-to-matrix load transfer, and the material hardening owing to the difference between the thermal-expansion coefficients of the particles and the aluminum matrix.

5.1. Hall–Petch Relationship

The contribution of grain refinement to the improvement of the mechanical properties of the material can be calculated by the Hall–Petch relationship [23]:

$$\sigma_{GR} = k_y(D^{-\frac{1}{2}} - D_0^{\frac{1}{2}}), \tag{1}$$

where D and D_0 are the grain size, and k_y is the Hall–Petch coefficient (~68 MPa).

In accordance with the Hall–Petch relationship, the addition of 0.5 and 1 wt.% Al_2O_3 enhances the mechanical properties of pure aluminum by 2.4 and 3.7 MPa, respectively.

5.2. Orowan Strengthening

The Orowan strengthening mechanism implies the direct influence of homogeneously distributed solid particles on precipitation hardening. These particles impede the smooth motion of dislocations through the matrix over its slip planes. When the particle size is greater than 1 µm, this effect can be neglected [1]. The increase in precipitation hardening is in inverse proportion to the distance between particles. Thus, the smaller the particles, the stronger the hardening. The latter can be achieved in an inverse proportion to the volume fraction of particles. The contribution of Orowan [24,25] strengthening to the mechanical properties of the alloy can be obtained from

$$\sigma_{OR} = \frac{0.13bG_m}{\lambda} ln \frac{d_p}{2b} \tag{2}$$

where λ is the geometry of the distance between particles. This value is calculated as

$$\lambda = d_p((\frac{1}{2V_p})^{\frac{1}{3}} - 1) \tag{3}$$

where b is the Burgers vector, G_m is the shear modulus, d_p is the mean diameter of nanoparticles, and V_p is the volume fraction of the particles.

In accordance with Orowan strengthening, the addition of 0.5 and 1 wt.% Al_2O_3 allows us to improve the mechanical properties of pure aluminum, respectively, by 19.7 and 36.4 MPa.

5.3. Particle-to-Matrix Load Transfer

The particle-to-matrix load transfer is caused by the direct particle contribution to the hardening process. This particle contribution depends on the amount in the alloy and can be written as [16]

$$\sigma_{load} = 0.5V_p\sigma_m, \tag{4}$$

where V_p is the volume fraction of Al_2O_3 nanoparticles, and σ_m is the yield stress of the matrix alloy (12 MPa).

The calculated load transfer is 2.16 and 4.32 MPa for 0.5 and 1 wt.% Al_2O_3, respectively.

5.4. Difference between the Thermal-Expansion Coefficients of the Particles and Aluminum Matrix

When the alloy temperature drops to room temperature, volume inconsistency between the matrix and strengthening particles can be produced by the difference between their thermal-expansion coefficients. This subsequently leads to the formation of dislocations around the strengthening particles. The length of the generated dislocation loop is determined by the value of πd_p (d_p is the particle size). Material hardening due to the difference in thermal-expansion coefficients can then be evaluated as [16]

$$\Delta\sigma_{CTE} = \beta Gb(\frac{12(\alpha_m - \alpha_p)\Delta TV_p}{bd_p(1 - V_p)})^{\frac{1}{2}} \tag{5}$$

where β is the constant (~1.25), α_m is the thermal-expansion coefficient of the Al matrix (23×10^{-6} 1/K), α_p is the thermal-expansion coefficient of the strengthening particles (7×10^{-6} 1/K), and ΔT is the difference between the synthesis (725 °C) and room (25 °C) temperatures. The value of d_p is used here as an average distance between the particles (~1.8 µm) calculated for the alloy sample with 1 wt.% Al_2O_3. This is shown in Figure 6.

$$G = 0.5E_m/(1 + \vartheta) \tag{6}$$

where E_m is the Young's modulus for aluminum (70 GPa), v is the Poisson ratio (0.33), and b is the Burgers vector (0.286 nm). The difference in the thermal-expansion coefficients calculated for the alloy sample with 1 wt.% Al_2O_3 is 20 MPa. Based on the results obtained, it can be concluded that in introducing Al_2O_3 nanoparticles to pure aluminum, the Orowan strengthening and the difference in a uniform thermal-expansion coefficients of the matrix and the particles dominate among other indicated mechanisms of precipitation hardening.

For pure aluminum, the picture is observed when the introduction of nonmetallic inclusions increases the yield strength, tensile strength, and decreases the plasticity, which corresponds to the results presented in [2,14,26,27]. Moreover, the introduction of nanoparticles, including aluminum oxide, into aluminum alloys can lead to a simultaneous increase in plasticity [21,28–30]. All this indicates that the alloying elements contained in aluminum alloys significantly change the deformation process. Nanoparticles can lead to atypical dependences of changes in mechanical properties alloys but not pure metals. In the future, it will be necessary to consider the contribution of alloying elements and nanoparticles to the formation of the mechanical properties of the alloy.

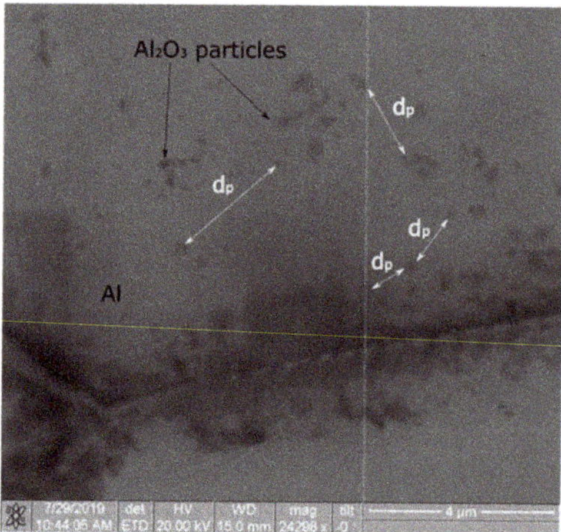

Figure 6. SEM image of the alloy sample with 1 wt.% Al_2O_3.

6. Conclusions

This work has clearly shown that the addition of Al_2O_3 nanoparticles to commercial-purity aluminum promoted its structural refinement and reduced the average grain size from 200 to 69 μm.

This grain refinement and the addition of Al_2O_3 nanoparticles allowed us to increase the alloy hardness from 19.36 to 21.73 HBW, the yield stress from 12.08 to 27.4 MPa, and the ultimate tensile strength from 48.75 to 79.1 MPa. It was shown that the mechanical properties of pure aluminum depend mostly on two hardening mechanisms: The Orowan strengthening and the difference in thermal-expansion coefficients of the matrix and the particles. The contribution of these two mechanisms to precipitation hardening of the alloy sample with 1 wt.% Al_2O_3 was 36 and 20 MPa, respectively.

Author Contributions: Data curation, I.A.Z.; formal analysis, A.A.K., A.P.K. and M.G.K.; funding acquisition, I.A.Z.; investigation, A.A.K.; methodology, A.P.K.; supervision, A.B.V.; visualization, N.I.K., E.N.M. and D.V.L.

Funding: This research was funded by Russian Foundation for Basic Research grant number 18-38-20081.

Acknowledgments: The reported study was funded by RFBR according to the research project No 18-38-20081.

Conflicts of Interest: The authors declare no conflict of interest.

References

1. Malaki, M.; Xu, W.; Kasar, A.K.; Menezes, P.L.; Dieringa, H.; Varma, R.S.; Gupta, M. Advanced Metal Matrix Nanocomposites. *Metals* **2019**, *9*, 330. [CrossRef]
2. Promakhov, V.V.; Khmeleva, M.G.; Zhukov, I.A.; Platov, V.V.; Khrustalyov, A.P.; Vorozhtsov, A.B. Influence of Vibration Treatment and Modification of A356 Aluminum Alloy on Its Structure and Mechanical Properties. *Metals* **2019**, *9*, 87. [CrossRef]
3. De Luca, A.; Dunand, D.C.; Seidman, D.N. Mechanical properties and optimization of the aging of a dilute Al-Sc-Er-Zr-Si alloy with a high Zr/Sc ratio. *Acta Mater.* **2016**, *119*, 35–42. [CrossRef]

4. Spierings, A.B.; Dawson, K.; Heeling, T.; Uggowitzer, P.J.; Schäublin, R.; Palm, F.; Wegener, K. Microstructural features of Sc-and Zr-modified Al-Mg alloys processed by selective laser melting. *Mater. Des.* **2017**, *115*, 52–63. [CrossRef]

5. Gao, T.; Ceguerra, A.; Breen, A.; Liu, X.; Wu, Y.; Ringer, S. Precipitation behaviors of cubic and tetragonal Zr–rich phase in Al–(Si–) Zr alloys. *J. Alloys Compd.* **2016**, *674*, 125–130. [CrossRef]

6. Shirvanimoghaddam, K.; Hamim, S.U.; Akbari, M.K.; Fakhrhoseini, S.M.; Khayyam, H.; Pakseresht, A.H.; Ghasali, E.; Zabet, M.; Munir, K.S.; Jia, S.; et al. Carbon fiber reinforced metal matrix composites: Fabrication processes and properties. *Compos. Part A Appl. Sci. Manuf.* **2017**, *92*, 70–96. [CrossRef]

7. Chen, B.; Shen, J.; Ye, X.; Jia, L.; Li, S.; Umeda, J.; Takahashi, M.; Kondoh, K. Length effect of carbon nanotubes on the strengthening mechanisms in metal matrix composites. *Acta Mater.* **2017**, *140*, 317–325. [CrossRef]

8. Maruyama, B.; Hunt, W.P. Discontinuously reinforced aluminum: Current status and future direction. *JOM* **1999**, *11*, 59–61. [CrossRef]

9. Umasankar, V.; Xavior, M.A.; Karthikeyan, S. Experimental evaluation of the influence of processing parameters on the mechanical properties of SiC particle reinforced AA6061 aluminium alloy matrix composite by powder processing. *J. Alloys Compd.* **2014**, *582*, 380–386. [CrossRef]

10. Khrustalyov, A.P.; Garkushin, G.V.; Zhukov, I.A.; Razorenov, S.V.; Vorozhtsov, A.B. Quasi-Static and Plate Impact Loading of Cast Magnesium Alloy ML5 Reinforced with Aluminum Nitride Nanoparticles. *Metals* **2019**, *9*, 715. [CrossRef]

11. Mirza, F.A.; Chen, D.L. A Unified Model for the Prediction of Yield Strength in Particulate-Reinforced Metal Matrix Nanocomposites. *Materials* **2015**, *8*, 5138–5153. [CrossRef] [PubMed]

12. Dieringa, H.; Katsarou, L.; Buzolin, R.; Szakács, G.; Horstmann, M.; Wolff, M.; Mendis, C.; Vorozhtsov, S.; StJohn, D. Ultrasound Assisted Casting of an AM60 Based Metal Matrix Nanocomposite, Its Properties, and Recyclability. *Metals* **2017**, *7*, 388. [CrossRef]

13. Casati, R.; Vedani, M. Metal Matrix Composites Reinforced by Nano-Particles—A Review. *Metals* **2014**, *4*, 65–83. [CrossRef]

14. Akbari, M.K.; Baharvandi, H.R.; Shirvanimoghaddam, K. Tensile and fracture behavior of nano/micro TiB2 particle reinforced casting A356 aluminum alloy composites. *Mater. Des.* **2015**, *66*, 150–161. [CrossRef]

15. Katsarou, L.; Mounib, M.; Lefebvre, W.; Vorozhtsov, S.; Pavese, M.; Badini, C.; Molina-Aldareguia, J.M.; Jimenez, C.C.; Pérez Prado, M.T.; Dieringa, H. Microstructure, mechanical properties and creep of magnesium alloy Elektron21 reinforced with AlN nanoparticles by ultrasound-assisted stirring. *Mater. Sci. Eng. A* **2016**, *659*, 84–92. [CrossRef]

16. Sreekumar, V.M.; Babu, N.H.; Eskin, D.G. Prospects of In-Situ α-Al$_2$O$_3$ as an Inoculant in Aluminum: A Feasibility Study. *J. Mater. Eng Perform.* **2017**, *26*, 4166–4176. [CrossRef]

17. Dieringa, H. Properties of magnesium alloys reinforced with nanoparticles and carbon nanotubes: A review. *J. Mater. Sci.* **2011**, *46*, 289–306. [CrossRef]

18. Puga, H.; Costa, S.; Barbosa, J.; Ribeiro, S.; Prokic, M. Influence of ultrasonic melttreatment on microstructure and mechanical properties of AlSi9Cu3 alloy. *J. Mater. Process. Technol.* **2011**, *211*, 1729–1735. [CrossRef]

19. Kudryashova, O.B.; Eskin, D.G.; Khrustalev, A.P.; Vorozhtsov, S.A. Ultrasonic effect on thepenetration of the metallic melt into submicron particles and their agglomerates. *Russ. J. Non-Ferr. Met.* **2017**, *58*, 427–433. [CrossRef]

20. Zhang, F.; Jacobi, A.M. Aluminum surface wettability changes by pool boiling of nanofluids. *Colloids Surf. A Physicochem. Eng. Asp.* **2016**, *506*, 438–444. [CrossRef]

21. Vorozhtsov, S.; Zhukov, I.; Promakhov, V.; Naydenkin, E.; Khrustalyov, A.; Vorozhtsov, A. The Influence of ScF$_3$ Nanoparticles on the Physical and Mechanical Properties of New Metal Matrix Composites Based on A356 Aluminum Alloy. *JOM* **2016**, *68*, 3101–3106. [CrossRef]

22. Zhukov, I.A.; Kozulin, A.A.; Khrustalyov, A.P.; Matveev, A.E.; Platov, V.V.; Vorozhtsov, A.B.; Zhukova, T.V.; Promakhov, V.V. The Impact of Particle Reinforcement with Al2O3, TiB2, and TiC and Severe Plastic Deformation Treatment on the Combination of Strength and Electrical Conductivity of Pure Aluminum. *Metals* **2019**, *9*, 65. [CrossRef]

23. Ramakrishnan, N. An analytical study on strengthening of particulate reinforced metal matrix composites. *Acta Mater.* **1996**, *44*, 69–77. [CrossRef]

24. Zhang, Z.; Chen, D.L. Consideration of Orowan strengthening effect in particulate-reinforced metal matrix nanocomposites: A model for predicting their yield strength. *Scr. Mater.* **2006**, *54*, 1321–1326. [CrossRef]

25. Zhang, Z.; Chen, D.L. Contribution of Orowan strengthening effect in particulate-reinforced metal matrix nanocomposites. *Mat. Sci. Eng. A* **2008**, *483–484*, 148–152. [CrossRef]
26. Shin, S.; Lee, D.; Lee, Y.-H.; Ko, S.; Park, H.; Lee, S.-B.; Cho, S.; Kim, Y.; Lee, S.-K.; Jo, I. High Temperature Mechanical Properties and Wear Performance of B4C/Al7075 Metal Matrix Composites. *Metals* **2019**, *9*, 1108. [CrossRef]
27. Khrustalyov, A.P.; Kozulin, A.A.; Zhukov, I.A.; Khmeleva, M.G.; Vorozhtsov, A.B.; Eskin, D.; Chankitmunkong, S.; Platov, V.V.; Vasilyev, S.V. Influence of Titanium Diboride Particle Size on Structure and Mechanical Properties of an Al-Mg Alloy. *Metals* **2019**, *9*, 1030. [CrossRef]
28. Belov, N.A. Effect of eutectic phases on the fracture behavior of high-strength castable aluminum alloys. *Met. Sci. Heat Treat.* **1995**, *37*, 237–242. [CrossRef]
29. Chen, Z.; Wang, T.; Zheng, Y.; Zhao, Y.; Kang, H.; Gao, L. Development of TiB2 reinforced aluminum foundry alloy based in situ composites—Part I: An improved halide salt route to fabricate Al–5 wt%TiB2 master composite. *Mat. Sci. Eng. A* **2014**, *605*, 301–309. [CrossRef]
30. Vorozhtsov, S.A.; Eskin, D.G.; Tamayo, J.; Vorozhtsov, A.B.; Promakhov, V.V.; Averin, A.A.; Khrustalyov, A.P. The application of external fields to the manufacturing of novel dense composite master alloys and aluminum-based nanocomposites. *Metal. Mater. Trans. A* **2015**, *46*, 2870–2875. [CrossRef]

Article

Influence of Titanium Diboride Particle Size on Structure and Mechanical Properties of an Al-Mg Alloy

Anton P. Khrustalyov [1,*], Alexander A. Kozulin [1], Ilya A. Zhukov [1], Marina G. Khmeleva [1], Alexander B. Vorozhtsov [1], Dmitry Eskin [1,2], Suwaree Chankitmunkong [2,3], Vladimir V. Platov [1] and Sergey V. Vasilyev [4]

[1] Faculty of Physics and Engineering, National Research Tomsk State University, 634050 Tomsk, Russia; kzln2015@yandex.ru (A.A.K.); gofra930@gmail.com (I.A.Z.); khmelmg@gmail.com (M.G.K.); abv1953@mail.ru (A.B.V.); vova.platov.85@mail.ru (V.V.P.); dmitry.eskin@brunel.ac.uk (D.E.)

[2] Brunel Centre for Advanced Solidification Technology, Brunel University London, Uxbridge, Middlesex UB8 3PH, UK; suwaree.03@mail.kmutt.ac.th

[3] Department of Production Engineering, King Mongkut's University of Technology Thonburi, Bangkok 10140, Thailand

[4] Joint-Stock Company "Scientific-production concern" Mechanical engineering, 125212 Moscow, Russia; ecamo@mail.ru

* Correspondence: tofik0014@mail.ru; Tel.: +79-52-155-5568

Received: 7 August 2019; Accepted: 20 September 2019; Published: 23 September 2019

Abstract: In the present study, aluminum alloys of the Al-Mg system with titanium diboride particles of different size distribution were obtained. The introduction of particles in the alloy was carried out using master alloys obtained through self-propagating high-temperature synthesis (SHS) process. The master alloys consisted of the intermetallic matrix Al-Ti with distributed TiB_2 particles. The master alloys with TiB_2 particles of different size distribution were introduced in the melt with simultaneous ultrasonic treatment, which allowed the grain refining of the aluminum alloy during subsequent solidification. It was found that the introduction of micro- and nanoparticles TiB_2 increased the yield strength, tensile strength, and plasticity of as-cast aluminum alloys. After pass rolling the castings and subsequent annealing, the effect of the presence of particles on the increase of strength properties is much less felt, as compared with the initial alloy. The recrystallization of the structure after pass rolling and annealing was the major contributor to hardening that minimized the effect of dispersion hardening due to the low content of nanosized titanium diboride.

Keywords: aluminum alloy; titanium diboride; master alloy; structure; mechanical properties

1. Introduction

At present, a wrought AA5056 alloy is widely used in aircraft engineering, maritime transport, and pipeline design due to high corrosion resistance and good weldability by traditional methods [1,2]. The AA5056 alloy is mainly used as sheets. The highest mechanical properties of this alloy are achieved by dispersion hardening with the introduction of elements such as zirconium or scandium [3–5]. The main disadvantage of dispersion hardening is the high cost which leads to a significant increase in the cost of products. Under the production of the rolled metal, additional deformation treatment affects the formation of the internal structure of an alloy that directly influences the change in its mechanical properties. In addition to dispersion hardening and deformation, there are methods to obtain high physical and mechanical properties of aluminum alloys, such as the modification of the structure by grain refining during melt solidification and the hardening of the metal matrix by introducing submicron non-metallic particles [6–9].

To refine or modify the structure, a chemical inoculant is usually introduced into an aluminum alloy. Due to the close parameters of the crystal structure and specific size, inoculants can act as centers of heterogeneous nucleation upon undercooling the liquid metal during its solidification [10]. Titanium diboride (TiB$_2$) is the most widely used grain refiner for aluminum alloys. TiB$_2$ modifier is introduced by the "ex-situ" method using Al-5Ti-1B master alloys [10–12] (containing Al$_3$Ti and TiB$_2$ particles in the aluminum matrix), or is synthesized by the "in-situ" method using K$_2$TiF$_6$ and KBF$_4$ [13]. As shown by recent studies, titanium diboride particles are not sufficiently active centers of solidification by themselves and require the formation of a two-dimensional compound Al$_3$Ti on their surface [10]. It has been shown in [14] that the optimal size of titanium diboride particles is from 1 to 5 μm for their use as inoculants in aluminum alloys upon conventional cooling rates and corresponding levels of melt undercooling. At the same time, it is known that for dispersion hardening of aluminum alloys, involving the Orowan mechanism, it is necessary to use particles with size less than 500 nm, such as aluminum oxide [15,16], silicon carbide [17], etc. These fine non-metallic particles become an obstacle for the moving dislocations in aluminum upon its deformation [18]. In this case, non-metallic particles should be uniformly distributed in the volume of the aluminum matrix, and have a good connection with the matrix which is retained during the dislocation motion. However, there are some problems with the introduction of particles into the metallic melt, related to their agglomeration and flotation due to poor wettability by the liquid metal. To solve these problems, ultrasonic (US) treatment can be used. US treatment allows for the degassing of the melt [19], improves the wettability and the distribution of nanoparticles [20] in the liquid volume. Master alloys with nanoparticle-reinforced aluminum matrix [16] are also used to improve the wettability of nanoparticles with the melt. Titanium diboride can be an effective obstacle to the motion of dislocations in aluminum due to its high hardness and stability [21]. Hence, the motivation for this study is the simultaneous use of titanium diboride particles for both the grain refinement and the dispersion hardening of the structure of aluminum alloys. This can be accomplished through the usage of the master alloys with a given chemical composition containing titanium diboride particles in a bimodal size distribution in nano- and micro-size ranges. For example, it has been shown in [22] that when modifying titanium diboride microparticles and aluminum oxide nanoparticles were introduced separately, the microstructure was refined and the aluminum matrix was reinforced, resulting in the increased strength and electrical conductivity.

One of the possible methods for obtaining the bimodal distribution of titanium diboride particles in a master alloy can be self-propagating high-temperature synthesis (SHS) [23]. In this case, titanium diboride particles in the master alloy are formed during combustion, and the size of titanium diboride particles and the phase composition of the matrix of obtained master alloys can be changed by controlling the process (speed, combustion temperature) of the initial powder system Al-Ti-B. Under the use of the exothermic reaction of stoichiometric ratios of titanium and boron at the production of the master alloys, the method specified can be considered as energy efficient since it does not require additional power sources to initiate and sustain the combustion.

Thus, the aim of this study is to investigate the effect of introducing master alloys containing titanium diboride particles on the structure and mechanical properties of aluminum alloys of the Al-Mg system in the as-cast condition and after deformation.

2. Materials and Methods

2.1. Master Alloys Production

Master alloys (MA) obtained by SHS process from the initial powder mixture of aluminum, titanium, and boron were used for the introduction of titanium diboride particles into the aluminum melt. Morphology, dispersion, chemical composition of powder materials, equipment for obtaining of alloys, and the procedure of SHS testing are described in detail in [22,24]. Here we give briefly some essential data.

The exothermic reaction of the powder systems is characterized by the interaction of titanium and boron, based on the formation of the intermetallic phase Ti_3Al:

$$4Ti + 2B + Al = TiB_2 + Ti_3Al + Q.$$

The SHS process is used to obtain composites. Each individual powder particle can be represented schematically as follows. SEM images of the obtained SHS materials can be found elsewhere [22]. The master alloy contains uniformly distributed particles of titanium diboride in the Ti-Al matrix (see Figure 1).

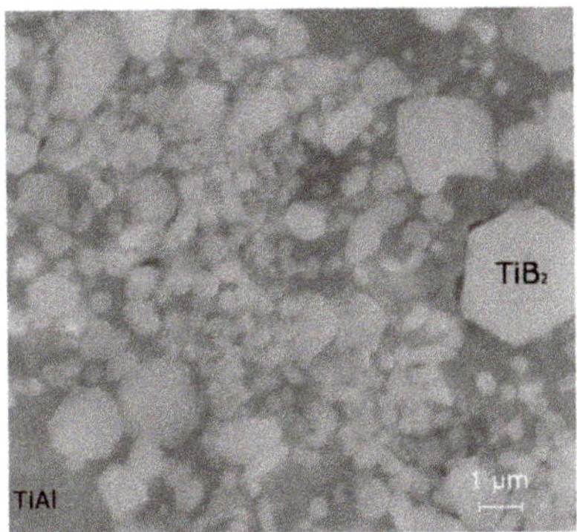

Figure 1. The microstructure of the master alloy [22]

The phase compositions of the obtained master alloys are listed in Table 1, and X-ray patterns are presented in Figures 2–4.

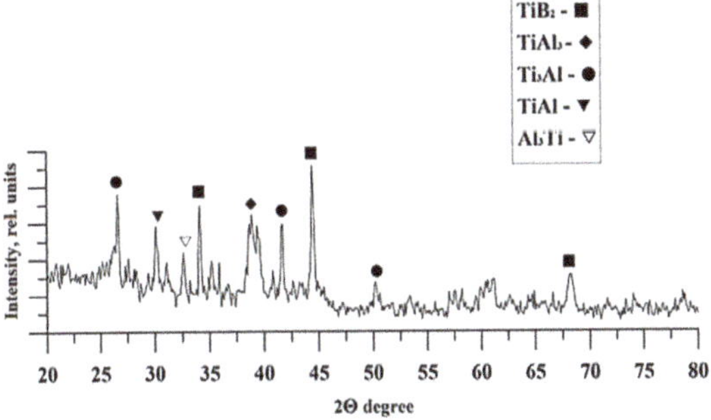

Figure 2. X-ray diffraction diagram of the MA1.

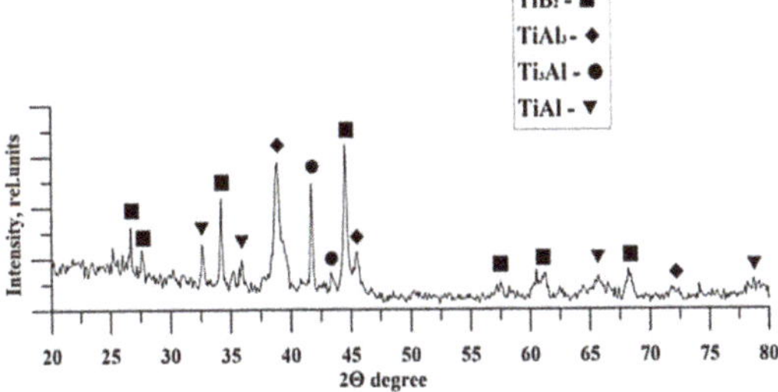

Figure 3. X-ray diffraction diagram of the MA2.

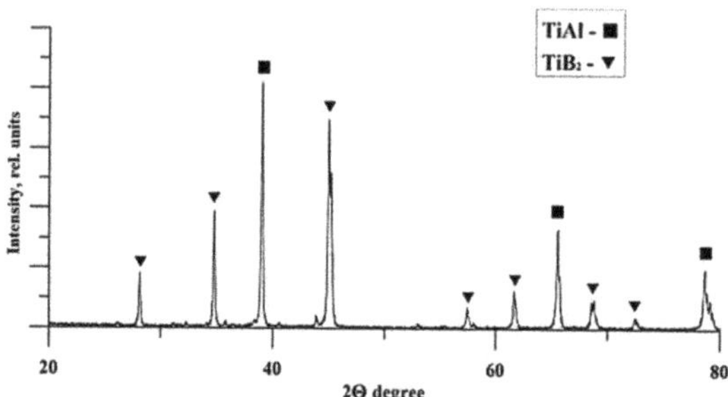

Figure 4. X-ray diffraction diagram of the MA3.

Table 1. Compositions of master alloys.

Master-Alloy	Phase	Phase Content, wt%	Lattice Parameter, Å
	TiB_2	30	a = 3.0296, c = 3.2260
1	Al_3Ti	26	a = 4.0123
	Ti_3Al	9	a = 5.6683, c = 4.5854
	TiAl	35	a = 4.0115
	TiB_2	30	a = 3.0293, c = 3.2257
2	Al_3Ti	40	a = 3.9484, c = 8.4989
	Ti_3Al	14	a = 5.6640, c = 4.6344
	TiAl	16	a = 4.0278
3	TiAl	57	a = 4.0319
	TiB_2	43	a = 3.0140, c = 3.2000

Changes in the lattice parameters of the components in the obtained master alloys were observed. This is due to the high temperature and flow rate of SHS, during which there is a distortion of the lattice and the change of lattice parameters. Due to the different composition of the initial components, the time and speed of the process change, which leads to different lattice parameters for the TiB_2, Al_3Ti, Ti_3Al, and TiAl phases in the master alloys.

The histograms of the particle size distribution for titanium diboride particles are presented in Figure 5. The particle is initially composite. After synthesis, the obtained materials were crushed and milled into powder. The particles of the obtained powder material consisted of an intermetallic matrix of type (Ti-Al) and TiB_2 particles were uniformly distributed in it. The matrix was then etched with a solution of H2O-20%HCl for 72 h. Further TiB_2 particles were mixed with aluminum powder. The resulting mixtures were pressed into tablets with a diameter of 23 mm. The resulting tablets were sintered in a vacuum furnace for 1 h at a temperature of 600 °C.

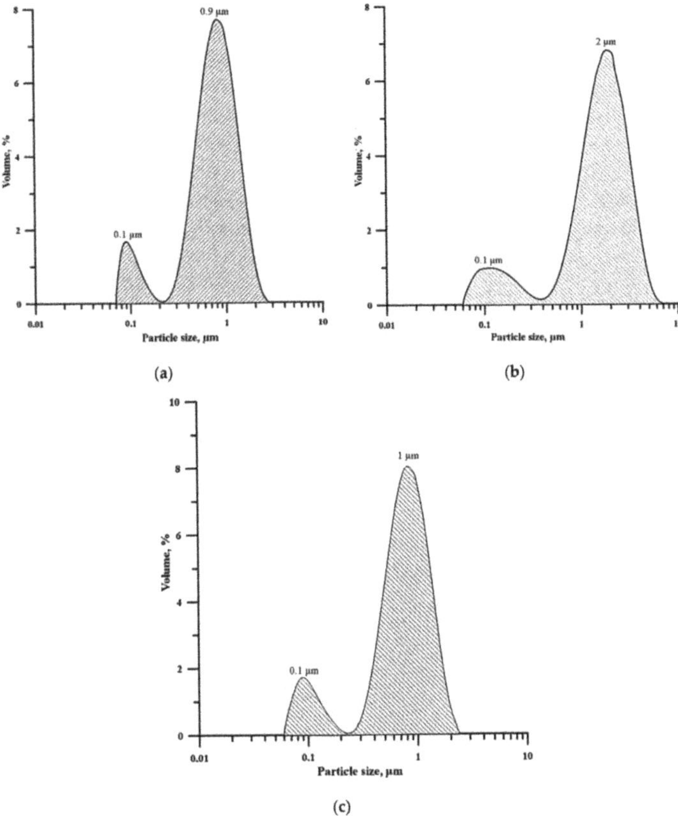

Figure 5. Histograms of the particle distribution in master alloys MA1 (**a**), MA2 (**b**), and MA3 (**c**) as determined in ANALYSETTE 22 MicroTec plus.

2.2. Obtaining the Alloys

An aluminum alloy AA5056 (91.9–94.68 wt% Al, 4.8–5.8 wt% Mg) was used as the starting material. A total of 1 kg of the AA5056 alloy was placed in a clay-graphite crucible, melted in an electric muffle furnace at a temperature of 780 °C, and kept for 2 h. Then the crucible was removed from the furnace by using a holding device, and a master alloy was introduced into the melt with simultaneous ultrasonic treatment at a temperature of 730 °C. Ultrasonic processing was carried out using a magnetostrictive water-cooled transducer at a power of 4.1 kW and a frequency of 17.6 kHz. After complete dissolution of the master alloys ultrasonic treatment continued for a further 2 min. The liquid metal was then placed in the furnace for 30 min, and then ultrasonic treatment was carried out for another 2 min. The melt was poured into a chill mold at a temperature of 720 °C. In addition, AA5056 + MA1 alloys without ultrasonic treatment and an AA5056 alloy with ultrasonic treatment were obtained as references. The data of obtained alloys are given in Table 2.

Table 2. A list of obtained alloys.

Alloy Matrix	Ultrasonic Treatment	Master Alloy	TiB$_2$ Particle Quantity in a Master Alloy, %	TiB$_2$ Particle Quantity in 1 kg of an Alloy
AA5056	+	-	-	-
AA5056	-	MA1	30	$4 \times 10^{20} \pm 6 \times 10^{10}$
AA5056	+	MA1	30	$4 \times 10^{20} \pm 6 \times 10^{10}$
AA5056	+	MA2	32	$4 \times 10^{20} \pm 6 \times 10^{10}$
AA5056	+	MA3	43	$4.5 \times 10^{20} \pm 7.3 \times 10^{10}$

Particle quantity for MA1 was 17 vol.% nano-83 vol.% micro, for MA2 14 vol.% nano-86 vol.% micro, for MA3 18% nano-82% micro.

2.3. Rolling Technique

Prismatic samples with a size of $11 \times 17 \times 40$ mm^3 were machined from the resulting castings. For rolling, a rolling mill with a roll diameter of 80 mm and a rotation speed of 24 rpm was used. Rolling of the aluminum alloys was carried out after preheating the samples in a muffle furnace at 300 °C for 30 min. Rolling was carried out to change the sample thickness from 11 to 2 mm in several cycles with intermediate heating for 15 min at 300 °C. Each cycle was divided into seven reversible passes. A single-pass rolling provided compression up to 4%. The number of passes per cycle was selected experimentally based on the sample temperature during cooling. Rolling at lower temperatures leads to the appearance of defects within the sample volume and the main crack growth during subsequent working cycles regardless of the temperature and deformation modes applied.

2.4. Methods of Analysis of Initial Materials and Alloys

The structures of the obtained materials were investigated through optical microscopy, Olympus GX71 (Olympus Scientific Solutions Americas, Waltham, MA, USA). Samples were subjected to preliminary mechanical polishing, electrolytical etching, and anodization. The electrochemical oxidation of the metallographic specimen surface in a 5% solution of hydrofluoric acid (HBF$_4$) at a voltage of 20 V and a current of 2 A was carried out to identify grain boundaries. Grain sizes were determined by a random linear intercept method from electronic images of the structure.

The phase composition and the structure of master alloys were performed using X-ray phase and X-ray diffraction methods. X-ray diffraction analysis of master alloys was performed using a SHIMADZU XRD 6000 diffractometer (Shimadzu, Tsukinowa, Japan). The phase composition analysis was carried out using PDF 4+ databases, as well as the POWDER CELL 2.4 full-profile analysis program. The particle dispersion was researched on ANALYSETTE 22 MicroTec plus (FRITSCH, Gamburg, Germany) by laser diffraction. Particle dispersion was measured in water. Mechanical tests were performed on a universal testing machine, Instron 3369 (Instron European Headquarters, High Wycombe, UK), at the speed of 0.2 mm/min. The samples were cut from castings and rolling products using electrical discharge machining (Scientific Industrial Corporation DELTA-TEST, Fryazino, Moscow Region, Russia). The samples are shaped as flat double-sided blades with a thickness of 2 mm, the ratio of the width of the working and holding parts is greater or equal to 1.5. Tests were conducted according to ASTM B557-15.

3. Results

The optical micrographs of the microstructures of the AA5056-based alloys are presented in Figure 6.

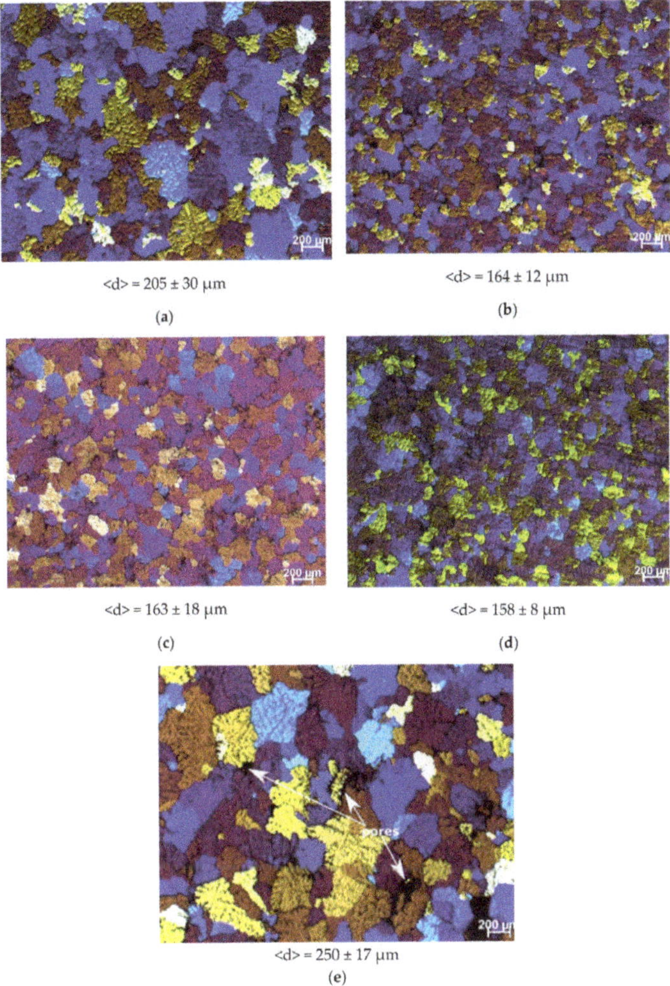

Figure 6. Microstructure of obtained alloys: AA5056 reference ultrasonic (US) (**a**), AA5056 + MA1 US (**b**), AA5056 + MA2 US (**c**), AA5056 + MA3 US (**d**), AA5056 + MA1 without US (**e**).

The average grain size of the initial alloy AA5056 after ultrasonic treatment (AA5056 reference US) was 205 ± 30 μm (Figure 6a). Ultrasonic treatment of the alloy AA5056 allows obtaining the microstructure with equiaxed grains. The introduction of the master alloy MA1 with ultrasonic treatment (AA5056 + MA1 US) reduced significantly the average grain size of the alloy AA5056 from 205 ± 30 to 164 ± 12 μm (Figure 6b). The addition of the master alloy MA2 (AA5056 + MA2 US) had a similar effect, reducing the average grain size to 163 ± 18 μm (Figure 6c), while the master alloy MA3 (AA5056 + MA3 US) reduced the grain size to 158 ± 8 μm (Figure 6d). Grain distribution histograms of the AA5056-based alloys are presented in Figure 7. At the same time, there are grains with size larger than 250 μm in the structure of the alloy AA5056 obtained without US (AA5056 + MA1 without US). The use of the master alloy without ultrasonic processing did not allow for introducing particles and distributing them in the volume of the ingot. Since the alloy had non-uniform distribution of particles, the grain of the alloy AA5056 cannot be refined, and its average size was 250 ± 17 μm and large pores were observed in the structure (see black inclusions in Figure 6e).

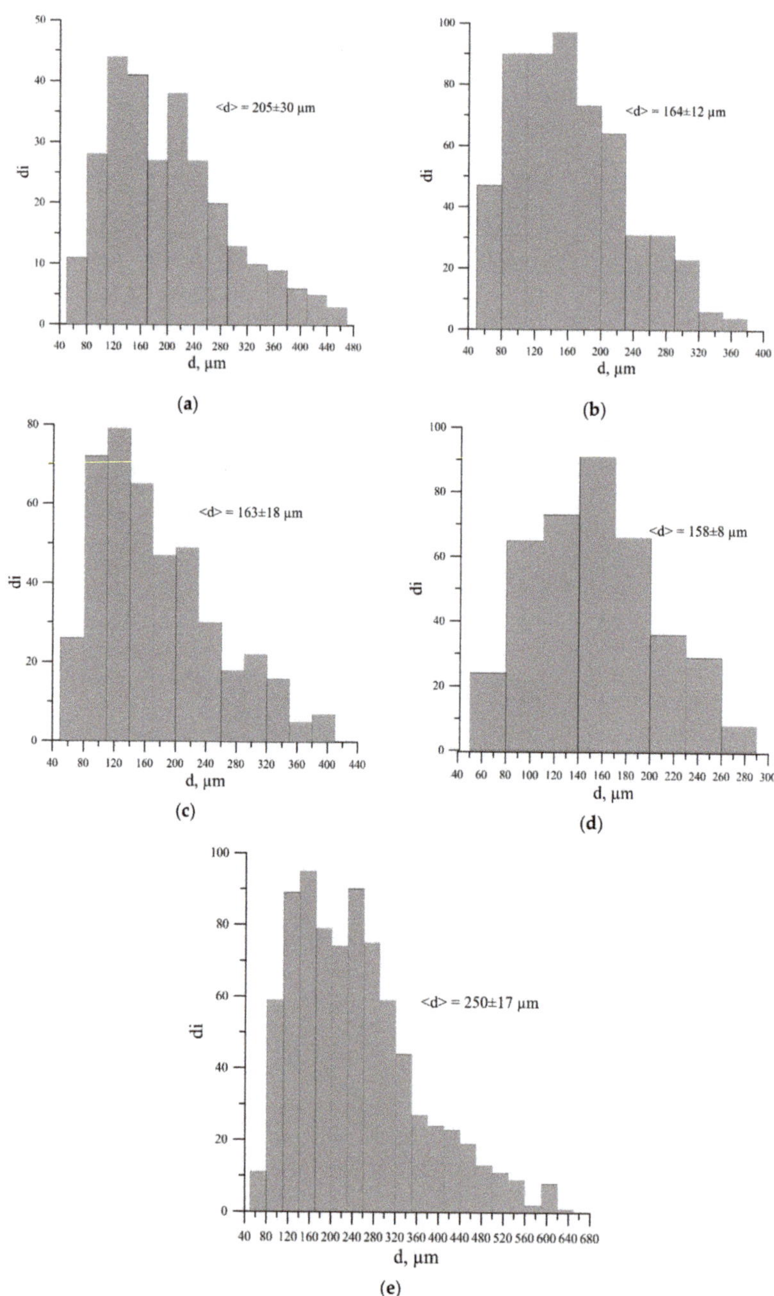

Figure 7. Grain distribution histograms of obtained alloys: AA5056 reference US (**a**), AA5056 + MA1 US (**b**), AA5056 + MA2 US (**c**), AA5056 + MA3 US (**d**), AA5056 + MA1 without US (**e**).

The stress-strain diagrams obtained under uniaxial tension of flat samples from tested AA5056-based aluminum alloys are given in Figure 8.

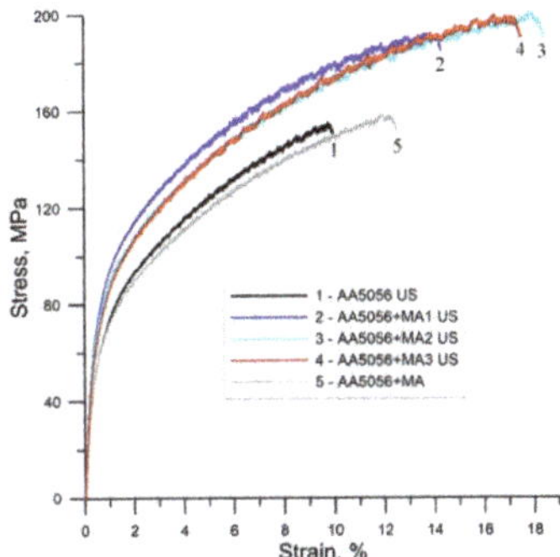

Figure 8. Diagrams of uniaxial tension of AA5056-based cast alloys.

The analysis of the diagrams allowed us to estimate the yield strength, tensile strength, and elongation of the cast AA5056 reference US alloy as 57 MPa, 155 MPa, and 11.5%, respectively (Table 3). After addition of MA1 (AA5056 + MA1 US), its mechanical properties increased significantly: yield strength from 57 to 74 MPa, tensile strength from 155 to 192 MPa, and elongation from 11.5% to 14.5% (Table 3). The introduction of particles from MA2 (AA5056 + MA2 US) also led to the increase in yield strength from 57 to 71 MPa, tensile strength from 155 to 201 MPa, and elongation from 11.5% to 18.8% (Table 3). Finally, the addition of MA3 (AA5056 + MA3 US) did not change the yield strength which amounted 69 MPa, increase tensile strength from 155 to 200 MPa, and elongation from 11.5% to 17.8% (Table 3). Without ultrasonic processing, the addition of MA1 (AA5056 + MA1 without US) into the base alloy decreased the yield strength from 57 to 53 MPa with negligible change in tensile strength from 155 to 159 MPa, and elongation from 11.5% to 12.9%.

Table 3. Mechanical properties of as-cast AA5056-based alloys.

Alloy	$\sigma_{0.2}$, MPa	σ_B, MPa	δ, %
AA5056 US	57 ± 4	155 ± 11	11.5 ± 0.8
AA5056 US + MA1	74 ± 7	192 ± 14	14.5 ± 0.4
AA5056 US + MA2	71 ± 6	201 ± 12	18.8 ± 0.6
AA5056 US + MA3	69 ± 8	200 ± 10	17.8 ± 0.5

The yield strength increases up to 345 MPa under the subsequent deformation by rolling of workpieces obtained from castings of the alloy AA5056 US, and it decreased to 328, 304, 330 MPa with the additions of master alloys MA1 (AA5056 US + MA1), MA2 (AA5056 US + MA2), and MA3 (AA5056 US + MA3), respectively (Table 4).

Table 4. The mechanical properties of deformed AA5056-based alloys.

Alloy	$\sigma_{0.2}$, MPa	σ_B, MPa	δ, %
	Pass rolling		
AA5056 US	345 ± 11	369 ± 16	9.6 ± 0.3
AA5056 US + MA1	328 ± 13	365 ± 17	12.3 ± 0.4
AA5056 US + MA2	304 ± 16	349 ± 19	8.4 ± 0.2
AA5056 US + MA3	330 ± 12	350 ± 17	10 ± 0.3
	Pass rolling + annealing		
AA5056 US	217 ± 9	311 ± 11	22.2 ± 0.3
AA5056 US + MA1	189 ± 10	309 ± 13	17.9 ± 0.1
AA5056 US + MA2	197 ± 7	286 ± 10	18.9 ± 0.2
AA5056 US + MA3	226 ± 8	316 ± 12	22.4 ± 0.1

The introduction of MA1 (AA5056 US + MA1) led to an increase in the elongation from 9.6% to 12.3%, but the strength did not change and equaled 365 MPa (Table 4, Figure 9). The addition of MA2 (AA5056 US + MA2) reduced the elongation to 8.4% and the strength, from 369 to 349 MPa (Table 4, Figure 9). The introduction of MA3 (AA5056 US + MA3) did not reduce the elongation but decreased the tensile strength from 369 to 350 MPa.

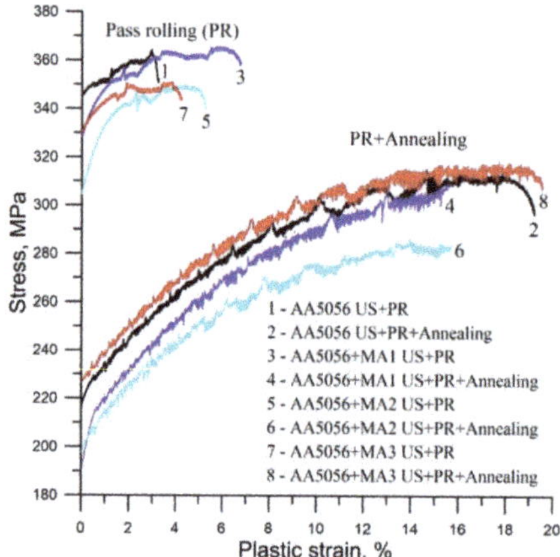

Figure 9. Stress-strain diagrams of aluminum alloys with master alloy additions MA1–MA3 after pass rolling and annealing.

After annealing, the yield strength, tensile strength, and elongation of the AA5056 US alloy were 217 MPa, 311 MPa and 22.2%, respectively. The additions of MA1 (AA5056 US + MA1) or MA2 (AA5056 US + MA2) did not increase mechanical properties after pass rolling and annealing (see Table 4 and Figure 9). The introduction of MA3 (AA5056 US + MA3) did not reduce the mechanical properties (yield strength, tensile strength, and plasticity: 226 MPa, 316 MPa, and 22.4%, respectively).

4. Discussion

The inhomogeneity in the alloy grain structure with the master alloy MA2 (manifested by a large scatter of the results, see Figure 7c) may be related to the average particle sizes in the master alloys (Figure 1a,b). In the MA1 master alloy the size of most particles is 0.9 µm, while in the MA2 it is 2 µm, that leads to a half decrease in the number of potential solidification centers if we assume that this size should be around 1 µm. This effect is confirmed by introducing the MA3 master alloy into the AA5056 alloy, where the average particle size was 1 µm (Figure 5d), and a more homogenous microstructure (see Figure 7c) was obtained as compared to the AA5056 US + MA2. A slightly better grain refinement effect by the MA3 master alloy may also be due to a higher concentration of particles per 1 kg of the alloy, which was 4.5×10^{20} compared to 4×10^{20} for the MA1 and MA2 master alloys (Table 2). The number of large particles in MA3 was 3.69×10^{20} compared with 3.44×10^{20} for MA2. The data obtained indicate that titanium diboride microparticles contained in the master alloys allowed the grain refinement of the AA5056 aluminum alloy.

The contribution to an increase of mechanical characteristics of the alloy with the master alloy additions may be due to the alloy grain size reduction (the Hall–Petch law) according to Equation (1).

$$\sigma_{GR} = k_y\left(D^{-\frac{1}{2}} - D_0^{\frac{1}{2}}\right),\tag{1}$$

where k_y is the Hall-Petch parameter (68 MPa), D is the grain size with a master alloy, D_0 is the grain size without a master alloy. These contributions can be estimated as 9.6–9.7 for the tested MA additions.

An additional contribution to an increase in mechanical properties is made by hardening the metal matrix of the AA5056 aluminum alloy with titanium diboride nanoparticles (0.1 μm) by the Orowan mechanism.

Titanium diboride nanoparticles can also provide load redistribution in the matrix, whose contribution according to Equation (2) is 8.4 MPa.

$$\sigma_{load} = 0.5 V_p \sigma_m, \tag{2}$$

where V_p is the volume content of particles, σ_m is the matrix yield strength.

The simultaneous increase in the yield strength, strength, and plasticity can be associated with the load redistribution in the matrix due to the introduction and distribution of nanoparticles, as previously suggested in [16,25].

Lower mechanical characteristics of the deformed alloys with titanium diboride particles in relation to the base AA5056 US alloy without particles can be associated with a decrease in the contribution according to the Hall-Petch law, since the deformation treatment allows one to significantly refine grains of aluminum alloys [9,26]. To determine the effect of particles on the mechanical properties and exclude the internal stresses in the matrix, annealing of the aluminum alloy samples obtained was performed. After annealing, there was a decrease in the mechanical properties of the aluminum alloys with the MA1 and MA2 master alloys (4×10^{20}). The lower characteristics of the alloy with MA2 can be associated with a smaller amount of titanium diboride nanoparticles about 14 vol.% compared with the MA1 master alloy of 17 vol.%. A larger number of titanium diboride particles (4.5×10^{20}) in the MA3 master alloy did not reduce the mechanical characteristics of the alloy, providing additional dispersion hardening. A lower amount of titanium diboride nanoparticles did not, apparently, allow one to sufficiently strengthen the aluminum matrix after pass rolling with majority of particles redistributed to the grain boundaries (Figure 10a).

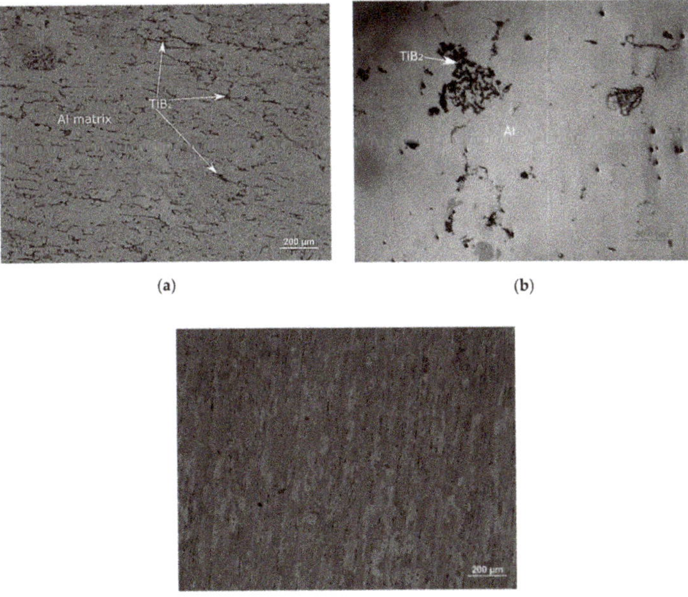

(a)

(b)

(c)

Figure 10. Microstructure of the AA5056 + MA3 US alloy after pass rolling (**a**) and cast AA5056 + MA3 US (**b**), AA5056 US (**c**).

According to optical microscopy data the cast AA5056 + MA3 US alloy (Figure 10b) demonstrates separate agglomerates of particles, but their number was insignificant relative to the total area of the surface studied. The deformation significantly increases the amount of titanium diboride particles and agglomerates at grain boundaries of the AA5056 + MA3 US alloy due to the transfer of the agglomerates and individual particles from the bulk volume of the grain to the boundaries. Apparently, the selected number of deformation cycles was too large for treating an alloy with titanium diboride particles of different sizes. During the deformation of the AA5056 + MA1 US and AA5056 + MA2 US alloys, the number of cycles was five, therefore, an estimated number of 0.8×10^{20} particles of titanium diboride might move to the grain boundary during each rolling pass. The uniformity of particle motion from the volume to the grain boundary requires additional evaluation on alloys with an intermediate number of cycles. For the given number of rolling passes, part of TiB_2 particles in the AA5056 + MA3 US alloy can be preserved inside the grains, but those would end up at the grain boundaries as well upon further deformation. It may be suggested that a reduction in the number of rolling cycles will allow the particles to remain in grain volume, and the effect of grain refinement and reinforcement on the mechanical properties of an AA5056 alloy can be retained after deformation.

5. Conclusions

It was found that the introduction of bi-modal sized titanium diboride particles (4–4.5×10^{20}) allows one to refine the grain structure of the cast aluminum alloy AA5056 from 205 to 158 μm, thereby increasing its yield strength, tensile strength, and ductility from 57 to 71 MPa, from 155 to 201 MPa, and from 11.5% to 18.8%, respectively. The greatest effect of the structure refinement is achieved by using alloys containing titanium diboride microparticles with a size of 1 μm. Nanoparticles present in bi-modal master alloys provide for additional hardening. After deformation the amount of titanium diboride nanoparticles is not enough for the dispersion hardening of the aluminum matrix and increasing the yield strength, tensile strength, and ductility of the AA5056 alloy after pass rolling, mostly due to their redistribution to grain boundaries. This effect can be controlled by the numbers of rolling passes.

Author Contributions: Conceptualization, I.A.Z. and A.P.K.; methodology, D.E., A.A.K., S.C. and S.V.V.; software, M.G.K.; A.P.K. and V.V.P.; resources, A.B.V. and D.E.; visualization, A.P.K. and A.A.K.; supervision, A.B.V. and D.E.

Funding: Russian Science Foundation: 17-13-01252; Council on grants of the President of the Russian Federation: 14.Y30.17.837-MK; EPSRC: EP/R011001/1, EP/R011044/1, EP/R011095/1; The Research and Researchers for Industry (RRi) under the Thailand Research Fund (TRF).

Acknowledgments: The research was performed with the financial support of Grant of Russian Science Foundation (Project No. 17-13-01252). Z.I.A. thank financial support of Presidential Grant MK-837.2017.8, contract number 14.Y30.17.837-MK in part of the work on the production of master alloys for aluminum alloys with TiB_2 different size distributions. D.E and S.C. thank financial support of EPSRC grant UltraMelt2 (EP/R011001/1, EP/R011044/1 and EP/R011095/1) and The Research and Researchers for Industry (RRi) under the Thailand Research Fund (TRF).

Conflicts of Interest: The authors declare no conflict of interest.

References

1. Kawazoe, M.; Shibata, T.; Mukai, T.; Higashi, K. Elevated temperature mechanical properties of A 5056 Al-Mg alloy processed by equal-channel-angular-extrusion. *Scr. Mater.* **1997**, *36*, 699–705. [CrossRef]
2. Jones, R.H. The influence of hydrogen on the stress-corrosion cracking of low-strength Al-Mg alloys. *JOM* **2003**, *55*, 42–46. [CrossRef]
3. Lee, S.; Utsunomiya, A.; Akamatsu, H.; Neishi, K.; Furukawa, M.; Horita, Z.; Langdon, T.G. Influence of scandium and zirconium on grain stability and superplastic ductilities in ultrafine-grained Al–Mg alloys. *Acta Mater.* **2002**, *50*, 553–564. [CrossRef]
4. Filatov, Y.A.; Yelagin, V.I.; Zakharov, V.V. New Al–Mg–Sc alloys. *Mater. Sci. Eng. A* **2000**, *280*, 97–101. [CrossRef]
5. Ahmad, Z. The properties and application of scandium-reinforced aluminum. *JOM* **2003**, *55*, 35–39. [CrossRef]

6. Yan, S.J.; Dai, S.L.; Zhang, X.Y.; Yang, C.; Hong, Q.H.; Chen, J.Z.; Lin, Z.M. Investigating aluminum alloy reinforced by graphene nanoflakes. *Mater. Sci. Eng. A* **2014**, *612*, 440–444. [CrossRef]

7. Vorozhtsov, S.; Minkov, L.; Dammer, V.; Khrustalyov, A.; Zhukov, I.; Promakhov, V.; Vorozhtsov, A.; Khmeleva, M. Ex situ introduction and distribution of nonmetallic particles in aluminum melt: Modeling and experiment. *JOM* **2017**, *69*, 2653–2657. [CrossRef]

8. Li, B.; Zhang, Z.; Shen, Y.; Hu, W.; Luo, L. Dissimilar friction stir welding of Ti–6Al–4V alloy and aluminum alloy employing a modified butt joint configuration: Influences of process variables on the weld interfaces and tensile properties. *Mater. Des.* **2014**, *53*, 838–848. [CrossRef]

9. Zhukov, I.; Promakhov, V.; Vorozhtsov, S.; Kozulin, A.; Khrustalyov, A.; Vorozhtsov, A. Influence of Dispersion Hardening and Severe Plastic Deformation on Structure, Strength and Ductility Behavior of an AA6082 Aluminum Alloy. *JOM* **2018**, *70*, 2731–2738. [CrossRef]

10. Fan, Z.; Wang, Y.; Zhang, Y.; Qin, T.; Zhou, X.R.; Thompson, G.E.; Pennycook, T.; Hashimoto, T. Grain refining mechanism in the Al/Al–Ti–B system. *Acta Mater.* **2015**, *84*, 292–304. [CrossRef]

11. Kotadia, H.R.; Qian, M.; Eskin, D.G.; Das, A. On the microstructural refinement in commercial purity Al and Al-10 wt % Cu alloy under ultrasonication during solidification. *Mater. Des.* **2017**, *132*, 266–274. [CrossRef]

12. Li, Y.; Bai, Q.L.; Liu, J.C.; Li, H.X.; Du, Q.; Zhang, J.S.; Zhuang, L.Z. The influences of grain size and morphology on the hot tearing susceptibility, contraction, and load behaviors of AA7050 alloy inoculated with Al-5Ti-1B master alloy. *Metal. Mater. Trans. A* **2016**, *47*, 4024–4037. [CrossRef]

13. Mahamani, A.; Jayasree, A.; Mounika, K.; Reddi, K.; Sakthivelan, N. Evaluation of mechanical properties of AA6061-TiB2/ZrB2 in-situ metal matrix composites fabricated by K2TiF6-KBF4-K2ZrF6 reaction system. *Int. J. Microstruct. Mater. Prop.* **2015**, *10*, 185–200.

14. Greer, A.L.; Bunn, A.M.; Tronche, A.; Evans, P.V.; Bristow, D.J. Modelling of inoculation of metallic melts: Application to grain refinement of aluminium by Al–Ti–B. *Acta Mater.* **2000**, *48*, 2823–2835. [CrossRef]

15. Ezatpour, H.R.; Torabi Parizi, M.; Sajjadi, S.A.; Ebrahimi, G.R.; Chaichi, A. Microstructure, mechanical analysis and optimal selection of 7075 aluminum alloy based composite reinforced with alumina nanoparticles. *Mater. Chem. Phys.* **2016**, *178*, 119–127. [CrossRef]

16. Vorozhtsov, S.A.; Eskin, D.G.; Tamayo, J.; Vorozhtsov, A.B.; Promakhov, V.V.; Averin, A.A.; Khrustalyov, A.P. The application of external fields to the manufacturing of novel dense composite master alloys and aluminum-based nanocomposites. *Metal. Mater. Trans. A* **2015**, *46*, 2870–2875. [CrossRef]

17. Mousavian, R.T.; Khosroshahi, R.A.; Yazdani, S.; Brabazon, D.; Boostani, A.F. Fabrication of aluminum matrix composites reinforced with nano-to micrometer-sized SiC particles. *Mater. Des.* **2016**, *89*, 58–70. [CrossRef]

18. Liu, H.; Gao, Y.; Qi, L.; Wang, Y.; Nie, J.F. Phase-field simulation of Orowan strengthening by coherent precipitate plates in an aluminum alloy. *Metal. Mater. Trans. A* **2015**, *46*, 3287–3301. [CrossRef]

19. Eskin, D.G.; Al-Helal, K.; Tzanakis, I. Application of a plate sonotrode to ultrasonic degassing of aluminum melt: Acoustic measurements and feasibility study. *J. Mater. Proc. Technol.* **2015**, *222*, 148–154. [CrossRef]

20. Xuan, Y.; Nastac, L. The role of ultrasonic cavitation in refining the microstructure of aluminum based nanocomposites during the solidification process. *Ultrasonics* **2018**, *83*, 94–102. [CrossRef]

21. Gao, Q.; Wu, S.; Lü, S.; Xiong, X.; Du, R.; An, P. Improvement of particles distribution of in-situ 5 vol % TiB2 particulates reinforced Al-4.5 Cu alloy matrix composites with ultrasonic vibration treatment. *J. Alloys Compd.* **2017**, *692*, 1–9. [CrossRef]

22. Promakhov, V.V.; Khmeleva, M.G.; Zhukov, I.A.; Platov, V.V.; Khrustalyov, A.P.; Vorozhtsov, A.B. The Impact of Particle Reinforcement with Al2O3, TiB2, and TiC and Severe Plastic Deformation Treatment on the Combination of Strength and Electrical Conductivity of Pure Aluminum. *Metals* **2019**, *65*, 9.

23. Zhukov, I.A.; Ziatdinov, M.K.; Vorozhtsov, A.B.; Zhukov, A.S.; Vorozhtsov, S.A.; Promakhov, V.V. Self-propagating high-temperature synthesis of Al and Ti borides. *Rus. Phys. J.* **2016**, *59*, 1324–1326. [CrossRef]

24. Zhukov, I.A.; Promakhov, V.V.; Matveev, A.E.; Platov, V.V.; Khrustalev, A.P.; Dubkova, Y.A.; Vorozhtsov, S.A.; Potekaev, A.I. Principles of Structure and Phase Composition Formation in Composite Master Alloys of the Al–Ti–B4C Systems Used for Aluminum Alloy Modification. *Rus. Phys. J.* **2018**, *60*, 2025–2031. [CrossRef]

25. Belov, N.A. Effect of eutectic phases on the fracture behavior of high-strength castable aluminum alloys. *Met. Sci. Heat Treat.* **1995**, *37*, 237–242. [CrossRef]
26. Sun, N.; Apelian, D. Friction stir processing of aluminum cast alloys for high performance applications. *JOM* **2011**, *63*, 44–50. [CrossRef]

Review

A Review of Friction Stir Processing of Structural Metallic Materials: Process, Properties, and Methods

Anna P. Zykova [1,*], **Sergei Yu. Tarasov** [1,2], **Andrey V. Chumaevskiy** [1] and **Evgeniy A. Kolubaev** [1]

[1] Institute of Strength Physics and Materials Science, Siberian Branch of Russian Academy of Sciences, 634055 Tomsk, Russia; tsy@ispms.ru (S.Y.T.); tch7av@gmail.com (A.V.C.); eak@ispms.ru (E.A.K.)

[2] Division for Materials Science, National Research Tomsk Polytechnic University; 634050 Tomsk, Russia

[*] Correspondence: zykovaap@mail.ru; Tel.: +7-(3822)28-68-63

Received: 22 May 2020; Accepted: 4 June 2020; Published: 9 June 2020

Abstract: Friction stir processing (FSP) has attracted much attention in the last decade and contributed significantly to the creation of functionally graded materials with both gradient structure and gradient mechanical properties. Subsurface gradient structures are formed in FSPed metallic materials due to ultrafine grained structure formation, surface modification and hardening with various reinforcing particles, fabrication of hybrid and in situ surfaces. This paper is a review of the latest achievements in FSP of non-ferrous metal alloys (aluminum, copper, titanium, and magnesium alloys). It describes the general formation mechanisms of subsurface gradient structures in metal alloys processed by FSP under various conditions. A summary of experimental data is given for the microstructure, mechanical, and tribological properties of non-ferrous metal alloys.

Keywords: friction stir processing; aluminum alloys; copper alloys; titanium alloys; magnesium alloys; subsurface gradient structures; surface modification; hardening with reinforcing particles; hybrid in situ surfaces

1. Introduction

Structural metal alloys have a long history of industrial applications and are still of great practical relevance for the manufacture of multifunctional products, components, and structures. These are aircraft fuselage and wing components, fuel and cryo tanks, rocket bodies, engine mounts, wheel disks, automobiles, aluminum bridges and pipelines, heat exchangers, air conditioners in construction engineering, railway car bodies, frames and bases of underground trains, and many others. The main feature of such components and structures is their long-term performance capabilities under given loads, which is largely determined by the choice of a suitable alloy to provide the desired properties. Along with the chemical composition of the alloy, the mechanical properties and performance of structures are also governed by an appropriately selected high-quality method of processing.

It is well known that the strength of metal parts can be improved by reinforcing them with alloying elements, metal fibers, or powders of various size and chemical composition [1–16]. This topic has been extensively studied since the 1980s. Over the past decade, much attention has been given to methods for the formation of subsurface gradient structures in metallic materials, such as plasma spraying [17–20], cold spraying [21–23], laser melting [24–26], ion implantation [27–32], and others. Unfortunately, the listed methods for the bulk and surface processing of metallic materials have many drawbacks, e.g., agglomeration of additive particles and their nonuniform distribution both in the bulk and on the surface of the alloy, formation of unwanted phases and interfacial reactions due to high processing temperatures, formation of numerous defects in ion implanted surface layers or formation of amorphous layers at high radiation doses, the need for thermal treatment or other additional processing methods, sophisticated processing equipment, low processing efficiency, and so on.

Friction stir processing technology is a good alternative to overcome the disadvantages mentioned above, because it is performed at temperatures below the melting temperature of base alloys [33–35]. This method is relatively new and is based on the physical principles of friction stir welding (FSW) [33]. Unlike FSW intended for joining together dissimilar solid materials, friction stir processing locally modifies the alloy microstructure to achieve the desired specific properties. It has clear advantages over other surface treatment methods for metallic materials:

(1) FSP is a solid-state, one-stage processing technique that provides grain refinement, strengthening, and structural homogeneity without changing the shape and size of the processed metallic material [33];

(2) the microstructure and mechanical properties of the processed parts can be easily controlled by varying the process parameters [33,35,36];

(3) the method is both environmentally friendly and energy efficient. FSP has greatly evolved over recent decades and have found many practical and scientific applications [33].

The growth of interest in FSP according to the Scopus database began in 2001 and continues to the present. In 2011–2015, the studies were mostly devoted to the surface processing of various metals and alloys. Since 2009 there has been increasing interest in the fabrication of particle-reinforced metal matrix composites, hybrid composites, and in situ composites on the basis of metals and alloys. Currently, the metallic materials produced by different FSP techniques can be conditionally classified into several main groups:

- materials with a subsurface gradient structure obtained through the formation of equiaxed nanograins and structural homogenization [37–57];
- materials with a compositional subsurface gradient structure formed by modifying and hardening the material surface with various reinforcements [32,58–77];
- in situ and hybrid composites [71,78–91].

The aim of this work is to review the latest progress in friction stir processing of non-ferrous metal alloys (aluminum, copper, titanium, magnesium alloys) in accordance with the proposed classification. We will discuss the mechanisms used in FSP for the formation of subsurface gradient structures under various conditions. Particular experimental results will be summarized to show the relationship between the FSP parameters and the resulting microstructure/mechanical properties.

2. Friction Stir Processing

2.1. Principles and Processes

The friction stir processing method evolved from the friction stir welding technology and involves similar processes and principles [92,93]. The friction-heated and plasticized metal is subjected to severe plastic deformation by stirring, which results in obtaining a homogeneous recrystallized fine-grained microstructure. The principle diagram of the FSP/FSW process is shown in Figure 1. The base metal (matrix) is mechanically stirred using a non-consumable rotating tool with a pin (Figure 1a). The tool rotates at a high rate and then is plunged into the workpiece under axial force until the tool shoulder contacts the workpiece surface. Then the tool is advanced over the workpiece along the processing direction. Friction between the tool and the workpiece produces a large amount of heat. As the temperature rises due to frictional heat, the base metal softens in the processing zone and undergoes severe plastic deformation while being entrained by the rotating and traversing pin. This is the basic principle of modifying metallic materials by FSP, resulting in the formation of a subsurface gradient structure in the material via grain refinement and microstructural homogenization. Some friction stir welding or processing techniques involve additional processes, e.g., application of ultrasound to the welding/processing zone (Figure 1c) [94] or multi-pass processing to harden the entire surface area (Figure 1b) [42].

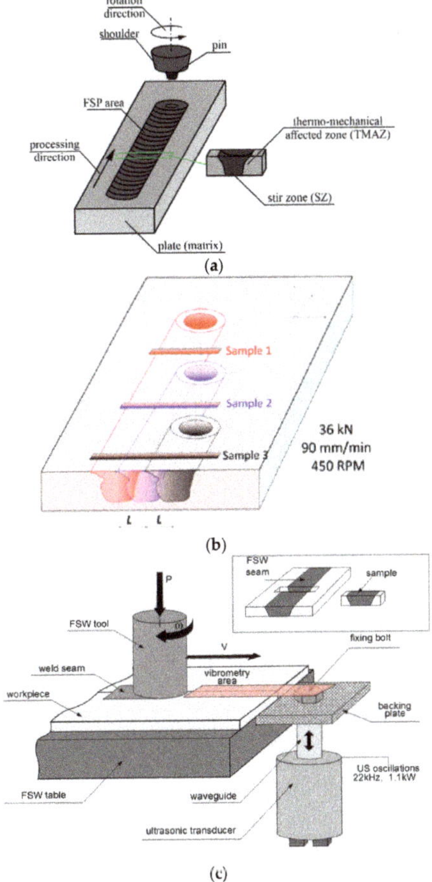

Figure 1. Schematic diagram of the FSP process: (**a**) multi-pass FSP process (**b**) (reproduced from [42], with permission from Springer, 2019) and FSW process with ultrasound assistance (**c**) (reproduced from [94], with permission from Springer, 2017).

2.2. FSP Process Parameters

The main FSP parameters are the tool rotation rate, traverse speed, tool tilt angle, pass time, tool geometry and size, and axial force on the tool.

The temperature in the processing zone during FSP ranges from 0.6 T_m to 0.9 T_m (where T_m is the metal melting temperature), and the strain rate is 1–10^{-3} s^{-1}. Together, they cause pronounced thermal effect, plastic deformation, and material stirring [36]. The most important parameters affecting the microstructure and mechanical properties of the processed material are the tool rotation rate, traverse speed, and axial force. By varying these parameters, FSP can be performed with different heat input conditions and material plastic flow regimes.

FSP allows healing the metal defects such as porosity, cracks, etc. and modifies the alloy microstructure by crushing large matrix grains, second phase particles, and dendrites in cast alloys. Similarly, FSP may crush and dissolve agglomerates of reinforcing particles introduced into metal matrix composites. In both cases, second phase or reinforcing particles are homogenized or uniformly distributed in the metal matrix. The structural homogenization and elimination of defects becomes more pronounced with either increasing rotational rate or decreasing traverse speed due to higher heat release, effective metal viscosity, and more intense flow of the plasticized material. At a lower rotational rate but higher traverse speed, the generated heat contributes to the grain refinement and corresponding metal strengthening [45,51]. A higher rotational rate but lower traverse speed lead, to less pin travel per revolution, producing larger amounts of heat, possibly resulting in grain coarsening and hardness deterioration [33,36,95]. A long-term thermal effect on the material can be favorable for in situ reactions because of the formation of larger amounts of hardening phases uniformly distributed in the matrix [86,96–99].

The microstructure and mechanical properties of metal alloys can also be modified by increasing the tool pass time, changing the tool rotation direction between the passes, or using multi-pass FSP; however, the results vary for different materials [42,44,45,47,93,100–104]. Multi-pass FSP is widely used to fabricate composites with a more homogeneous phase distribution than in single-pass FSPed materials as well as with more efficient in situ reactions during processing [78,86,87,105]. Multi-pass FSP is also used to obtain materials with a large processing area by making closely spaced tracks; the overlapping zone between two adjacent tracks exhibits a complex structure [42].

The tool size and the pin shape strongly influence the heat production and material flow during FSP [40,41,92,93]. At the very beginning of tool plunging, heating occurs mainly due to friction between the pin and the workpiece. Some additional heating results from material deformation. The tool is plunged into the workpiece until the shoulder contacts its surface. According to the studies on the role of pin geometry in heat generation at the plunge stage, the effective pin area has a direct effect on friction-induced deformation and heat production. This suggests that circular pins produce the lowest temperature during plunging [93]. The effect of the tool size and geometry on the microstructure and properties of materials is studied in detail in Refs. [36,40,41,92,93,106–108].

2.3. Microstructure in FSP

Microstructural changes in FSPed materials are caused by thermomechanical effects. As in the case of FSW, the FSP area has a stir zone (SZ), a thermomechanically affected zone (TMAZ), a heat affected zone (HAZ), and a base metal zone (BM) (Figure 2) [36,41,92]. The stir zone has a typical onion ring structure formed when layers of plasticized material flow in the direction from the advancing to the retreating side of the tool. The stir zone material is strongly heated in FSP due to friction and severe plastic deformation, leading to a dynamically recrystallized microstructure. That is why the stir zone consists mainly of uniform refined equiaxed grains much finer than those in the base metal [42]. The structure of these grains is in most cases characterized by a high proportion of high-angle boundaries [37,48,49,54,89,109,110]. FSP parameters such as the tool geometry, workpiece temperature, and axial pressure significantly affect the size of recrystallized grains in the stir zone.

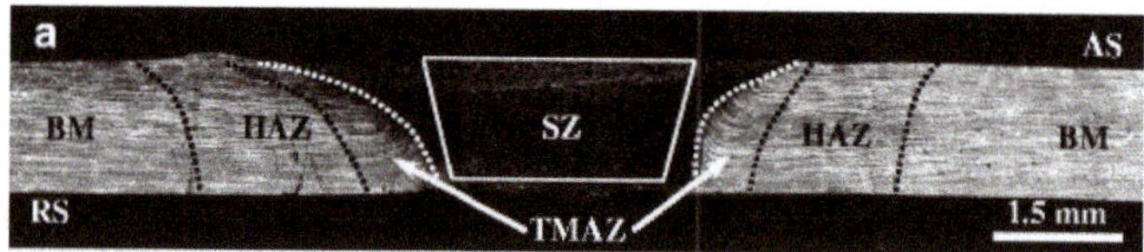

Figure 2. A typical macroimage of different microstructural zones in a FSWed material (reproduced from [94], with permission from Springer, 2017).

3. FSP Applications for Different Materials

3.1. FSP of Structural Alloys

Mechanical characteristics of crystalline structural materials are largely dependent on defects (porosity, shrinkage, cracks, etc.) and the coarse-grained structure of cast material. It is known that the strength of an alloy is enhanced by decreasing the grain size, according to the Hall-Petch equation [111]. To turn a coarse-grained alloy into a fine-grained crystalline material, the alloy can be subjected to severe plastic deformation in order to produce a high density of dislocations and then to rearrange these dislocations to form a grain boundary array. In friction stir processing of structural alloys, the material is heated up due to friction and severe plastic deformation, and as a result the stir zone is dynamically recrystallized. Givi and Asadi [112] proposed three types of dynamic recrystallization mechanism during FSP: (1) intermittent dynamic recrystallization occurring during the nucleation and growth of new grains; (2) continuous dynamic recrystallization involving the formation of low-angle boundary arrays and a gradual increase in boundary misorientation under hot deformation, which finally leads to the nucleation of new grains; and (3) geometric dynamic recrystallization resulting from collision with serrated grain boundaries formed when grains are highly elongated due to severe hot deformation. Thus, FSP forms a subsurface gradient structure with fine equiaxed recrystallized grains of uniform size, due to which the alloy strength and hardness increase. According to the literature data, FSP was applied to aluminum [37–44], copper [46,48–50,107,110], titanium [51–55,113], magnesium alloys [56,57], steels [114–119] and high-entropy alloys [120–125]. The FSP efficiency depends on the tool rotation rate, traverse speed and the number of passes, and is determined for each type of alloy differently. It was shown below for light structural alloys basing on Al, Cu, Ti, Mg.

3.2. FSP of Aluminum Alloys

A review of experimental studies shows that single-pass FSP with low tool rotation rates may reduce the average grain size in aluminum alloys by 85–96%. Figure 3 illustrates a typical macro- and microstructure of aluminum alloys before and after FSP.

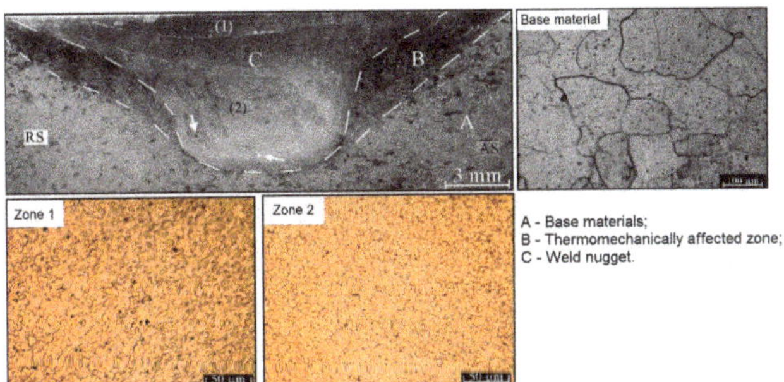

Figure 3. Macro- and microstructure of single-pass FSPed 6063 aluminum alloy (reproduced from [39], with permission from Elsevier, 2019).

By way of example, Table 1 gives the experimental data for FSPed aluminum alloys, showing the effect of FSP on the grain structure and strength of the alloys. The choice of parameters such as the number of FSP passes, tool rotation rate, and traverse speed depends on the aluminum alloy grade and leads to ambiguous results.

A study of single- and multi-pass FSPed A356 aluminum alloy [37] revealed that an increase in the number of passes leads to porosity elimination, refinement of primary silicon particles (from 188 to 2.5–1.6 µm) and α-Al grains (up to 0.4–0.51 µm), as well as to higher hardness. The α-Al grains demonstrate mainly high-angle boundaries and various stages of recovered substructures and dislocation densities. The first FSP pass may produce subgrains and low-angle boundaries by migration of dislocations. During the second and third passes, the formed second phase particles impede the grain boundary migration and thereby limit the recrystallization front, leading to the formation of submicron grains [37]. After six passes, second phase particles are coarsened and cannot provide a sufficient Zener pinning-type effect. That is why no noticeable refinement of subgrains was observed compared to other passes [37]. Similar results were obtained in Ref. [38] (Table 1).

Table 1. Experimental data on FSP of aluminum alloys.

Material	Tool Rotation Rate, rpm	Traverse Speed, mm/min	Number of Passes	Average grain size of the Base Alloy/Average Grain Size after FSP, μm	Mechanical Properties	Ref. No.
A356	350	16	1	-/0.74	MH: 68 HV	[37]
			2	-/0.58	MH: 92 HV	
			3	-/0.45	MH: 113 HV	
			6	-/0.51	MH: 133 HV	
Al-12Si	1400	28	1	25/-	MH: ↑ 20.9% UTS: ↑ 15.1% Elong.: ↑ 3.7 times	[38]
Al5052	1120	80	1	243/16.5	MH: ↑ 13.3%	[59]
AA5005-H34	490	127	1	-/10.7	MH: 42.6 HV UTS: 135.3 MPa Elong.: 34.4%	[40]
	970			-/18.5	MH: 38.9 HV UTS: 118.7 MPa Elong.: 37.3%	
	1200			-/20.4	MH: 37.9 HV UTS: 119.3 MPa Elong.: 41.4%	
6063	300		1	134/5.3	UTS: ↓ 6% Elong.: ↓ 42%	[39]
			2	134/8.6	UTS: ↓ 21% Elong.: ↓ 40%	
	500		1	134/5.5	UTS: no change Elong.: ↓ 28%	
			2	134/9.6	UTS: ↓ 10% Elong.: ↓ 29%	
	700		1	134/7.5	UTS: ↑ 15% Elong.: ↓ 36%	
			2	134/9.7	UTS: ↑ 5% Elong.: ↓ 36%	
	1000		1	134/8	-	
	1200		1	134/7.8	-	

Table 1. *Cont.*

Material	Tool Rotation Rate, rpm	Traverse Speed, mm/min	Number of Passes	Average grain size of the Base Alloy/Average Grain Size after FSP, µm	Mechanical Properties	Ref. No.
5086	1025	30	1	48/7	MH: ↑ 8.6% UTS: ↑ 3.8% Elong.: ↑ 30.7%	[43]
		80		48/10.5	MH: ↑ 8.6% UTS: ↑ 9.6% Elong.: ↑ 23%	
		150		48/3.8	MH: ↑ 10% UTS: ↑ 1.9% Elong.: ↑ 19.2%	
		30	12 (intermittent)	48/8	MH: ↑ 6.9% UTS: ↑ 5.7% Elong.: ↑ 40.3%	
		80		48/13.5	MH: ↑ 5.7% UTS: ↓ 19.2% Elong.: ↑ 19.2%	
		150		48/4	MH: ↑ 5.6% UTS: ↓ 3.8% Elong.: ↑ 15.3%	
		30	12 (continuous)	48/10.5	MH: ↓ 4.3% UTS: ↑ 1.9% Elong.: ↑ 32.7%	
		80		48/15	MH: ↑ 1.4% UTS: ↓ 30.8% Elong.: ↑ 3.8%	
		150		48/6	MH: ↑ 4.3% UTS: no change Elong.: ↑ 7.6%	
AA1050	1600	20	1	42.85/10.58	MH: ↑ 47.6% CF: ↓ 13.8%	[62]

A higher tool rotation rate may serve to increase the mean grain size and reduce the strength. Zhao et al. [39] who studied the influence of the tool rotation rate ranging from 300 to 1200 rpm and showed a 16–26-fold reduction in the mean grain size in the stir zones of single- and multi-pass FSPed 6063 aluminum alloy. As the tool rotation rate increased from 300 to 1200 rpm, the grain size increased from ~5 to ~8 μm and from ~8.5 to ~9.7 μm after one and two passes, respectively [39]. However, despite the significant grain refinement after FSP, some mechanical properties deteriorated in comparison with those of the base alloy (Table 1). This is explained by the fact that the base material contains a high density of needle-shaped precipitates, the amount of which decreases after FSP, or the precipitates undergo strain-induced dissolution at high temperature during processing. The reduced density of needle-shaped fine precipitates in the stir zone is the reason for its lower mechanical properties than those of the base material. Similar results were reported in Ref. [40]. Another study by Ramesh et al. [43] performed on 5086 aluminum alloy subjected to discontinuous and continuous single- and twelve-pass FSP at a constant rotation rate of 1025 rpm but varying traverse speed (30–150 mm/min) revealed a growth of the mean grain size and a corresponding decrease in strength. It was shown that the best structure and properties of the alloy are achieved in single- and multi-pass FSP with the traverse speed of 30 mm/min. With further increasing traverse speed, the average grain size first increases and then decreases, the ductility is enhanced, and the strength of 5086 alloy is reduced (Table 1).

3.3. FSP of Copper Alloys

Pure copper is widely used in optical and electronic applications owing to its high electrical conductivity, thermal conductivity, and corrosion resistance. High purity copper alloys have low strength and wear resistance; therefore they are not popular in applications that require high strength properties. FSPed copper alloys (Cu 99.9%) demonstrate high ductility (up to $\delta = 70\%$) and relatively high strength (up to $\sigma_B = 330$ MPa), because their average grain size is reduced by about 51–99% when varying FSP parameters and performing additional passes [46–49,110]. The effect of FSP on pure copper was investigated using various tool pin geometries (plain cylindrical, threaded cylindrical, triflute, triangle, square, and hexagonal) [107,126] with fixed rotational rate and traverse speed to provide low heat input. The tool with the threaded cylindrical pin profile was found to be more effective in bringing about the desired mechanical modification of pure copper than other pin profiles used under low heat input conditions [107].

A typical macro- and microstructure of FSPed copper alloy is presented in Figure 4. As one can see, the macrostructure of the processed area exhibits the typical zones described in Section 2.3. Table 2 lists the properties of copper alloy samples, indicating the presence of fine equiaxed grains with a large fraction of high-angle boundaries which contribute to higher strength of copper alloys.

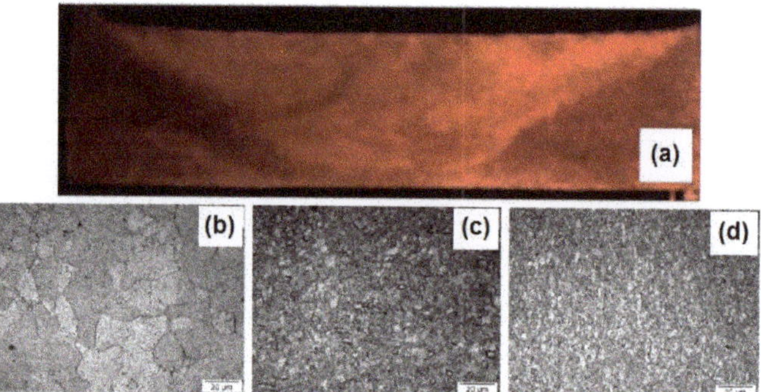

Figure 4. Optical images of single-pass FSPed copper alloy (reproduced from [47], with permission from Elsevier, 2011) (**a**) macrostructure; (**b**) base material; (**c**) nugget regions of specimens processed at 50 and (**d**) 250 mm/min.

Table 2. Experimental data on FSP of copper alloys.

Material	Tool Rotation Rate, rpm	Traverse Speed, mm/min	Number of Passes	Average Grain Size of the Base Alloy/Average Grain Size after FSP, μm	Mechanical Properties	Ref. No.
Cu (99.86%)	300	50	1	19/9.3	MH: ↑ 20% UTS: ↑ 18.1% Elong.: ↑ 9%	[47]
		100	1	19/6.1	MH: ↑ 21% UTS: ↑ 19.2% Elong.: ↑ 4.5%	
		150	1	19/5.9	MH: ↑ 32% UTS: ↑ 19.6% Elong.: ↑ 4.5%	
		200	1	19/3.6	MH: ↑ 33% UTS: ↑ 19.6% Elong.: ↑ 4.5%	
		250	1	19/3.0	MH: ↑ 34% UTS: ↑ 21.4% Elong.: ↑ 4.5%	
Cu (99.99%)	630	40	1	50–60/7.5	UTS: ↑ 30% Elong.: ↑ 2.9 times	[110]
			4	50–60/0.7–0.8	UTS: ↑ 43.3% Elong.: ↑ 1.8 times	
	630	315	1	50–60/2.5	UTS: ↑ 43.3% Elong.: ↑ 2.4 times	
			4	50–60/4–5	UTS: ↑ 43.3% Elong.: ↑ 2.4 times	
	1600	40	1	50–60/6	UTS: ↑ 46.7% Elong.: ↑ 3.9 times	
			4	50–60/2	UTS: ↑ 33.3% Elong.: ↑ 4.2 times	
Cu (99.95%)	400 600 800 1200	20	1	15/0.156 15/0.265 15/0.126 15/0.109	-	[48]

Table 2. *Cont.*

Material	Tool Rotation Rate, rpm	Traverse Speed, mm/min	Number of Passes	Average Grain Size of the Base Alloy/Average Grain Size after FSP, µm	Mechanical Properties	Ref. No.
Cu (99.98%)	250	50	1	35/5-20	MH: ↑ 18.2% UTS: - Elong.: -	[49]
	350				MH: ↑ 13.4% UTS: ↓ 18.2% Elong.: ↓ 1.7 times	
	500				MH: ↑ 7.3% UTS: ↓ 17.9% Elong.: ↑ 1.4 times	
Cu-0.18wt%Zr	600	50 100 150 200	1	40.5/9.7 40.5/6.6 40.5/4.9 40.5/4.6	-	[50]

Analysis of experimental data shows that FSP of copper alloys (Cu 99.9%) is efficient under low heat input conditions, and the efficiency increases with increasing traverse speed. For example, Surekha and Els-Botes [47] synthesized a high strength and high-conductivity copper using FSP with low heat input by varying the traverse speed from 50 to 250 mm/min (Figure 4) at a constant rotational rate of 300 rpm. By increasing the traverse speed from 50 to 250 mm/min, they refined the grains in the stir zone from 9 to 3 μm and simultaneously increased the hardness from 102 to 114 HV [47]. The grain refinement achieved at a constant rotational rate but increased traverse speed led to the improvement of mechanical characteristics, according to the Hall–Petch relationship [47]. Similar results were obtained in Refs. [50,110]. Barmouz et al. [45] investigated single-pass FSP on a pure copper plate by applying the traverse speeds of 40 and 315 mm/min while keeping the tool rotation rate at 630 rpm. FSP at higher traverse speed resulted in higher strength as demonstrated in Table 2. Four-pass FSPed specimens of the same material had an ultrafine-grained structure with a mean grain size of up to 800 nm [45].

Using higher tool rotation rate during copper alloy processing may cause the formation of both ultrafine and nano-sized grains with high-angle boundaries [48,49,110], as well as consequent improvement of the tensile strength [110]. Cartigueyen et al. [49] studied the effect of the FSP heat release on the mechanical properties of FSPed copper. The results showed that the temperatures reached during FSP strongly affected the microstructure and properties of the processed copper. It was found that the peak temperatures for the characteristic FSP zones range between 320 °C and 445 °C (about 0.3–0.42 T_m), indicating the achievement of low heat input conditions. The peak temperatures were higher on the advancing side of the FSP track as compared to those in the middle of the stir zone and on the retreating side. A fine and homogeneous grain structure was produced with various FSP tool rotation rates (Table 2). The authors observed the formation of a tunnel defect at 250 rpm, which was caused by insufficient heat input and significantly impaired the mechanical properties of the processed metal. The hardness of the FSPed copper was strongly dependent on the tool rotation rate (Table 2). The minimum rotational rate for performing efficient FSP under low heat input conditions was found to be equal to 350 rpm [49].

3.4. FSP of Titanium Alloys

Titanium and its alloys are widely used in aerospace, chemical, and biomedical industries due to high specific strength, corrosion resistance, and good biocompatibility. Biomedical and aerospace applications often require only surface hardening of titanium alloy products, while retaining its original structure and composition in the bulk. Since the surface layer hardness determines the wear resistance, surface hardening is performed to improve the surface of soft pure titanium. FSP can be used to increase the sliding wear resistance and surface hardness of alloys by changing the surface microstructural characteristics such as grain refinement and strain hardening [51–54]. However, the issue of wear resistance is too complex and cannot be reduced only to increasing the hardness. As far as nanostructured metals are concerned, there is no unambiguous opinion on whether their wear resistance is higher or lower as compared to that of coarse-grained ones. Many nanostructured metals and alloys lose their ductility and therefore become prone to subsurface fracture during sliding friction. Also, the abundance of grain boundaries adds to a higher amount of dangling bonds and therefore, a higher probability of adhesion bonding to the counter body. Zhang et al. [54] produced an ultrafine microstructure in FSPed Ti-6Al-4V alloy, which consisted of α grains (~0.51 μm) and a small amount of β-phase with a high fraction of high-angle grain boundaries (89.3%). Mironov et al. [52] found that the stress state in FSPed pure titanium is close to simple shear, where the shear plane resembles a truncated cone with a diameter close to the tool shoulder diameter in the upper part of the stir zone and close to the pin diameter in its lower part. The authors [52] demonstrated that the material flow arises mainly from the prism slip and leads to a pronounced P-fiber {hkil} $\langle 11\bar{2}0 \rangle$ shear texture in the stir zone. The texture evolution governs the development of deformation-induced grain boundaries in this zone. The macro- and microstructure of pure Ti after single-pass FSP is shown in Figure 5 [51]. Table 3 gives the experimental data on the properties of FSPed titanium alloys.

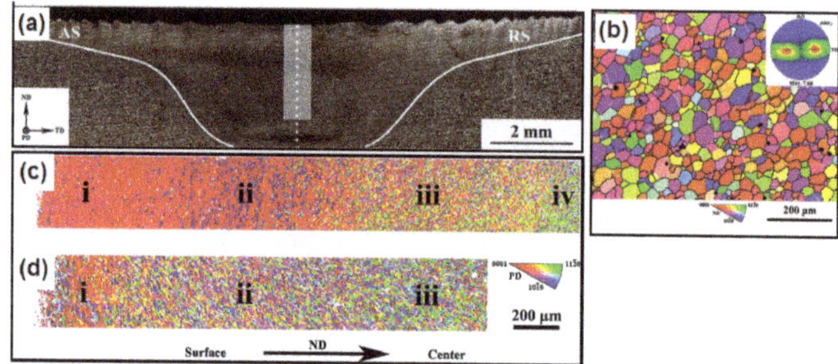

Figure 5. Macrostructure of single-pass FSPed titanium (reproduced from [51], with permission from Elsevier, 2019): (**a**) region processed at 180 rpm; (**b**) EBSD map of pure titanium sheet; EBSD maps of the specimen cross sections processed at 180 rpm (**c**) and 270 rpm (**d**).

Table 3. Experimental data on FSP of titanium alloys.

Material	Tool Rotation Rate, rpm	Traverse Speed, mm/min	Number of Passes	Average Grain Size of the Base Alloy/Average Grain Size after FSP, µm	Mechanical Properties	Ref. No.
α-Ti (99.6%)	180	25	1	33.1/5.8	MH: ↑ 27% YS: ↑ 71.7% UTS: ↑ 35.1%	[51]
α-Ti (99.85%)	250 300 350	75	1	42/7	MH: ↑ 18.4% UTS: 382–384 MPa	[53]
Ti-6Al-4V	120	30	1	-/0.51	-	[54]
Ti grade 2	1400	14	1 2 3	-	MH: ↑ 15% FC: ↓ 31% MH: ↑ 34.6% FC: ↓ 66% MH: ↑ 55.4% FC: ↓ 88.8%	[55]

Efficient FSP of pure titanium can be performed at both high (> 250 rpm) [53,55] and low tool rotation rates (< 250 rpm) [43]. At 180 rpm, the grain size in the stir zone decreases by 82% (from 33.1 to 5.8 µm), the microhardness increases by 27%, and the yield strength increases by 71.7% [51].

A study of the multi-pass FSP effect on the assessment of the microstructure and wear resistance of pure titanium showed that higher wear resistance and microhardness of specimens after 3 passes correlate with a smaller grain size [55] (Table 3).

3.5. FSP of Magnesium Alloys

The attractive properties of magnesium and its alloys include reduced weight, electromagnetic shielding, high specific strength, and so on. However, the alloys have limited formability, especially at ambient temperature, which significantly limits their industrial application. It is believed that grain refinement and texture weakening are effective ways to improve the ductility of magnesium. This can be achieved by FSP that can change the alloy microstructure and thus significantly increase its ductility without the tensile strength loss [127–129]. Typical macro- and microstructures of a magnesium alloy before and after FSP are shown in Figure 6. The experimental data on friction stir processing of particle-reinforced structural magnesium alloys are analyzed in Table 4.

Table 4. Experimental data on FSP of magnesium alloys.

Material	Tool Rotation Rate, rpm	Traverse Speed, mm/min	Number of Passes	Average Grain Size of the Base Alloy/Average Grain Size after FSP, μm	Mechanical Properties	Ref. No.
Al-Cu-Mg	450	-	1	137 × 22.2/9.1 × 6.4	MH: ↑ 15% UTS: ↑ 9%	[56]
AZ31	200	50	1	-/-	UTS: ↑ 4% Elong.: ↑ 9.5%	[129]
Mg-6Zn-1Y-0.5Zr	800	20	1	-/3.20	UTS: ↑ 32.6% Elong.: ↑ 146.7%	[128]
		80	1	-/2.37	UTS: ↑ 37.7% Elong.: ↑ 183.4%	
		200	1	-/1.65	UTS: ↑ 53% Elong.: ↑ 151.4%	
AZ31	400	50	1	16–300/6.6–3.5	MH: ↑ 22.2% UTS: ↑ 2 times Elong.: ↑ 1.5 times	[127]
AZ31	600	20	1	-/-	MH: ↑ 17.8% MH: ↑ 24.3% MH: ↑ 38% MH: ↑ 44.6% MH: ↑ 48.7% MH: ↑ 53.7%	[113]
	600	30				
	600	40				
	800	20				
	800	30				
	800	40				
AZ61	1000	37	2	75/0.04–0.2	MH: ↑ 3 times	[130]
AZ80	375	118	1 (in air)	-/7.1	MH: 69.4 HV	[131]
			1 (under water)	-/2.7	MH: 75.3 HV	
AE42	950	75	1	81/7.4	MH: ↓ 19.1% UTS: ↑ 22.9% Elong.: ↑ 2.7 times	[132]
QE22	800	100	1	38/0.88	UTS: ↓ 13.5% Elong.: ↑ 3 times	[133]
	800	100	1	38/0.63	UTS: ↓ 1.9% Elong.: ↑ 3.4 times	
	600	100	2			
	800	100	1	38/2.30	UTS: ↓ 30.7% Elong.: ↑ 1.7 times	
	600	100	2			

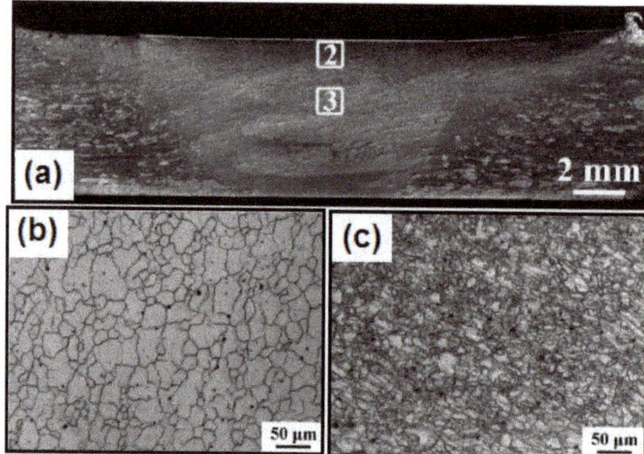

Figure 6. Macro- and microstructure of single-pass FSPed AZ31 alloy (reproduced from [129], with permission from Elsevier, 2019): (**a**) specimen macrostructure in the region processed at 200 rpm; (**b**) base alloy microstructure; (**c**) stir zone microstructure (2).

According to Wang et al. [128], FSP of a Mg-6Zn-1Y-0.5Zr casting resulted in dissolution and dispersion of the intergranular eutectic I-phase (Mg_3Zn_6Y). Hot deformation by FSP led to $I{\rightarrow}W$ ($Mg_3Zn_3Y_2$) phase transformation. An increase in the traverse speed caused significant grain refinement and the formation of a large fraction of fine particles, which greatly improved the yield strength (93.1%), tensile strength (53.0%), and relative elongation (151.4%) in comparison with those of the cast material [128].

A change in the phase composition after FSP was also observed in the cast alloy AE42 [132]. The β-$Mg_{17}Al_{12}$ and $Al_{11}RE_3$ phases dissolved after single-pass FSP, with the formation of a new Al_2RE phase. The factors affecting the strength of the cast magnesium alloy AE42 were found to be secondary phases (most influential), texture, and grain size [132].

As reported by Du and Wu [130] for AZ61 Mg alloy, a nano-grained structure can be produced by double-pass FSP under the condition of rapid heat removal by means of using an additional liquid nitrogen cooling system. The proposed processing technique allows reducing the mean grain size to <100 nm, thus increasing the alloy microhardness to 155 HV. The authors described the nanostructure evolution process as follows: (1) in the first FSP pass, submicron-sized grains are formed in the processed sheet by continuous dynamic recrystallization; (2) in the second pass, numerous nuclei are formed by discontinuous dynamic recrystallization due to the presence of submicron-sized grains, subgrains, and a high density of dislocation walls; (3) the growth of recrystallized grains is limited by effective liquid nitrogen cooling. Similar effects of remarkable grain structure refinement and improvement of mechanical properties by the above scenario are described in Refs. [128,132,133].

4. Friction Stir Processing of Particle-Reinforced Structural Alloys

The last decade showed a growing interest in friction stir processing of particle-reinforced metallic materials. Such materials are referred to in the literature as metal matrix composites [59,60,63,71], composite materials [58,61,66], hybrid composites [69,73–75], and others. This processing method is used for fabricating surface composite coatings with an average thickness from 50 to 600 μm on the basis of aluminum, copper, titanium, and magnesium alloys. The reinforcing additives for the surface composites can be in the form of powder, fibers, or platelets, which are most commonly filled into especially milled grooves [59,61,63–65,69,73,78,80,81,86,93] or drilled holes [69,73–75] (Figure 7). A typical subsurface macro- and microstructure of the stir zone with introduced particles is shown in Figure 8.

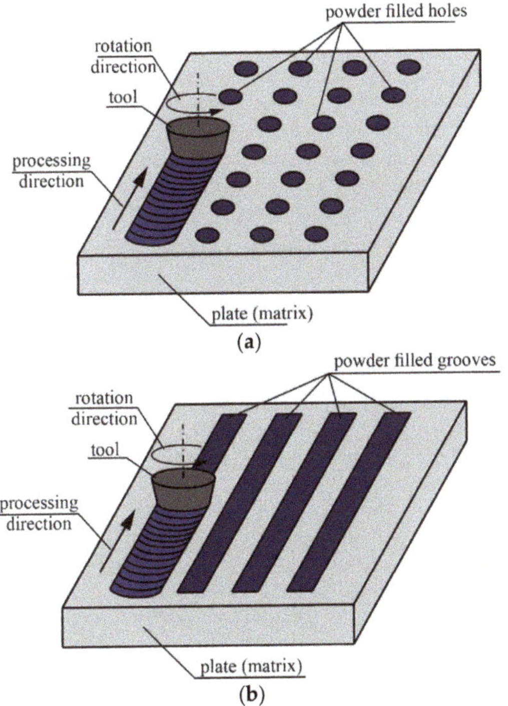

Figure 7. Schematic of the FSP process with the addition of reinforcements filled (**a**) into holes in the substrate and (**b**) into a groove.

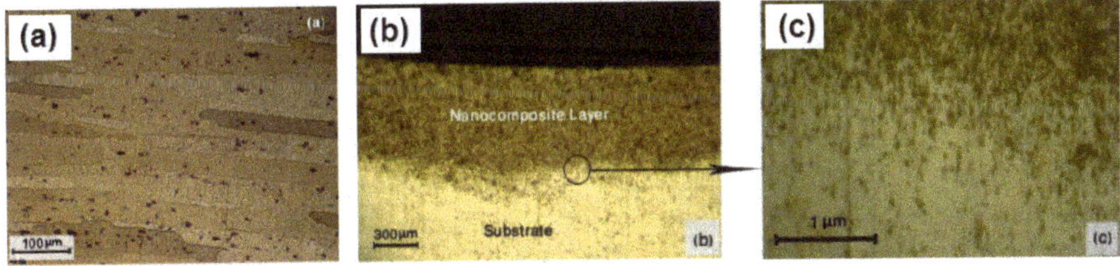

Figure 8. Microstructure of Al2024 (reproduced from [65], with permission from Elsevier, 2010): (**a**) microstructure of the base Al2024; (**b**) cross-sectional microstructure with Al_2O_3 particles after single-pass FSP; (**c**) interfacial microstructure with Al_2O_3 particles.

Hard fine-grained particles can be admixed to the substrate during FSP by the following mechanism. The heat generated by the friction of the tool shoulder and the pin plasticizes the metal matrix around and under the tool. Its rotational and translational motion entrains the plasticized metal matrix material from the advancing side to the retreating side. The flow of the matrix material breaks the grooves (or holes) and admixes the compacted particles to the plasticized metal matrix material. The tool rotation rate and traverse speed determine the stirring intensity and provide the formation of a composite. Analysis of experimental data shows that all types of reinforcing particles can be stirred with the plasticized metal matrix to form a composite. This fact is clearly demonstrated by Dinaharan et al. [71] who synthesized copper matrix composite layers reinforced with various ceramic particles. The authors showed that the type of ceramic particles does not affect the particle distribution pattern in the composite. Neither the density gradient nor the wettability of ceramic particles by the copper matrix lead to a nonuniform particle distribution. It is also noted that merging of the material flows caused separately by the tool shoulder and the pin leads to the formation of layers with a high and low volume fraction of ceramic particles due to the temperature gradient along the depth of the plate. The tool penetration depth is not equal to the total plate thickness in FSP. The absence of "onion rings" indicates that the temperature gradient along the pin penetration depth is negligible [71].

Severe plastic deformation and dynamic recrystallization during FSP result in a fine equiaxed grain microstructure in the stir zone, in which reinforcing particles are located both inside the grains and at the grain boundaries [61,63,71,76]. When introducing reinforcement particles into the matrix by FSP, no interfacial reactions were observed; there is a distinct boundary between the matrix and the introduced particles, e.g., as shown in Figure 9 for AA6063 alloy FSPed with the addition of vanadium particles [63]. The composite image in Figure 9 demonstrates a sharp boundary between the vanadium particles and the aluminum matrix. In Figure 9b, there are no reaction layers that would show contrast other than those of the matrix and the particle. This is confirmed by the EDX line scan (see inset in Figure 9b) indicating a sharp change in EDX counts in the narrow transition zone at the particle-matrix interface.

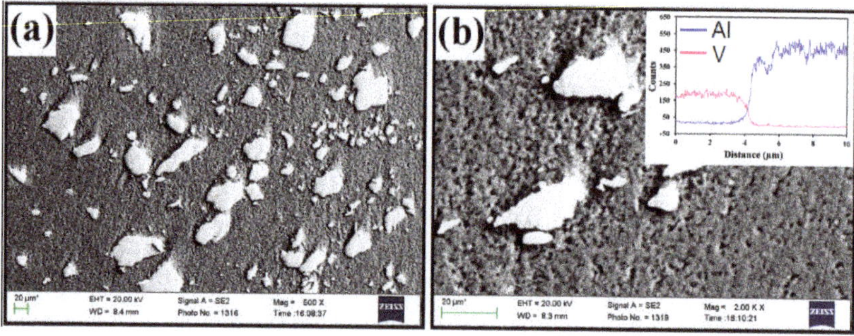

Figure 9. Microstructure of FSPed AA6063 with 12 vol. % V content at magnification: (**a**) 500× and (**b**) 2000× (the insert shows the EDX line scan along the particle interface) (reproduced from [63], with permission from Elsevier, 2019).

However, as noted in Refs. [59,62,64], the single-pass FSPed particle-reinforced alloy surface layers exhibit heterogeneous grain structure, nonuniform particle distribution and tensile properties. Multi-pass FSP allows producing a composite with more homogeneous particle distribution and grain structure, and with better tensile strength (Table 5) [59,62,64,68,70,72]. The mechanical and performance characteristics of composite metallic materials are also greatly affected by the content/volume fraction of the particles introduced. As shown for cast aluminum alloy A356 with a dendritic structure and a small number of pores [61], its processing with the addition of different volume fractions of Ti_3AlC_2 causes considerable elimination of coarse needle-like silicon particles and large primary aluminum dendrites, and produces a uniform distribution of fine Si and Ti_3AlC_2 particles in the matrix. A356 and Ti_3AlC_2 do not react during FSP, because the process time and temperature are too low to initiate mechanochemical or diffusion-controlled phase transformations. After FSP and an increase in the Ti_3AlC_2 particle volume fraction from 2.5 vol. % to 7 vol. %, tensile tests revealed a 2-fold increase in microhardness and mechanical properties (Table 5) [61]. A 3-fold improvement of the mechanical characteristics with increasing volume fraction of reinforcement particles during FSP was observed in Refs. [68,76]. Surface composites with different volume fractions of reinforcing particles (25 vol. % B_4C-75 vol. % TiB_2, 50 vol. % B_4C-50 vol. % TiB_2, and 75 vol. % B_4C-25 vol. % TiB_2) were synthesized by FSP in AA7005 alloy by Pol et al. [70]. They found that the hardness of the base alloy and Al7005-25 vol. % TiB_2-75 vol. % B_4C composite were 90 HV and 150 HV, respectively. The microhardness of surface composites with different volume fractions of the introduced particles were almost the same, which might be due to the same powder particle sizes. The synthesized surface composites of aluminum alloys demonstrated better ballistic resistance. The penetration depth of a steel projectile into the base alloy and composites 25 vol. % B$_4$C-75 vol. % TiB$_2$, 50 vol. % B$_4$C-50 vol. % TiB$_2$, and 75 vol. % B$_4$C-25 vol. % TiB$_2$ was 37 mm, 26 mm, 24 mm, and 20 mm, respectively, which is explained by the presence of hard reinforcing ceramic particles in the surface composite and by a hard core of the matrix [70].

Of particular interest are carbon materials (graphene, SWCNTs, MWCNTs, fibers, etc.) as high strength (30 GPa) reinforcement agents [67,68,134,135] for next-generation automotive and aerospace materials. S. Zhang et al. [68] demonstrated the microstructure and mechanical properties of nanocomposites are closely related to the energy input. In the cited work, different energy inputs led to different dispersion of CNTs in a CNTs/Al nanocomposite. Better CNT dispersion and higher tensile strength of a CNTs/Al nanocomposite was obtained at higher energy input (Table 5). The highest energy input led to a 53.8% higher maximum tensile strength of the nanocomposite than that of unreinforced aluminum. Moreover, nanocomposites showed a good improvement of ductility from 25% to 33% [68].

For an AA6061-graphene-TiB2 hybrid nanocomposite synthesized by Nazari et al. [69], it was shown that the simultaneous addition of graphene and TiB$_2$ particles during FSP led to a significant grain structure refinement in the stir zone; the average grain size was reduced to < 1 μm. Both graphene and TiB$_2$ particles retained their structure while being high-speed stirred into the aluminum alloy matrix. The hardness of the aluminum alloy increased to ~102 HV, mainly at the cost of TiB$_2$ particles introduced together with graphene with an optimal hybrid ratio of 1 wt. % graphene-20 wt. % TiB$_2$ [69]. With the same ratio of components, the processed hybrid nanocomposites demonstrated the best combination of tensile properties, namely three times higher yield strength and ~70% higher ultimate tensile strength (Table 5) [69].

There are also experiments on the fabrication and single-/multi-pass FSP of cast metal matrix composites [75,108,136]. The FSPed cast metal matrix composites exhibit a gradient structure represented by the bulk-reinforced matrix and the FSP-hardened surface layers. Arokiasamy and Ronald [75] described the process of stir casting a magnesium-based hybrid composite at the melting temperature of 700 °C with the introduction of SiC and Al$_2$O$_3$. It was shown that additional FSP of the cast composite increases its microhardness by 17.5%. Single-pass FSP led to a considerable grain refinement in the cast magnesium composite (Table 5). Microstructural studies revealed uniform distribution of SiC and Al$_2$O$_3$ particles both in the bulk of the material and in the stir zone [75].

Hardening of composite aluminum alloy surfaces by FSP is performed using fine powders of the following chemical composition: Al6061-SiO$_2$ [58], Al-Al$_2$O$_3$ [64,65], AA6016-(Al$_2$O$_3$ + AlN) [91], CaCO$_3$ [66], Al-SiC [59,137–139], Al-Ti$_3$AlC$_2$ [61], Al-TiO$_2$ [62], Al-B$_4$C-TiB$_2$ [70], Al-NbC [76], Al-V [63], Al-graphene/carbon SWCNTs/MWCNTs [67,68,134,135], and Al-TiB$_2$ [69]. The following compositions are used to synthesize copper alloy matrix composites by FSP: Cu-SiC [71], Cu-B$_4$C [71], Cu-TiC [71], Cu-Al$_2$O$_3$ [71], Cu-TiO$_2$ [72], and Cu-AlN-BN [73]. Titanium alloy matrix composites are fabricated by FSP using SiC [140] and Al$_2$O$_3$ [141]. The most frequently synthesized magnesium metal matrix composites are AZ31B-MWCNT-graphene [74], Mg-NiTi [77] and Mg-SiC-Al$_2$O$_3$ [75]. An analysis of the experimental data on FSP of structural alloys with the addition of various reinforcing particles is presented in Table 5.

Table 5. Experimental data on FSP of particle-reinforced structural alloys.

Material	FSP Parameters	Particle Introduction Method	Reinforcing Particles (size)	Average grain Size of the Base Alloy/Average Grain Size after FSP, μm	Mechanical Properties	Ref. No.
			Aluminum alloys			
Al6061	1150 rpm, 31.5 mm/min 1 pass	V-shaped grooves	SiO_2 (d_{av} = 20 nm)	-/15,53	CR: ↑ 78% ↑ MH, UTS, Elong.	[58]
Al5052	1120 rpm, 80 mm/min 1 pass	Groove (depth 2 mm, width 1 mm)	SiC (d_{av} = 5 μm)	243/5.4	MH: ↑ 29.3%	[59]
	1120 rpm, 80 mm/min 4 passes		SiC (d_{av} = 5 μm)	243/4.2	MH: ↑ 42.6%	
	1120 rpm, 80 mm/min 4 passes		SiC (d_{av} = 50 nm)	243/0.9	MH: ↑ 54.6%	
Al6061-T651	1000 rpm 72 mm/min 1 pass	Slot in the butt end of the plate	SiC (d_{av} = 3–6 μm)	-/-	UTS: ↓ 28.8% Elong.: ↓ 8.3%	[60]
	1000 rpm 72 mm/min 2 passes			-/-	UTS: ↓ 23% Elong.: ↑ 59.3%	
A356	1000 rpm 112 mm/min 1 pass	Groove	2.5 vol. % Ti_3AlC_2	-/-	MH: ↑ 18.4% UTS: ↑ 9% Elong.: ↑ 1.4 times	[61]
			5 vol. % Ti_3AlC_2	-/-	MH: ↑ 27.6% UTS: ↑ 14.2% Elong.: ↑ 1.5 times	
			7 vol. % Ti_3AlC_2	-/-	MH: ↑ 33.8% UTS: ↑ 19.4% Elong.: ↑ 1.7 times	

Table 5. *Cont.*

Material	FSP Parameters	Particle Introduction Method	Reinforcing Particles (size)	Average grain Size of the Base Alloy/Average Grain Size after FSP, μm	Mechanical Properties	Ref. No.
AA1050	1600 rpm 20 mm/min 1 pass	Holes (diameter 2.5 mm, spacing 3 mm)	TiO_2	42.85/5	MH: ↑ 61.9% CF: ↓ 19.2%	[62]
	1600 rpm 20 mm/min 2 passes		TiO_2	42.85/5	MH: ↑ 80.9% CF: ↓ 29.2%	
AA6063	1600 rpm 60 mm/min 1 pass	Grooves $(1.2 \times 5.5 \times 100 \text{ mm}^3)$	12 vol. % V $(d_{av} = 18 \text{ μm})$	72/7.6	UTS: ↑ 24.6% Elong.: ↑ 1.2 times	[63]
AA1050	1180 rpm 80 mm/min 1 pass	Groove (width 1 mm, depth 3 mm)	Al_2O_3	128/29	-	[64]
	1180 rpm 80 mm/min 2 passes			128/23	WT: ↑ 1.8 times	
Al2024	800 rpm 25 mm/min 1 pass	Groove	Al–10 vol. % Al_2O_3 powders $(d_{av} = 50{-}150 \text{ μm})$	$250 \times 8/4$	MH: ↑ 2.5 times WT: ↑ 3 times	[65]
AA6082	1250 rpm 40 mm/min 1 pass	-	-	141/15–20	MH: ↑ 43.5% WT: ↑ 1.2 times	[66]
		Groove (width 2 mm, depth 2 mm)	$CaCO_3$ $(d_{av} = 3{-}5 \text{ μm})$	141/10–12	MH: ↑ 35.9% WT: ↑ 1.6 times	
AA7075	1200 rpm 1 pass	Groove	5 vol. % NbC $(d_{av} = 10{-}20 \text{ μm})$	50/40	UTS: ↑ 13.6% Elong.: ↓ 20% MH: ↑ 17.3%	[76]
			10 vol. % NbC $(d_{av} = 10{-}20 \text{ μm})$	50/26	UTS: ↑ 36.3% Elong.: ↓ 30% MH: ↑ 37.7%	
			15 vol. % NbC $(d_{av} = 10{-}20 \text{ μm})$	50/16	UTS: ↑ 47.7% Elong.: ↓ 65% MH: ↑ 53%	

Table 5. *Cont.*

Material	FSP Parameters	Particle Introduction Method	Reinforcing Particles (size)	Average grain Size of the Base Alloy/Average Grain Size after FSP, μm	Mechanical Properties	Ref. No.
AA7075	800 rpm 60 mm/min 1 pass	Hole (diameter 2 mm, depth 3 mm)	- MWCNT (diameter 15–20 nm, length 5 μm) Cu (d_{av} = 10–20 μm) SiC (d_{av} = 15–20 μm)	82.70/2.98 82.70/2.88 82.70/2.57 82.70/2.53	MH: ↓ 11.8% IT: ↓ 37.9% MH: ↑ 2.8% IT: ↓ 29.7% MH: ↑ 6.9% IT: ↓ 8.8% MH: ↑ 2.8% IT: ↓ 6.3%	[67]
AA1060	950 rpm 30 mm/min 3 passes 950 rpm 150 mm/min 3 passes 600 rpm 95 mm/min 3 passes 750 rpm 30 mm/min 3 passes	3 plates, groove in the middle plate (length 150 mm, depth 1.5 mm)	1.6 vol. % CNT (diameter 12.1 nm, length 1 μm)	-/4.8 -/- -/- -/-	UTS: 90.2 MPa Elong.: 36.8% UTS: 102.3 MPa Elong.: 25.3% UTS: 103.4 MPa Elong.: 33.4% UTS: 110.9 MPa Elong.: 32.3%	[68]
	600 rpm 150 mm/min 3 passes 750 rpm 95 mm/min 3 passes 950 rpm 30 mm/min 3 passes	3 plates, groove in the middle plate (length 150 mm, depth 2 mm)	3.2 vol. % CNT (diameter 12.1 nm, length 1 μm)	-/1.9 -/2.1 -/3.3	UTS: 93.6 MPa Elong.: 32.1% UTS: 127 MPa Elong.: 23.3% UTS: 138.8 MPa Elong.: 31.2%	

Table 5. *Cont.*

Material	FSP Parameters	Particle Introduction Method	Reinforcing Particles (size)	Average grain Size of the Base Alloy/Average Grain Size after FSP, μm	Mechanical Properties	Ref. No.
AA6061	1000 rpm 340 mm/min 1 pass	-	-	70/20	MH: ↑ 15.4% UTS: ↑ 10% Elong.: ↑ 20.8% CF: ↓ 8.7%	[69]
	1000 rpm 340 mm/min 4 passes	Groove 2×3 mm^2	Micro-sized TiB$_2$ particles and nano-sized graphene platelets: 10 wt. % TiB$_2$–0 wt. % graphene	70/< 1 μm	MH: ↑ 31.6% UTS: ↑ 18.1% Elong.: ↓ 16.6% CF: ↓ 14%	
			20 wt. % TiB$_2$–0 wt. % graphene		MH: ↑ 48.2% UTS: ↑ 31.3% Elong.: ↓ 25% CF: ↓ 26.3%	
			30 wt. % TiB$_2$–0 wt. % graphene		MH: ↑ 45% UTS: ↑ 45% Elong.: ↓ 50% CF: ↓ 7%	
			0 wt. % TiB$_2$–0.5 wt. % graphene		MH: ↑ 22.1% UTS: ↑ 37.5% Elong.: ↑ 4.2% CF: ↓ 12%	
			0 wt. % TiB$_2$–1 wt. % graphene		MH: ↑ 37.5% UTS: ↑ 54.4% Elong.: ↑ 20.8% CF: ↓ 24.5%	
			0 wt. % TiB$_2$–2 wt. % graphene		MH: ↑ 38.5% UTS: ↑ 59.4% Elong.: ↓ 16.7% CF: ↓ 3.5%	
			20 wt. % TiB$_2$–0.5 wt. % graphene		MH: ↑ 54.4% UTS: ↑ 61.9% Elong.: ↓ 8.3% CF: ↓ 29.8%	
			20 wt. % TiB$_2$–1 wt. % graphene		MH: ↑ 66.5% UTS: ↑ 69.4% Elong.: ↓ 4.2% CF: ↓ 29.6%	
			20 wt. % TiB$_2$–2 wt. % graphene		MH: ↑ 62.8% UTS: ↑ 75.6% Elong.: ↓ 62.5% CF: ↓ 1.7%	

Table 5. *Cont.*

Material	FSP Parameters	Particle Introduction Method	Reinforcing Particles (size)	Average grain Size of the Base Alloy/Average Grain Size after FSP, μm	Mechanical Properties	Ref. No.
Al7005	750 rpm 50 mm/min 2 passes	- Holes (diameter 1.5 mm, depth 3 mm)	- 50% B_4C + 50% TiB_2 75% B_4C + 25% TiB_2 25% B_4C + 75% TiB_2	-/- -/- -/- -/-	MH: ↑ 33.3% MH: ↑ 66.6% MH: ↑ 64.4% MH: ↑ 61.1%	[70]
Copper alloys						
Cu (99.9%)	1000 rpm 40 mm/min 1 pass	Groove (depth 2.5 mm, width 0.7 mm)	12 vol. % SiC 12 vol. % Al_2O_3 12 vol. % B_4C 12 vol. % TiC	35/6 35/3 35/5 35/4	MH: ↑ 54.6% MH: ↑ 58.6% MH: ↑ 80% MH: ↑ 68%	[71]
Cu (99.9%)	710 rpm 20 mm/min 1 pass	Holes (depth 3 mm, length 2 mm, spacing 4 mm)	TiO_2 (d_{av} = 41 nm)	30/21	MH: ↑ 8% UTS: ↑ 2.5% Elong.: ↓ 1.9 time CF: ↓ 14%	[72]
	710 rpm 20 mm/min 1 pass			30/9.3	MH: ↑ 28.3% UTS: ↑ 22.6% Elong.: ↓ 2.7 time CF: ↓ 48.4%	
	710 rpm 20 mm/min 2 passes			30/6.4	MH: ↑ 50% UTS: ↑ 27.6% Elong.: ↓ 3.5 time CF: ↓ 60.9%	
	710 rpm 20 mm/min 4 passes			30/2.4	MH: ↑ 77% UTS: ↑ 33% Elong.: ↓ 2.4 time CF: ↓ 75%	
Cu	1000 rpm 30 mm/min 1 pass	Groove	AlN (d_{av} = 10 μm), BN (d_{av} = 1 μm): 5 vol. % (25 mass. % AlN + 75 mass. % BN) AlN (d_{av} = 10 μm), BN (d_{av} = 1 μm): 10 vol. % (25 mass. % AlN + 75 mass. % BN) AlN (d_{av} = 10 μm), BN (d_{av} = 1 μm): 15 vol. % (25 mass. % AlN + 75 mass. % BN)	-/- -/- -/-	MH: ↑ 25% UTS: ↓ 26.6% Elong.: ↓ 1.4 times MH: ↑ 28.3% UTS: ↓ 19.7% Elong.: ↓ 1.5 times MH: ↑ 29.2% UTS: ↓ 19.7% Elong.: ↓ 2.3 time	[73]

Table 5. *Cont.*

Material	FSP Parameters	Particle Introduction Method	Reinforcing Particles (size)	Average grain Size of the Base Alloy/Average Grain Size after FSP, μm	Mechanical Properties	Ref. No.
Titanium alloys						
CP-Ti	800 rpm 45 mm/min 3 passes	-	-	75/4	MH: ↑ 56.2%	[140]
	500 rpm 50 mm/min 4 passes	Grooves (width 2 mm, depth 2 mm)	β-SiC powder (d_{av} = 50 nm)	75/0.4	MH: ↑ 228%	
CP-Ti grade 2	500 rpm 50 mm/min 1 pass	-	-	28/4.4	-	[141]
	500 rpm 50 mm/min 4 passes	-	-	28/2.6	-	
	500 rpm 50 mm/min 1 pass	Groove (width 1 mm, depth 3 mm)	~1.8 vol. % Al_2O_3 (d_{av} = 80 nm)	28/1.14	-	
Magnesium alloys						
AZ31B	1000 rpm 40 mm/min 3 passes	Groove 2×4 mm^2	1.6 vol. % MWCNT (d_{av} = 10–30 nm) + 0.3 vol. % graphene (d_{av} = 5–10 nm)	-/-		[74]
	1200 rpm 40 mm/min 3 passes			-/-		
	1400 rpm 40 mm/min 3 passes			-/-		

Table 5. *Cont.*

Material	FSP Parameters	Particle Introduction Method	Reinforcing Particles (size)	Average grain Size of the Base Alloy/Average Grain Size after FSP, μm	Mechanical Properties	Ref. No.
Mg + 5 wt. % (SiC + Al$_2$O$_3$)	SiC and Al$_2$O$_3$ hybrid particles were added to molten metal at 700 °C. The mixture was stirred for 20 min at 400 rpm with a stirrer, followed by pouring into a permanent mould	Casting	5 wt. % (SiC + Al$_2$O$_3$)	82	MH: 59.3 HV	[75]
	220 rpm 10 mm/min 1 pass			82/15	MH: ↑ 13.9%	
	340 rpm 20 mm/min 1 pass			82/11	MH: ↑ 15.5%	
	560 rpm 30 mm/min 1 pass			82/7	MH: ↑ 17.5%	

FSP of structural alloys allows the formation of gradient composite structures with the hardness increased by 13–80%, tensile strength by 2.5–75%, compressive strength by 70%, and wear resistance by 14–26% compared to the base metal (Table 5). As can be seen from Table 5, the tensile ductility values of many particle-reinforced structural alloys are lower than those of the base metals. The tensile characteristics depend on many microstructural factors such as the interaction between the base metal matrix and reinforcing particles, the particle size distribution in the processed area, and the dislocation density. The main reason for the deteriorated tensile strength of simply processed and reinforced materials as compared to the base metals is the residual stresses induced by the enormous heat release in FSP [74]. In addition, the presence of hard reinforcing particles inside the grains and at the grain boundaries causes high stress concentrations in zones with harder particles prone to crack initiation and growth, as a result of which the material ductility is reduced [74]. An analysis of Table 5 also shows increased hardness for all FSPed particle-reinforced structural alloys, despite some cases when both ductility and tensile strength are reduced. In most experimental studies, the hardness enhancement is attributed to grain refinement and the presence of fine reinforcing particles in accordance to the Hall-Petch and Orowan mechanisms, respectively [68,72,138,142]. In addition, as a result of increasing dispersion of reinforcement particles the distance between them is reduced and therefore the free run length of dislocations is restricted. The restriction of dislocation motion also contributes to higher microhardness of surface composites.

5. Friction Stir Processing of Structural Alloys for Fabricating In Situ Hybrid Surfaces

Of greatest interest in the last decade is the fabrication of hybrid composites by in situ reactions during FSP. The given FSP technique provides almost complete mixing of the introduced powder with the plasticized substrate metal due to a complex quasi-viscous material flow at temperatures below the melting point. The in situ hybrid composite FSP method has several advantages over other FSP methods used for composite fabrication: (1) more thermodynamically stable matrix reinforcement [143], (2) coherent/semi-coherent bonding at the particle/matrix interfaces (Figure 10) [60,79,136], and (3) formation of finer reinforcing particles uniformly distributed in the matrix [82]. The interfacial characteristics, including the interfacial bonding structure, intermediate phase formation, and thermal expansion difference, are also fundamental and depend on the chemical composition of both the introduced particles and the matrix. The complexity of interfacial reactions affects the adhesion between particles and the matrix, which has an additional effect on the mechanical properties of in situ hybrid composites. The high strain rate and friction during FSP produce a large amount of heat, the material temperature rises, resulting in a higher diffusion rate and shorter diffusion distances. All these factors accelerate the in situ exothermic reactions between the metal matrix atoms and the introduced particles. Since the reactions are exothermic, there is additional heat release that also contributes to the temperature rise and reaction acceleration. High strains and temperatures reached during FSP cause fragmentation and dissolution of the reinforcing particles, which leads to further precipitation of smaller intermetallic particles and their more uniform distribution in the matrix.

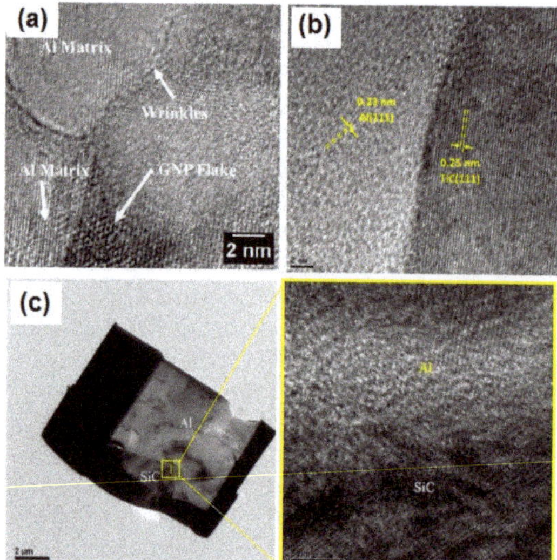

Figure 10. Coherent and semi-coherent interfaces between the matrix and reinforcing particles: (**a**) TEM-BF images of FSPed A356 reinforced with graphene nanoplatelets (GNP) showing the interface between the matrix and encapsulated GNP flakes (reproduced from [136], with permission from Elsevier, 2020); (**b**) HRTEM micrograph of the particle/matrix interface in the six-pass FSPed composite (reproduced from [79], with permission from Elsevier,2019); (**c**) HRTEM micrographs of the stir zone containing bare SiC reinforcement (reproduced from [60], with permission from authors, 2018).

As noted by Zhang et al. [143], the heat release of the metal/metal oxide reaction is much higher than that of the metal/transition metal reaction. Therefore, a reaction with enhanced formation kinetics is expected for the metal/metal oxide system. Moreover, the formation of nano-sized reaction products with coherent or semi-coherent interfaces can improve the mechanical properties. Experimental studies showed that in situ hybrid composites can be fabricated by FSP using the systems Al-CeO$_2$ [96], Al-TiO$_2$ [143], Al + Mg + CuO [98], and Al-Al$_{13}$Fe$_4$-Al$_2$O$_3$ [82]. In many oxide/aluminum substitution reactions, the reduced metal can exothermically react with Al to form an intermetallic compound, due to which the system temperature rises [82]. As was shown for an aluminum-based in situ composite synthesized from an Al-Mg-CuO powder mixture by FSP, the use of the Mg/metal oxide substitution reaction instead of the Al/metal oxide one has a positive effect on the synthesized aluminum-based in situ nanocomposites [98]. The nano-sized MgO and Al$_2$Cu particle-reinforced composite exhibits an excellent Young's modulus (88 GPa) and yield strength (350 MPa in tension and 436 MPa in compression) [98].

In the work by Azimi-Roeen et al. [82], pre-milled powder mixture (Al$_{13}$Fe$_4$ + Al$_2$O$_3$) was introduced into the stir zone formed in a 1050 aluminum alloy sheet by FSP. The homogeneous and active mixture reacted with plasticized aluminum to form Al$_{13}$Fe$_4$ + Al$_2$O$_3$ particles. The intermetallic Al$_{13}$Fe$_4$ was represented by elliptical particles of ~100 nm in size, and nano-sized Al$_2$O$_3$ precipitated in the form of flocculated particles with the remnants of iron oxide particles. With increasing milling time (1–3 h) of the introduced powder mixture, the volume fraction of Al$_{13}$Fe$_4$ + Al$_2$O$_3$ increased in the fabricated composite. The hardness and tensile strength of the nanocomposites varied from 54.5 HV to 75 HV and from 139 MPa to 159 MPa, respectively (Table 6) [82].

Table 6. Experimental data on FSP of in situ hybrid composites.

Material	FSP Parameters	Particle Introduction Method	Introduced Particles (Size)	Average Grain Size of the Base Alloy/Average Grain Size after FSP, μm	Formation of Additional/Intermetallic Phases	Mechanical Properties	Ref. No.
Aluminum alloys							
7075	1200 rpm 30 mm/min 1 pass	Groove (width 3 mm, depth 3 mm)	Ti-6Al-4V (d_{av} = 35 nm)	-/-	Al_3Ti AlTi $AlTi_3$	MH: ↑ 3.3% UTS: ↑ 7.4% Elong.: ↑ 1.2 times FC: 0.7	[78]
	1200 rpm 30 mm/min 2 passes		Ti-6Al-4V (d_{av} = 35 nm)	-/-	Al_3Ti AlTi $AlTi_3$	MH: ↑ 28.3% UTS: ↑ 23.5% Elong.: ↑ 2 times FC: 0.58	
	1200 rpm 30 mm/min 3 passes		Ti-6Al-4V (d_{av} = 35 nm)	-/-	Al_3Ti AlTi $AlTi_3$	MH: ↑ 60% UTS: ↑ 38.8% Elong.: ↑ 2.1 times FC: 0.32	
AA1050	1400 rpm 40 mm/min 2 passes	Holes (diameter 2 mm, depth 3 mm)	Ni (≤ 20 μm), Ti (40-60 μm), C (50 μm). Powder mixture Ni-32 mass. % Ti-8 mass % C. Preliminary planetary ball milling	-/-	Al_3Ni TiC	-	[79]
	1400 rpm 40 mm/min 4 passes			-/-	Al_3Ni TiC	-	
	1400 rpm 40 mm/min 6 passes			-/-	Al_3Ni TiC	MH: ↑ 214%	
Al6061-T651	1000 rpm 72 mm/min 1 pass	Slot in the butt end of the plate	SiC (d_{av} = 3-6 μm) with 1.3–1.8 μm thick copper coating	-/-	Al_2Cu Al_4Cu_9	MH: ↓ 11% UTS: ↓ 24.6% Elong.: ↑ 18.7%	[60]
	1000 rpm 72 mm/min 2 passes		SiC (d_{av} = 3–6 μm) with 1.3–1.8 μm thick copper coating	-/-	Al_2Cu Al_4Cu_9	MH: ↑ 16.6% UTS: ↓ 15% Elong.: ↑ 29.6%	

Table 6. *Cont.*

Material	FSP Parameters	Particle Introduction Method	Introduced Particles (Size)	Average Grain Size of the Base Alloy/Average Grain Size after FSP, μm	Formation of Additional/Intermetallic Phases	Mechanical Properties	Ref. No.
A356	1600 rpm 50 mm/min 1 pass	Groove (width 0.6 mm, depth 3.5 mm)	Powder mixture SiCp (d_{av} = 30 μm) − MoS_2 (d_{av} = 5 μm)	Destruction of needle-like Si and Al dendrites	SiCp and MoS_2 particles (d_{av} ~10 μm)	MH: ↑ 45.4% FC: ↓ 2 times	[80]
	1600 rpm 50 mm/min 1 pass	Groove (width 0.6 mm, depth 3.5 mm)	SiCp (d_{av} = 30 μm)	Destruction of needle-like Si and Al dendrites	SiCp particles (d_{av} ~10 μm)	MH: ↑ 54.5%	
A6061	1600 rpm 60 mm/min 2 passes	Groove dimensions correspond to 18 vol.% of reinforcing particles	18 vol. % fly ash (d_{av} = 5 μm)	76.85/5.61	Uniform distribution of fly ash particles independently of the metal matrix type	MH: ↑ 2 times	[81]
1050	1120 rpm 125 mm/min 4 passes	Groove (depth 3.5 mm, width 1.4 mm)	Powder mixture Fe_2O_3 (d_{av} = 1 μm) − Al (d_{av} = 100 μm), pre-mixed and pre-ground	−/~2–3	$Al_{13}Fe_4$ (~100 nm) α-Al_2O_3, Fe_3O_4	MH: ↑ 27.3%	[82]
Al-1050-H24	750 rpm 99.4 mm/min	Groove (width 3 mm, depth 1.5 mm)	Cu powder (d_{av} = 5 μm)	−/−	$CuAl_2$ Al-Cu Al_4Cu_9	MH: ↑ 4 times	[83]
	750 rpm 49.7 mm/min			−/−		MH: ↑ 5 times	
A413	2000 rpm 8 mm/min 1 pass 2000 rpm 8 mm/min 3 passes	Groove 2 × 3 mm²	Ni powder (d_{av} = 1–3 μm)	Si: 40.6/4.58 Si: 40.6/2.8	Al_3Ni	MH: ↑ 18.8% CF: ↓ 1.5 times MH: ↑ 26.5% CF: ↓ 1.5 times	[84]

Table 6. *Cont.*

Material	FSP Parameters	Particle Introduction Method	Introduced Particles (Size)	Average Grain Size of the Base Alloy/Average Grain Size after FSP, μm	Formation of Additional/Intermetallic Phases	Mechanical Properties	Ref. No.
Al1100	1180 rpm 60 mm/min 2 passes	Groove (width 3 mm, depth 5 mm)	Ni powder (d_{av} = 25–38 μm)	-/-	Nonuniform distribution of a small amount of Al_3Ni particles	MH: ↑ 1.8 times UTS: ↑ 1.5 times Elong.: ↓ 1.9 times	[85]
	1180 rpm 60 mm/min 4 passes			-/-	More uniform distribution of Al_3Ni	MH: ↑ 2.5 times UTS: ↑ 1.8 times Elong.: ↓ 3.5 times	
	1180 rpm 60 mm/min 6 passes			-/-	Uniform Al_3Ni distribution (d_{av} ≤ 1 μm)	MH: ↑ 2.7 times UTS: ↑ 1.9 times Elong.: ↓ 3.9 times	
Al1050	1600 rpm 20 mm/min 2 passes	Groove $1 \times 2 \times 160$ mm^3	Nb powder (d = 1–10 μm)	60/23	Al_3Nb Al_3Nb Al_3Nb	MH: ↑ 13.6% UTS: ↑ 13.3% Elong.: ↓ 2.5 times	[86]
	1600 rpm 20 mm/min 4 passes	Groove $1 \times 2 \times 160$ mm^3		60/6.5		MH: ↑ 54.5% UTS: ↑ 33.3% Elong.: ↓ 1.6 times	
	1600 rpm 20 mm/min 4 passes	Groove $2 \times 2 \times 160$ mm^3		60/4		MH: ↑ 100% UTS: ↑ 33.3% Elong.: ↑ 2 times	
AA5052	1250 rpm 25 mm/min 1 pass	-	-	10.7/9.7	-	MH: ↓ 9% UTS: ↑ 14.4% Elong.: ↑ 3.4%	[87]
	1200 rpm 100 mm/min 5 passes	Groove (width 1.2 mm, depth 3.5 mm)	Powders of graphene nancplatelets (diameter 2 μm, thickness 1–20 nm)	10.7/2.1	$(Fe,Mn,Cr)_3SiAl_{12}$ particles (d_{av} ≤ 1 μm), Al_4C_3 particles	MH: ↑ 52.7% UTS: ↑ 35.7% Elong.: ↓ 31.8%	

Table 6. *Cont.*

Material	FSP Parameters	Particle Introduction Method	Introduced Particles (Size)	Average Grain Size of the Base Alloy/Average Grain Size after FSP, μm	Formation of Additional/Intermetallic Phases	Mechanical Properties	Ref. No.
Copper alloys							
Cu plate (99.9% pure)	1000 rpm 40 mm/min 2 passes	Groove dimensions correspond to 18 vol. % of reinforcing particles	18 vol. % fly ash ($d_{av} = 5$ μm)	35.43/2.79	Uniform distribution of fly ash particles independently of the metal matrix type	MH: ↑ 2.13 times	[81]
Titanium alloys							
Ti-6Al-4V	800 rpm 25 mm/min 1 pass	-	-	-/-	-	MH: ↑ 5.4% CS: ↑ 1.7%	[88]
		Holes (diameter 1.2 mm, depth 3.8 mm, spacing 2.5 mm)	B_4C ($d_{av} = 10$ μm)	-/-	TiB, TiB$_2$, TiC	MH: ↑ 68% CS: ↑ 47.9%	
Ti	1200 rpm 50 mm/min 1 pass	-	-	92.2/~2	-	MH: ↑ 25.5% UTS: ↑ 28.8% Elong.: ↓ 33.7%	[89]
		Groove (length 210 mm, width 1.2 mm, depth 3.5 mm)	Hydroxy-apatite powder Ca$_{10}$(PO$_4$)$_6$(OH)$_2$ ($d_{av} = 120$ nm)	92.2/1.4–14.8	Decomposition products in the form of elemental calcium (Ca) and phosphide (PO$_3$)	MH: ↑ 34.8% UTS: ↓ 41.6% Elong.: ↓ 52.2%	
Magnesium alloys							
AZ31	1200 rpm 40 mm/min 2 passes	Groove dimensions correspond to 18 vol.% of reinforcing particles	18 vol.% fly ash ($d_{av} = 5$ μm)	66.35/6.09	Uniform distribution of fly ash particles independently of the metal matrix type	MH: ↑ 1.75 times	[81]

In the case of an incoherent bonding interface between particles and the metal matrix, the surface characteristics of the particles can be modified by additional processing, e.g., by plating. Huang and Aoh [60] performed electroless plating to deposit a copper coating on the surface of SiC ceramic particles to change their surface characteristics. The preliminary processing of the particles provided interphase coherence through the formation of Al_2Cu and Al_4Cu_9 intermetallic compounds at the interphase boundary. Double-pass FSP with the Cu-coated reinforcement increased the composite hardness and ductility by about 20% (Figure 11, Table 6).

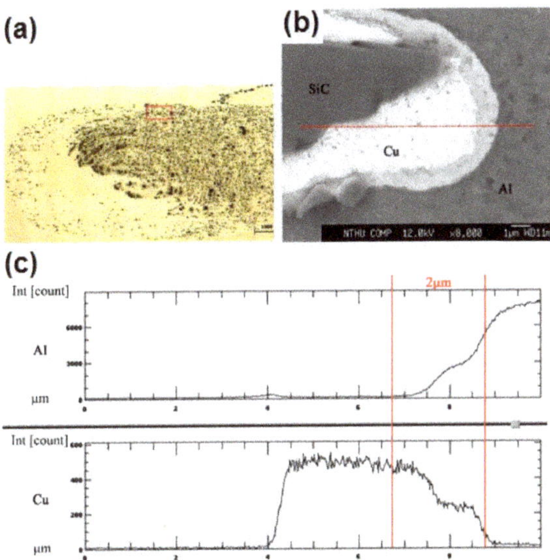

Figure 11. Micrographs of copper-coated SiC particles embedded in the matrix (**a**,**b**); Al-SiC/Cu reinforcement with EPMA line scan across Cu-coated SiC and Al matrix showing Al and Cu distribution (**c**) (reproduced from [60], with permission from authors, 2018).

The FSP method is also used to fabricate in situ metal matrix composites of the compositions Al-Al_2Cu [60], Al-Al_3Ti [78], Al-$Al_{13}Fe_4$ [82], Al-Al_3Ni [79,84,85] with the formation of intermetallic phases. Such composites are mainly synthesized using powder mixtures subjected to pre-processing and special preparation. For an FSPed A413 alloy reinforced by Ni powder, Golmohammadi et al. [88] observed the destruction of needle-like Si particles and in situ formation of uniformly distributed intermetallic Al_3Ni particles. An increase in the number of FSP passes led to less agglomeration, finer and more uniform dispersion of reinforcing particles, as well as to an increase in the intermetallic phase length. The authors showed that the wear resistance of the Al-Al3Ni composite is higher by approximately a factor of 2 than that of the base alloy (Table 6) [84].

Experiments showed that the addition of carbon-based solid lubricating particles (graphene particles and platelets, nanotubes, fibers, etc.) together with hard particles improves the tribological behavior of in situ composites under sliding wear conditions [69,80,87]. Dixit et al. [144] synthesized new multi-layer graphene-reinforced aluminum composites by exfoliating cheap graphite into graphene using friction stir alloying, and observed a twofold increase in strength. This method opened up new possibilities for the efficient and scalable production of graphene-based metal matrix nanocomposites [144].

Experimental studies were performed for FSPed in situ composites on the basis of aluminum alloys: Al7075-Ti-6Al-4V [78], Al1050-Ni-Ti-C [79], Al-SiC [60,80], Al6061-fly ash [81,90], Al1050-Fe_2O_3-Al [82], Al-1050-Cu [83], Al-Ni [84,85], Al-Nb [86], Al-graphene [69,80,87]; copper alloys: Cu-fly ash [81]; titanium alloys: Ti-6Al-4V-B_4C [81], Ti-hydroxyapatite powder [89]; magnesium alloys: AZ31-fly ash [81]. A review of the experimental data on FSP of in situ hybrid composite materials is given in Table 6.

6. Conclusions

This paper summarizes the latest progress in the study of friction stir processing of aluminum, copper, titanium, and magnesium alloys. Severe plastic deformation and thermal effects during FSP cause the destruction of large dendrites and second phase particles, grain refinement in the matrix, elimination of porosity, as well as the formation of a homogeneous fine-grained structure. It was shown that FSP can be applied to fabricate metallic materials:

(1)   with a subsurface gradient structure obtained through the formation of equiaxed nanograins and structural homogenization;
(2)   with a compositional subsurface gradient structure formed by modifying and hardening the material surface with reinforcing particles;
(3)   in situ composites.

FSP of structural alloys proves to be the most energy efficient, environmentally friendly, and versatile method that allows local controlled modification of the subsurface microstructure in the processed structural materials.

However, the literature contains a wide scatter of experimental results on the properties of FSPed metallic materials, indicating the necessity of further research in this relevant area. A reason for the large data scatter can be the physical nature of the friction stir process based on the phenomenon of adhesion friction, which is of highly inhomogeneous nature as compared to lubricated friction. In view of the frictional inhomogeneity, it is possible to fabricate materials with markedly different properties by using slightly different FSP tool geometries at the same processing parameters. Despite the presence of unresolved issues concerning the FSP of structural alloys, the given method shows much promise for commercial applications.

Author Contributions: Resources, writing part 1–3, 6, A.P.Z.; writing part 4, A.V.C.; writing part 5, S.Y.T.; guidance and corrections to the article, E.A.K. All authors have read and agreed to the published version of the manuscript.

Funding: A review of references of friction stir processing of alloys, hardening with various reinforcing particles, fabrication of hybrid and in situ surfaces was funded by Government research assignment for ISPMS SB RAS, project No. III.23.2.11. A review of references of friction stir processing of aluminum and copper alloys was performed with financial support from RSF Grant No. 19-79-00136.

Conflicts of Interest: The authors declare no conflict of interest.

Nomenclature

FSP	friction stir processing
FSW	friction stir welding
MH	microhardness
UTS	ultimate tensile strength
Elong	elongation
WT	wear testing
CR	corrosion resistance
IT	impact toughness
SWCNTs	single-walled carbon nanotubes
MWCNT	multi-walled carbon nanotubes
CNTs	carbon nanotubes
CS	compressive strength
HRTEM	high resolution transmission electron microscopy

References

1. Hekimoğlu, A.P.; Çalış, M.; Ayata, G. Effect of Strontium and Magnesium Additions on the Microstructure and Mechanical Properties of Al–12Si Alloys. *Met. Mater. Int.* **2019**, *25*, 1488–1499. [CrossRef]
2. He, T.; Chen, W.; Wang, W.; Ren, F.; Stock, H.R. Effect of different Cu contents on the microstructure and hydrogen production of Al–Cu-Ga-In-Sn alloys for dissolvable materials. *J. Alloys Compd.* **2020**, *821*, 153489. [CrossRef]
3. Trudonoshyn, O.; Rehm, S.; Randelzhofer, P. Körner Improvement of the high-pressure die casting alloy Al-5.7Mg-2.6Si-0.7Mn with Zn addition. *Mater. Charact.* **2019**, *158*, 109959. [CrossRef]
4. Prach, O.; Trudonoshyn, O.; Randelzhofer, P.; Körner, C.; Durst, K. Effect of Zr, Cr and Sc on the Al–Mg–Si–Mn high-pressure die casting alloys. *Mater. Sci. Eng. A* **2019**, *759*, 603–612. [CrossRef]
5. Fabrizi, A.; Ferraro, S.; Timelli, G. The influence of Sr, Mg and Cu addition on the microstructural properties of a secondary AlSi9Cu3(Fe) die casting alloy. *Mater. Charact.* **2013**, *85*, 13–25. [CrossRef]
6. Seifeddine, S.; Johansson, S.; Svensson, I.L. The influence of cooling rate and manganese content on the β-Al$_5$FeSi phase formation and mechanical properties of Al–Si-based alloys. *Mater. Sci. Eng. A* **2008**, *490*, 385–390. [CrossRef]
7. Rao, Y.; Yan, H.; Hu, Z. Modification of eutectic silicon and β-Al5FeSi phases in as-cast ADC12 alloys by using samarium addition. *J. Rare Earths* **2013**, *31*, 916–922. [CrossRef]
8. Cinkilic, E.; Ridgeway, C.D.; Yan, X.; Luo, A.A. A Formation Map of Iron-Containing Intermetallic Phases in Recycled Cast Aluminum Alloys. *Metall. Mater. Trans. A Phys. Metall. Mater. Sci.* **2019**, *50*, 5945–5956. [CrossRef]
9. Irizalp, S.G.; Saklakoglu, N. Effect of Fe-rich intermetallics on the microstructure and mechanical properties of thixoformed A380 aluminum alloy. *Eng. Sci. Technol. Int. J.* **2014**, *17*, 58–62. [CrossRef]
10. Mohamed, A.M.A.; Samuel, A.M.; Samuel, F.H.; Doty, H.W. Influence of additives on the microstructure and tensile properties of near-eutectic Al–10.8%Si cast alloy. *Mater. Des.* **2009**, *30*, 3943–3957. [CrossRef]
11. Li, Q.; Xia, T.; Lan, Y.; Zhao, W.; Fan, L.; Li, P. Effect of in situ γ-Al2O3 particles on the microstructure of hypereutectic Al–20%Si alloy. *J. Alloys Compd.* **2013**, *577*, 232–236. [CrossRef]
12. Huang, X.; Yan, H. Effect of trace la addition on the microstructure and mechanical property of as-cast ADC12 Al-Alloy. *J. Wuhan Univ. Technol. Mater. Sci. Ed.* **2013**, *28*, 202–205. [CrossRef]
13. Li, J.H.; Wang, X.D.; Ludwig, T.H.; Tsunekawa, Y.; Arnberg, L.; Jiang, J.Z.; Schumacher, P. Modification of eutectic Si in Al–Si alloys with Eu addition. *Acta Mater.* **2015**, *84*, 153–163. [CrossRef]
14. Czerwinski, F. Cerium in aluminum alloys. *J. Mater. Sci.* **2020**, *55*, 24–72. [CrossRef]
15. Niu, G.; Mao, J.; Wang, J. Effect of Ce Addition on Fluidity of Casting Aluminum Alloy A356. *Metall. Mater. Trans. A Phys. Metall. Mater. Sci.* **2019**, *50*, 5935–5944. [CrossRef]
16. Ravi, M.; Pillai, U.T.S.; Pai, B.C.; Damodaran, A.D.; Dwarakadasa, E.S. The effect of mischmetal addition on the structure and mechanical properties of a cast Al-7Si-0.3Mg alloy containing excess iron (up to 0.6 pct). *Metall. Mater. Trans. A Phys. Metall. Mater. Sci.* **2002**, *33*, 391–400. [CrossRef]
17. Bijalwan, P.; Pandey, K.K.; Mukherjee, B.; Islam, A.; Pathak, A.; Dutta, M.; Keshri, A.K. Tailoring the bimodal zone in plasma sprayed CNT reinforced YSZ coating and its impact on mechanical and tribological properties. *Surf. Coat. Technol.* **2019**, *377*, 124870. [CrossRef]
18. Gui, M.; Kang, S.B.; Euh, K. Influence of spraying conditions on microstructures of Al-SiCpcomposites by plasma spraying. *Metall. Mater. Trans. A* **2005**, *36*, 2471–2480. [CrossRef]
19. Maharajan, S.; Ravindran, D.; Rajakarunakaran, S.; Khan, M.A. Analysis of surface properties of tungsten carbide (WC) coating over austenitic stainless steel (SS316) using plasma spray process. *Mater. Today Proc.* **2019**. [CrossRef]
20. Xi, H.; He, P.; Wang, H.; Liu, M.; Chen, S.; Xing, Z.; Ma, G.; Lv, Z. Microstructure and mechanical properties of Mo coating deposited by supersonic plasma spraying. *Int. J. Refract. Met. Hard Mater.* **2020**, *86*, 105095. [CrossRef]
21. Padmini, B.V.; Mathapati, M.; Niranjan, H.B.; Sampathkumaran, P.; Seetharamu, S.; Ramesh, M.R.; Mohan, N. High temperature tribological studies of cold sprayed nickel based alloy on low carbon steels. *Mater. Today Proc.* **2019**. [CrossRef]

22. Liu, Z.; Wang, H.; Haché, M.; Irissou, E.; Zou, Y. Formation of refined grains below 10 nm in size and nanoscale interlocking in the particle–particle interfacial regions of cold sprayed pure aluminum. *Scr. Mater.* **2020**, *177*, 96–100. [CrossRef]

23. Dabney, T.; Johnson, G.; Yeom, H.; Maier, B.; Walters, J.; Sridharan, K. Experimental evaluation of cold spray FeCrAl alloys coated zirconium-alloy for potential accident tolerant fuel cladding. *Nucl. Mater. Energy* **2019**, *21*, 100715. [CrossRef]

24. Hu, C.; Baker, T.N. A new aluminium silicon carbide formed in laser processing. *J. Mater. Sci.* **1997**, *32*, 5047–5051. [CrossRef]

25. Pantelis, D.; Tissandier, A.; Manolatos, P.; Ponthiaux, P. Formation of wear resistant Al–SiC surface composite by laser melt–particle injection process. *Mater. Sci. Technol.* **1995**, *11*, 299–303. [CrossRef]

26. Katipelli, L.R.; Dahotre, N.B. Mechanism of high temperature oxidation of laser surface engineered TiC/Al alloy 'composite' coating on 6061 aluminium alloy. *Mater. Sci. Technol.* **2001**, *17*, 1061–1068. [CrossRef]

27. Dong, M.; Cui, X.; Lu, B.; Jin, G.; Cai, Z.; Feng, X.; Liu, Z.; Wang, H. Effect of Ti+N and Zr+N ions implantation on mechanical and corrosion performance of carburized layer. *Thin Solid Films* **2019**, *692*, 137597. [CrossRef]

28. Chen, T.; Castanon, E.; Gigax, J.G.; Kim, H.; Balerio, R.; Fan, J.; Garner, F.A.; Shao, L. Nitrogen ion implantation into pure iron for formation of surface nitride layer. *Nucl. Instruments Methods Phys. Res. Sect. B Beam Interact. Mater. Atoms* **2019**, *451*, 10–13. [CrossRef]

29. Acciari, H.A.; Palma, D.P.S.; Codaro, E.N.; Zhou, Q.; Wang, J.; Ling, Y.; Zhang, J.; Zhang, Z. Surface modifications by both anodic oxidation and ion beam implantation on electropolished titanium substrates. *Appl. Surf. Sci.* **2019**, *487*, 1111–1120. [CrossRef]

30. Titov, A.I.; Karaseov, P.A.; Karabeshkin, K.V.; Ermolaeva, G.M.; Shilov, V.B. Effect of monatomic and molecular ion irradiation on time resolved photoluminescence decay in GaN. *Nucl. Instruments Methods Phys. Res. Sect. B Beam Interact. Mater. Atoms* **2019**, *458*, 164–168. [CrossRef]

31. Stepanov, A.L.; Vorobev, V.V.; Rogov, A.M.; Nuzhdin, V.I.; Valeev, V.F. Sputtering of silicon surface by silver-ion implantation. *Nucl. Instruments Methods Phys. Res. Sect. B Beam Interact. Mater. Atoms* **2019**, *457*, 1–3. [CrossRef]

32. Kalashnikov, M.P.; Fedorischeva, M.V.; Sergeev, V.P.; Neyfeld, V.V.; Popova, N.A. Features of surface layer structure of VT23 titanium alloy under bombardment with copper ions. *AIP Conf. Proc.* **2015**, *1683*. [CrossRef]

33. Ma, Z.Y. Friction Stir Processing Technology: A Review. *Metall. Mater. Trans. A* **2008**, *39*, 642–658. [CrossRef]

34. Kumar, R.A.; Kumar, R.G.A.; Ahamed, K.A.; Alstyn, B.D.; Vignesh, V. Review of Friction Stir Processing of Aluminium Alloys. *Mater. Today Proc.* **2019**, *16*, 1048–1054. [CrossRef]

35. Węglowski, M.S. Friction stir processing – State of the art. *Arch. Civ. Mech. Eng.* **2018**, *18*, 114–129. [CrossRef]

36. Padhy, G.K.; Wu, C.S.; Gao, S. Friction stir based welding and processing technologies - processes, parameters, microstructures and applications: A review. *J. Mater. Sci. Technol.* **2018**, *34*, 1–38. [CrossRef]

37. Rao, A.G.; Ravi, K.R.; Ramakrishnarao, B.; Deshmukh, V.P.; Sharma, A.; Prabhu, N.; Kashyap, B.P. Recrystallization Phenomena During Friction Stir Processing of Hypereutectic Aluminum-Silicon Alloy. *Metall. Mater. Trans. A* **2013**, *44*, 1519–1529. [CrossRef]

38. Sun, H.; Yang, S.; Jin, D. Improvement of Microstructure, Mechanical Properties and Corrosion Resistance of Cast Al–12Si Alloy by Friction Stir Processing. *Trans. Indian Inst. Met.* **2018**, *71*, 985–991. [CrossRef]

39. Zhao, H.; Pan, Q.; Qin, Q.; Wu, Y.; Su, X. Effect of the processing parameters of friction stir processing on the microstructure and mechanical properties of 6063 aluminum alloy. *Mater. Sci. Eng. A* **2019**, *751*, 70–79. [CrossRef]

40. Abrahams, R.; Mikhail, J.; Fasihi, P. Effect of friction stir process parameters on the mechanical properties of 5005-H34 and 7075-T651 aluminium alloys. *Mater. Sci. Eng. A* **2019**, *751*, 363–373. [CrossRef]

41. Kalashnikov, K.N.; Vorontsov, A.V.; Kalashnikova, T.A.; Chumaevskii, A.V. Changes in the structure and properties of aluminum alloys during friction stir processing by different types of tools. *AIP Conf. Proc.* **2018**, *2053*, 40038. [CrossRef]

42. Kalashnikov, K.N.; Tarasov, S.Y.; Chumaevskii, A.V.; Fortuna, S.V.; Eliseev, A.A.; Ivanov, A.N. Towards aging in a multipass friction stir–processed AA2024. *Int. J. Adv. Manuf. Technol.* **2019**, *103*, 2121–2132. [CrossRef]

43. Ramesh, K.N.; Pradeep, S.; Pancholi, V. Multipass Friction-Stir Processing and its Effect on Mechanical Properties of Aluminum Alloy 5086. *Metall. Mater. Trans. A* **2012**, *43*, 4311–4319. [CrossRef]

44. Senthilkumar, R.; Prakash, M.; Arun, N.; Jeyakumar, A.A. The effect of the number of passes in friction stir processing of aluminum alloy (AA6082) and its failure analysis. *Appl. Surf. Sci.* **2019**, *491*, 420–431. [CrossRef]

45. Barmouz, M.; Givi, M.K.B.; Seyfi, J. On the role of processing parameters in producing Cu/SiC metal matrix composites via friction stir processing: Investigating microstructure, microhardness, wear and tensile behavior. *Mater. Charact.* **2011**, *62*, 108–117. [CrossRef]

46. Cartigueyen, S.; Mahadevan, K. Role of Friction Stir Processing on Copper and Copper based Particle Reinforced Composites—A Review. *J. Mater. Sci. Surf. Eng.* **2015**, *2*, 133–145.

47. Surekha, K.; Els-Botes, A. Development of high strength, high conductivity copper by friction stir processing. *Mater. Des.* **2011**, *32*, 911–916. [CrossRef]

48. Wang, Y.; Fu, R.; Jing, L.; Li, Y.; Sang, D. Grain refinement and nanostructure formation in pure copper during cryogenic friction stir processing. *Mater. Sci. Eng. A* **2017**, *703*, 470–476. [CrossRef]

49. Cartigueyen, S.; Mahadevan, K. Influence of rotational speed on the formation of friction stir processed zone in pure copper at low-heat input conditions. *J. Manuf. Process.* **2015**, *18*, 124–130. [CrossRef]

50. Bheekya Naik, R.; Venkateswara Reddy, K.; Madhusudhan Reddy, G.; Arockia Kumar, R. Development of High-Strength and High-Electrical Conductivity Cu–Zr Alloy Through Friction Stir Processing. *Trans. Indian Inst. Met.* **2019**, *72*, 1431–1435. [CrossRef]

51. Jiang, L.; Huang, W.; Liu, C.; Chai, L.; Yang, X.; Xu, Q. Microstructure, texture evolution and mechanical properties of pure Ti by friction stir processing with slow rotation speed. *Mater. Charact.* **2019**, *148*, 1–8. [CrossRef]

52. Mironov, S.; Sato, Y.S.; Kokawa, H. Development of grain structure during friction stir welding of pure titanium. *Acta Mater.* **2009**, *57*, 4519–4528. [CrossRef]

53. Liu, F.C.; Liao, J.; Gao, Y.; Nakata, K. Influence of texture on strain localization in stir zone of friction stir welded titanium. *J. Alloys Compd.* **2015**, *626*, 304–308. [CrossRef]

54. Zhang, W.; Ding, H.; Cai, M.; Yang, W.; Li, J. Ultra-grain refinement and enhanced low-temperature superplasticity in a friction stir-processed Ti-6Al-4V alloy. *Mater. Sci. Eng. A* **2018**, *727*, 90–96. [CrossRef]

55. Vakili-Azghandi, M.; Roknian, M.; Szpunar, J.A.; Mousavizade, S.M. Surface modification of pure titanium via friction stir processing: Microstructure evolution and dry sliding wear performance. *J. Alloys Compd.* **2020**, *816*, 152557. [CrossRef]

56. Kalashnikov, K.N.; Kalashnikova, T.A.; Chumaevskii, A.V.; Tarasov, S.Y.; Rubtsov, V.E.; Ivanov, A.N.; Kolubaev, E.A. High-strength friction stir processed dispersion hardened Al-Cu-Mg alloy. *AIP Conf. Proc.* **2017**, *1909*, 1–5. [CrossRef]

57. Kalashnikov, K.N.; Kalashnikova, T.A.; Chumaevskii, A.V.; Ivanov, A.N.; Tarasov, S.Y.; Rubtsov, V.E.; Kolubaev, E.A. Friction-stir processed ultrafine grain high-strength Al-Mg alloy material. *AIP Conf. Proc.* **2017**, *1909*, 1–6. [CrossRef]

58. Barati, M.; Abbasi, M.; Abedini, M. The effects of friction stir processing and friction stir vibration processing on mechanical, wear and corrosion characteristics of Al6061/SiO2 surface composite. *J. Manuf. Process.* **2019**, *45*, 491–497. [CrossRef]

59. Dolatkhah, A.; Golbabaei, P.; Givi, M.K.B.; Molaiekiya, F. Investigating effects of process parameters on microstructural and mechanical properties of Al5052/SiC metal matrix composite fabricated via friction stir processing. *Mater. Des.* **2012**, *37*, 458–464. [CrossRef]

60. Huang, C.W.; Aoh, J.N. Friction stir processing of copper-coated SiC particulate-reinforced aluminum matrix composite. *Materials* **2018**, *11*, 599. [CrossRef]

61. Manochehrian, A.; Heidarpour, A.; Mazaheri, Y.; Ghasemi, S. On the surface reinforcing of A356 aluminum alloy by nanolayered Ti3AlC2 MAX phase via friction stir processing. *Surf. Coat. Technol.* **2019**, *377*, 124884. [CrossRef]

62. Jain, V.K.S.; Varghese, J.; Muthukumaran, S. Effect of First and Second Passes on Microstructure and Wear Properties of Titanium Dioxide-Reinforced Aluminum Surface Composite via Friction Stir Processing. *Arab. J. Sci. Eng.* **2019**, *44*, 949–957. [CrossRef]

63. Abraham, S.J.; Dinaharan, I.; Selvam, J.D.R.; Akinlabi, E.T. Microstructural characterization of vanadium particles reinforced AA6063 aluminum matrix composites via friction stir processing with improved tensile strength and appreciable ductility. *Compos. Commun.* **2019**, *12*, 54–58. [CrossRef]

64. Bourkhani, R.D.; Eivani, A.R.; Nateghi, H.R. Through-thickness inhomogeneity in microstructure and tensile properties and tribological performance of friction stir processed AA1050-Al2O3 nanocomposite. *Compos. Part B Eng.* **2019**, *174*, 107061. [CrossRef]

65. Zahmatkesh, B.; Enayati, M.H. A novel approach for development of surface nanocomposite by friction stir processing. *Mater. Sci. Eng. A* **2010**, *527*, 6734–6740. [CrossRef]

66. Prabhu, M.S.; Perumal, A.E.; Arulvel, S.; Issac, R.F. Friction and wear measurements of friction stir processed aluminium alloy 6082/CaCO$_3$ composite. *Measurement* **2019**, *142*, 10–20. [CrossRef]

67. Deore, H.A.; Mishra, J.; Rao, A.G.; Mehtani, H.; Hiwarkar, V.D. Effect of filler material and post process ageing treatment on microstructure, mechanical properties and wear behaviour of friction stir processed AA 7075 surface composites. *Surf. Coat. Technol.* **2019**, *374*, 52–64. [CrossRef]

68. Zhang, S.; Chen, G.; Wei, J.; Liu, Y.; Xie, R.; Liu, Q.; Zeng, S.; Zhang, G.; Shi, Q. Effects of energy input during friction stir processing on microstructures and mechanical properties of aluminum/carbon nanotubes nanocomposites. *J. Alloys Compd.* **2019**, *798*, 523–530. [CrossRef]

69. Nazari, M.; Eskandari, H.; Khodabakhshi, F. Production and characterization of an advanced AA6061-Graphene-TiB2 hybrid surface nanocomposite by multi-pass friction stir processing. *Surf. Coat. Technol.* **2019**, *377*, 124914. [CrossRef]

70. Pol, N.; Verma, G.; Pandey, R.P.; Shanmugasundaram, T. Fabrication of AA7005/TiB2-B4C surface composite by friction stir processing: Evaluation of ballistic behaviour. *Def. Technol.* **2019**, *15*, 363–368. [CrossRef]

71. Dinaharan, I.; Sathiskumar, R.; Murugan, N. Effect of ceramic particulate type on microstructure and properties of copper matrix composites synthesized by friction stir processing. *J. Mater. Res. Technol.* **2016**, *5*, 302–316. [CrossRef]

72. Heidarpour, A.; Mazaheri, Y.; Roknian, M.; Ghasemi, S. Development of Cu-TiO2 surface nanocomposite by friction stir processing: Effect of pass number on microstructure, mechanical properties, tribological and corrosion behavior. *J. Alloys Compd.* **2019**, *783*, 886–897. [CrossRef]

73. Thankachan, T.; Prakash, K.S.; Kavimani, V. Investigating the effects of hybrid reinforcement particles on the microstructural, mechanical and tribological properties of friction stir processed copper surface composites. *Compos. Part B Eng.* **2019**, *174*, 107057. [CrossRef]

74. Sharma, S.; Handa, A.; Singh, S.S.; Verma, D. Influence of tool rotation speeds on mechanical and morphological properties of friction stir processed nano hybrid composite of MWCNT-Graphene-AZ31 magnesium. *J. Magnes. Alloy.* **2019**, *7*, 487–500. [CrossRef]

75. Arokiasamy, S.; Ronald, B.A. Enhanced properties of Magnesium based metal matrix composites via Friction Stir Processing. *Mater. Today Proc.* **2018**, *5*, 6934–6939. [CrossRef]

76. Satish Kumar, T.; Suganya Priyadharshini, G.; Shalini, S.; Krishna Kumar, K.; Subramanian, R. Characterization of NbC-Reinforced AA7075 Alloy Composites Produced Using Friction Stir Processing. *Trans. Indian Inst. Met.* **2019**, *72*, 1593–1596. [CrossRef]

77. Gangil, N.; Nagar, H.; Kumar, R.; Singh, D. Shape memory alloy reinforced magnesium matrix composite fabricated via friction stir processing. *Mater. Today Proc.* **2020**. [CrossRef]

78. Adetunla, A.; Akinlabi, E. Fabrication of Aluminum Matrix Composites for Automotive Industry Via Multipass Friction Stir Processing Technique. *Int. J. Automot. Technol.* **2019**, *20*, 1079–1088. [CrossRef]

79. Fotoohi, H.; Lotfi, B.; Sadeghian, Z.; Byeon, J. Microstructural characterization and properties of in situ Al-Al3Ni/TiC hybrid composite fabricated by friction stir processing using reactive powder. *Mater. Charact.* **2019**, *149*, 124–132. [CrossRef]

80. Alidokht, S.A.; Abdollah-zadeh, A.; Soleymani, S.; Assadi, H. Microstructure and tribological performance of an aluminium alloy based hybrid composite produced by friction stir processing. *Mater. Des.* **2011**, *32*, 2727–2733. [CrossRef]

81. Dinaharan, I.; Akinlabi, E.T. Low cost metal matrix composites based on aluminum, magnesium and copper reinforced with fly ash prepared using friction stir processing. *Compos. Commun.* **2018**, *9*, 22–26. [CrossRef]

82. Azimi-Roeen, G.; Kashani-Bozorg, S.F.; Nosko, M.; Nagy, Š.; Matko, I. Correction to: Formation of Al/(Al$_{13}$Fe$_4$ + Al$_2$O$_3$) Nano-composites via Mechanical Alloying and Friction Stir Processing. *J. Mater. Eng. Perform.* **2018**, *27*, 6800. [CrossRef]

83. Mahmoud, E.R.I.; Al-qozaim, A.M.A. Fabrication of In-Situ Al–Cu Intermetallics on Aluminum Surface by Friction Stir Processing. *Arab. J. Sci. Eng.* **2016**, *41*, 1757–1769. [CrossRef]

84. Golmohammadi, M.; Atapour, M.; Ashrafi, A. Fabrication and wear characterization of an A413/Ni surface metal matrix composite fabricated via friction stir processing. *Mater. Des.* **2015**, *85*, 471–482. [CrossRef]

85. Qian, J.; Li, J.; Xiong, J.; Zhang, F.; Lin, X. In situ synthesizing Al3Ni for fabrication of intermetallic-reinforced aluminum alloy composites by friction stir processing. *Mater. Sci. Eng. A* **2012**, *550*, 279–285. [CrossRef]

86. Zeidabadi, S.R.H.; Daneshmanesh, H. Fabrication and characterization of in-situ Al/Nb metal/intermetallic surface composite by friction stir processing. *Mater. Sci. Eng. A* **2017**, *702*, 189–195. [CrossRef]

87. Khodabakhshi, F.; Arab, S.M.; Švec, P.; Gerlich, A.P. Fabrication of a new Al-Mg/graphene nanocomposite by multi-pass friction-stir processing: Dispersion, microstructure, stability, and strengthening. *Mater. Charact.* **2017**, *132*, 92–107. [CrossRef]

88. Wang, T.; Gwalani, B.; Shukla, S.; Frank, M.; Mishra, R.S. Development of in situ composites via reactive friction stir processing of Ti–B4C system. *Compos. Part B Eng.* **2019**, *172*, 54–60. [CrossRef]

89. Khodabakhshi, F.; Rahmati, R.; Nosko, M.; Orovčík, L.; Nagy, Š.; Gerlich, A.P. Orientation structural mapping and textural characterization of a CP-Ti/HA surface nanocomposite produced by friction-stir processing. *Surf. Coat. Technol.* **2019**, *374*, 460–475. [CrossRef]

90. Kumar, H.; Prasad, R.; Kumar, P.; Tewari, S.P.; Singh, J.K. Mechanical and tribological characterization of industrial wastes reinforced aluminum alloy composites fabricated via friction stir processing. *J. Alloys Compd.* **2020**, *831*, 154832. [CrossRef]

91. Kumar, K.N.; Aravindkumar, N.; Eswaramoorthi, K. Fabrication of AA6016/(Al$_2$O$_3$ + AlN) hybrid surface composite using friction stir processing. *Mater. Today Proc.* **2020**. [CrossRef]

92. Mishra, R.S.; Ma, Z.Y. Friction stir welding and processing. *Mater. Sci. Eng. R Reports* **2005**, *50*, 1–78. [CrossRef]

93. Li, K.; Liu, X.; Zhao, Y. Research Status and Prospect of Friction Stir Processing Technology. *Coatings* **2019**, *9*, 129. [CrossRef]

94. Tarasov, S.Y.; Rubtsov, V.E.; Fortuna, S.V.; Eliseev, A.A.; Chumaevsky, A.V.; Kalashnikova, T.A.; Kolubaev, E.A. Ultrasonic-assisted aging in friction stir welding on Al-Cu-Li-Mg aluminum alloy. *Weld. World* **2017**, *61*, 679–690. [CrossRef]

95. Sathiskumar, R.; Murugan, N.; Dinaharan, I.; Vijay, S.J. Role of friction stir processing parameters on microstructure and microhardness of boron carbide particulate reinforced copper surface composites. *Sadhana* **2013**, *38*, 1433–1450. [CrossRef]

96. Chen, C.F.; Kao, P.W.; Chang, L.W.; Ho, N.J. Effect of processing parameters on microstructure and mechanical properties of an Al-Al11Ce3-Al2O3 in-situ composite produced by friction stir processing. *Metall. Mater. Trans. A Phys. Metall. Mater. Sci.* **2010**, *41*, 513–522. [CrossRef]

97. Shahi, A.; Sohi, M.H.; Ahmadkhaniha, D.; Ghambari, M. In situ formation of Al–Al3Ni composites on commercially pure aluminium by friction stir processing. *Int. J. Adv. Manuf. Technol.* **2014**, *75*, 1331–1337. [CrossRef]

98. You, G.L.; Ho, N.J.; Kao, P.W. Aluminum based in situ nanocomposite produced from Al–Mg–CuO powder mixture by using friction stir processing. *Mater. Lett.* **2013**, *100*, 219–222. [CrossRef]

99. Hsu, C.J.; Kao, P.W.; Ho, N.J. Ultrafine-grained Al–Al2Cu composite produced in situ by friction stir processing. *Scr. Mater.* **2005**, *53*, 341–345. [CrossRef]

100. Fattah-alhosseini, A.; Vakili-Azghandi, M.; Sheikhi, M.; Keshavarz, M.K. Passive and electrochemical response of friction stir processed pure Titanium. *J. Alloys Compd.* **2017**, *704*, 499–508. [CrossRef]

101. Al-Fadhalah, K.J.; Almazrouee, A.I.; Aloraier, A.S. Microstructure and mechanical properties of multi-pass friction stir processed aluminum alloy 6063. *Mater. Des.* **2014**, *53*, 550–560. [CrossRef]

102. Nakata, K.; Kim, Y.G.; Fujii, H.; Tsumura, T.; Komazaki, T. Improvement of mechanical properties of aluminum die casting alloy by multi-pass friction stir processing. *Mater. Sci. Eng. A* **2006**, *437*, 274–280. [CrossRef]

103. El-Rayes, M.M.; El-Danaf, E.A. The influence of multi-pass friction stir processing on the microstructural and mechanical properties of Aluminum Alloy 6082. *J. Mater. Process. Technol.* **2012**, *212*, 1157–1168. [CrossRef]

104. Girish, G.; Anandakrishnan, V. Determination of friction stir processing window for AA7075. *Mater. Today Proc.* **2020**, *21*, 557–562. [CrossRef]

105. Yang, R.; Zhang, Z.; Zhao, Y.; Chen, G.; Guo, Y.; Liu, M.; Zhang, J. Effect of multi-pass friction stir processing on microstructure and mechanical properties of Al3Ti/A356 composites. *Mater. Charact.* **2015**, *106*, 62–69. [CrossRef]

106. John, J.; Shanmughanatan, S.P.; Kiran, M.B. Effect of tool geometry on microstructure and mechanical Properties of friction stir processed AA2024-T351 aluminium alloy. *Mater. Today Proc.* **2018**, *5*, 2965–2979. [CrossRef]

107. Cartigueyen, S.; Mahadevan, K. Study of friction stir processed zone under different tool pin profiles in pure copper. *IOSR J. Mech. Civ. Eng.* **2014**, *11*, 6–12. [CrossRef]

108. Vijayavel, P.; Balasubramanian, V. Effect of pin profile volume ratio on microstructure and tensile properties of friction stir processed aluminum based metal matrix composites. *J. Alloys Compd.* **2017**, *729*, 828–842. [CrossRef]

109. Surekha, K.; Murty, B.S.; Rao, K.P. Microstructural characterization and corrosion behavior of multipass friction stir processed AA2219 aluminium alloy. *Surf. Coat. Technol.* **2008**, *202*, 4057–4068. [CrossRef]

110. Barmouz, M.; Besharati Givi, M.K.; Jafari, J. Evaluation of Tensile Deformation Properties of Friction Stir Processed Pure Copper: Effect of Processing Parameters and Pass Number. *J. Mater. Eng. Perform.* **2014**, *23*, 101–107. [CrossRef]

111. Luo, P.; McDonald, D.T.; Xu, W.; Palanisamy, S.; Dargusch, M.S.; Xia, K. A modified Hall–Petch relationship in ultrafine-grained titanium recycled from chips by equal channel angular pressing. *Scr. Mater.* **2012**, *66*, 785–788. [CrossRef]

112. Besharati Givi, M.K.; Asadi, P.; Bag, S.; Yaduwanshi, D.; Pal, S.; Heidarzadeh, A.; Mudani, S.; Kazemi-Choobi, K.; Hanifian, H.; Braga, D.; et al. *Advances in Friction-Stir Welding and Processing*; Elsevier: Amsterdam, The Netherlands, 2014; ISBN 9780857094544.

113. Babu, J.; Anjaiah, M.; Mathew, A. Experimental studies on Friction stir processing of AZ31 Magnesium alloy. *Mater. Today Proc.* **2018**, *5*, 4515–4522. [CrossRef]

114. Pan, L.; Kwok, C.T.; Lo, K.H. Friction-stir processing of AISI 440C high-carbon martensitic stainless steel for improving hardness and corrosion resistance. *J. Mater. Process. Technol.* **2020**, *277*, 116448. [CrossRef]

115. Lashgari, H.R.; Kong, C.; Asnavandi, M.; Zangeneh, S. The effect of friction stir processing (FSP) on the microstructure, nanomechanical and corrosion properties of low carbon CoCr28Mo5 alloy. *Surf. Coat. Technol.* **2018**, *354*, 390–404. [CrossRef]

116. Jiang, X.; Overman, N.; Canfield, N.; Ross, K. Friction stir processing of dual certified 304/304L austenitic stainless steel for improved cavitation erosion resistance. *Appl. Surf. Sci.* **2019**, *471*, 387–393. [CrossRef]

117. Grewal, H.S.; Arora, H.S.; Singh, H.; Agrawal, A. Surface modification of hydroturbine steel using friction stir processing. *Appl. Surf. Sci.* **2013**, *268*, 547–555. [CrossRef]

118. Yasavol, N.; Jafari, H. Microstructure, Mechanical and Corrosion Properties of Friction Stir-Processed AISI D2 Tool Steel. *J. Mater. Eng. Perform.* **2015**, *24*, 2151–2157. [CrossRef]

119. Singh, S.; Kaur, M.; Saravanan, I. Enhanced microstructure and mechanical properties of boiler steel via Friction Stir Processing. *Mater. Today Proc.* **2020**, *22*, 482–486. [CrossRef]

120. Wang, T.; Shukla, S.; Komarasamy, M.; Liu, K.; Mishra, R.S. Towards heterogeneous AlxCoCrFeNi high entropy alloy via friction stir processing. *Mater. Lett.* **2019**, *236*, 472–475. [CrossRef]

121. Yang, X.; Yan, Z.; Dong, P.; Cheng, B.; Zhang, J.; Zhang, T.; Zhang, H.; Wang, W. Surface modification of aluminum alloy by incorporation of AlCoCrFeNi high entropy alloy particles via underwater friction stir processing. *Surf. Coat. Technol.* **2020**, *385*, 125438. [CrossRef]

122. Wang, T.; Komarasamy, M.; Shukla, S.; Mishra, R.S. Simultaneous enhancement of strength and ductility in an AlCoCrFeNi2.1 eutectic high-entropy alloy via friction stir processing. *J. Alloys Compd.* **2018**, *766*, 312–317. [CrossRef]

123. Yang, X.; Dong, P.; Yan, Z.; Cheng, B.; Zhai, X.; Chen, H.; Zhang, H.; Wang, W. AlCoCrFeNi high-entropy alloy particle reinforced 5083Al matrix composites with fine grain structure fabricated by submerged friction stir processing. *J. Alloys Compd.* **2020**, *836*, 155411. [CrossRef]

124. Sinha, S.; Nene, S.S.; Frank, M.; Liu, K.; Lebensohn, R.A.; Mishra, R.S. Deformation mechanisms and ductile fracture characteristics of a friction stir processed transformative high entropy alloy. *Acta Mater.* **2020**, *184*, 164–178. [CrossRef]

125. Shukla, S.; Wang, T.; Frank, M.; Agrawal, P.; Sinha, S.; Mirshams, R.A.; Mishra, R.S. Friction stir gradient alloying: A novel solid-state high throughput screening technique for high entropy alloys. *Mater. Today Commun.* **2020**, *23*, 100869. [CrossRef]

126. Leal, R.M.; Galvão, I.; Loureiro, A.; Rodrigues, D.M. Effect of friction stir processing parameters on the microstructural and electrical properties of copper. *Int. J. Adv. Manuf. Technol.* **2015**, *80*, 1655–1663. [CrossRef]

127. Peng, J.; Zhang, Z.; Huang, J.; Guo, P.; Li, Y.; Zhou, W.; Wu, Y. The effect of the inhomogeneous microstructure and texture on the mechanical properties of AZ31 Mg alloys processed by friction stir processing. *J. Alloys Compd.* **2019**, *792*, 16–24. [CrossRef]

128. Wang, Y.; Huang, Y.; Meng, X.; Wan, L.; Feng, J. Microstructural evolution and mechanical properties of MgZnYZr alloy during friction stir processing. *J. Alloys Compd.* **2017**, *696*, 875–883. [CrossRef]

129. Shang, Q.; Ni, D.R.; Xue, P.; Xiao, B.L.; Wang, K.S.; Ma, Z.Y. An approach to enhancement of Mg alloy joint performance by additional pass of friction stir processing. *J. Mater. Process. Technol.* **2019**, *264*, 336–345. [CrossRef]

130. Du, X.; Wu, B. Using two-pass friction stir processing to produce nanocrystalline microstructure in AZ61 magnesium alloy. *Sci. China Ser. E Technol. Sci.* **2009**, *52*, 1751–1755. [CrossRef]

131. Huang, L.; Wang, K.; Wang, W.; Yuan, J.; Qiao, K.; Yang, T.; Peng, P.; Li, T. Effects of grain size and texture on stress corrosion cracking of friction stir processed AZ80 magnesium alloy. *Eng. Fail. Anal.* **2018**, *92*, 392–404. [CrossRef]

132. Jin, Y.; Wang, K.; Wang, W.; Peng, P.; Zhou, S.; Huang, L.; Yang, T.; Qiao, K.; Zhang, B.; Cai, J.; et al. Microstructure and mechanical properties of AE42 rare earth-containing magnesium alloy prepared by friction stir processing. *Mater. Charact.* **2019**, *150*, 52–61. [CrossRef]

133. MD, F.K.; Karthik, G.M.; Panigrahi, S.K.; Ram, G.D.J. Friction stir processing of QE22 magnesium alloy to achieve ultrafine-grained microstructure with enhanced room temperature ductility and texture weakening. *Mater. Charact.* **2019**, *147*, 365–378. [CrossRef]

134. Johannes, L.B.; Yowell, L.L.; Sosa, E.; Arepalli, S.; Mishra, R.S. Survivability of single-walled carbon nanotubes during friction stir processing. *Nanotechnology* **2006**, *17*, 3081–3084. [CrossRef]

135. Reddy, K.V.; Naik, R.B.; Rao, G.R.; Reddy, G.M.; Kumar, R.A. Microstructure and damping capacity of AA6061/graphite surface composites produced through friction stir processing. *Compos. Commun.* **2020**, *20*, 100352. [CrossRef]

136. Kumar, P.A.; Madhu, H.C.; Pariyar, A.; Perugu, C.S.; Kailas, S.V.; Garg, U.; Rohatgi, P. Friction stir processing of squeeze cast A356 with surface compacted graphene nanoplatelets (GNPs) for the synthesis of metal matrix composites. *Mater. Sci. Eng. A* **2020**, *769*, 138517. [CrossRef]

137. Mishra, R.S.; Ma, Z.Y.; Charit, I. Friction stir processing: A novel technique for fabrication of surface composite. *Mater. Sci. Eng. A* **2003**, *341*, 307–310. [CrossRef]

138. Khorrami, M.S.; Saito, N.; Miyashita, Y.; Kondo, M. Texture variations and mechanical properties of aluminum during severe plastic deformation and friction stir processing with SiC nanoparticles. *Mater. Sci. Eng. A* **2019**, *744*, 349–364. [CrossRef]

139. Sivanesh Prabhu, M.; Elaya Perumal, A.; Arulvel, S. Development of multi-pass processed AA6082/SiCp surface composite using friction stir processing and its mechanical and tribology characterization. *Surf. Coat. Technol.* **2020**, *394*, 125900. [CrossRef]

140. Shamsipur, A.; Kashani-Bozorg, S.F.; Zarei-Hanzaki, A. The effects of friction-stir process parameters on the fabrication of Ti/SiC nano-composite surface layer. *Surf. Coat. Technol.* **2011**, *206*, 1372–1381. [CrossRef]

141. Shafiei-Zarghani, A.; Kashani-Bozorg, S.F.; Gerlich, A.P. Texture Analyses of Ti/Al2O3 Nanocomposite Produced Using Friction Stir Processing. *Metall. Mater. Trans. A Phys. Metall. Mater. Sci.* **2016**, *47*, 5618–5629. [CrossRef]

142. Shang, J.; Ke, L.; Liu, F.; Lv, F.; Xing, L. Aging behavior of nano SiC particles reinforced AZ91D composite fabricated via friction stir processing. *J. Alloys Compd.* **2019**, *797*, 1240–1248. [CrossRef]
143. Zhang, Q.; Xiao, B.L.; Wang, Q.Z.; Ma, Z.Y. In situ Al3Ti and Al2O3 nanoparticles reinforced Al composites produced by friction stir processing in an Al-TiO2 system. *Mater. Lett.* **2011**, *65*, 2070–2072. [CrossRef]
144. Dixit, S.; Mahata, A.; Mahapatra, D.R.; Kailas, S.V.; Chattopadhyay, K. Multi-layer graphene reinforced aluminum – Manufacturing of high strength composite by friction stir alloying. *Compos. Part B Eng.* **2018**, *136*, 63–71. [CrossRef]

MDPI

St. Alban-Anlage 66

4052 Basel

Switzerland

Tel. +41 61 683 77 34

Fax +41 61 302 89 18

www.mdpi.com

Metals Editorial Office

E-mail: metals@mdpi.com

www.mdpi.com/journal/metals

www.ingramcontent.com/pod-product-compliance
Lightning Source LLC
LaVergne TN
LVHW071029170726
843515LV00006B/1237